Edible Oil Processing

Edible Oil Processing

Second Edition

Edited by

Wolf Hamm
Harpenden, UK

Richard J. Hamilton
Liverpool John Moores University, Liverpool, UK

Gijs Calliauw
Desmet Ballestra Oils and Fats, Zaventem, Belgium

WILEY-BLACKWELL
A John Wiley & Sons, Ltd., Publication

This edition first published 2013

Wiley-Blackwell is an imprint of John Wiley & Sons, formed by the merger of Wiley's global Scientific, Technical and Medical business with Blackwell Publishing.

Registered office:
John Wiley & Sons, Ltd, The Atrium, Southern Gate, Chichester, West Sussex, PO19 8SQ, UK

Editorial offices:
9600 Garsington Road, Oxford, OX4 2DQ, UK
The Atrium, Southern Gate, Chichester, West Sussex, PO19 8SQ, UK
111 River Street, Hoboken, NJ 07030-5774, USA

For details of our global editorial offices, for customer services and for information about how to apply for permission to reuse the copyright material in this book please see our website at www.wiley.com/wiley-blackwell.

Library of Congress Cataloging-in-Publication Data has been applied for

ISBN 978-1-4443-3684-9 (hardback)

A catalogue record for this book is available from the British Library.

Wiley also publishes its books in a variety of electronic formats. Some content that appears in print may not be available in electronic books.

Cover image: Main image © Desmet Ballestra
Oil and water © Thomas Vogel/istockphoto.com
Oilseed rape © Matthew Dixon/istockphoto.com
Cover design by Meaden Creative

Set in 10.5/13pt Times by Laserwords Private Limited, Chennai, India

1 2013

Contents

List of Contributors

Dr Arjen Bot, Unilever R&D Vlaardingen, Vlaardingen, The Netherlands

Dr Gijs Calliauw, Development Manager Modification, Desmet Ballestra Oils and Fats, Zaventem, Belgium

Dr David Cowan, CS Application Scientist/Global Coordinator, Novozymes, Chesham, UK

Dr Wim De Greyt, R&D Manager, Desmet Ballestra Oils and Fats, Zaventem, Belgium

Gerrit den Dekker, Retired, Unilever R&D Vlaardingen, Vlaardingen, The Netherlands

Professor Eckhard Flöter, Technical University Berlin, Berlin, Germany

Frank D. Gunstone, Professor emeritus, St Andrews University, St Andrews, UK

Wolf Hamm, Retired, Harpenden, UK

Dr Marc Kellens, Group Technical Director, Desmet Ballestra Oils and Fats, Zaventem, Belgium

Timothy G. Kemper, Global Technical Director, Solvent Extraction, Desmet Ballestra, Marietta, GA, USA

Philippe van Doosselaere, Retired (formerly Product Manager, Crushing, Desmet Ballestra Oils and Fats), Brussels, Belgium

Dr Gerrit van Duijn, Maas Refinery, Rotterdam, The Netherlands

Mar Verhoeff, Laboratory Dr A. Verwey B.V., Rotterdam, The Netherlands

List of Abbreviations

ADI	Acceptable Daily Intake
ARfD	Acute Reference Dose
AMF	Anhydrous Milk Fat
ALARA	As Low As Reasonably Achievable
ATEX	Atmospheres Explosive
AES	Atomic Emission Spectroscopy
Barg	Bar gauge
CBE	Cocoa Butter Equivalents
CBI	Cocoa Butter Improvers
CBS	Cocoa Butter Substitutes
DSC	Differential Scanning Calorimetry
DHA	4,7,10,13,16,19-Docosahexaenoic acid
DACC	Donor Accepted Column Chromatographic
DOBI	Deterioration of Bleachability Index
EPA	5,8,11,14,17-Eicosapentaenoic Acid
ECD	Electron Capture Detection
EDTA	Ethylene Diamine Tetra-acetic Acid
EU-27	European Union – 27
FOSFA	Federation of Oils, Seeds and Fats Associations
FID	Flame Ionisation Detection
FFA	Free Fatty Acids
GCFID	Gas Chromatography–Flame Ionisation Detection
GC-MS	Gas Chromatography–Mass Spectrometry
GPC	Gel Permeation Chromatography
HACCP	Hazard Analysis and Critical Control Points
HAZOP	Hazard and Operational Study
HPLC	High-Performance Liquid Chromatography
ICP	Inductively Coupled Plasma
IMO	International Maritime Organization

MARPOL	International Convention for the Prevention of Pollution from Ships
ISO	International Organization for Standardization
LOD	Limit of Determination
LDL	Low-Density Lipoprotein
MRL	Maximum Residue Limits (s)
3-MCPD	3-Monochloropropane-diol
NIOP	National Institute of Oilseeds Products
NORES	Neutral Oil Recovery System
NPD	Nitrogen Phosphorus Detection
BOB	2-Oleo-1.3-dibehenin
EO	Operational Efficiency
PFAD	Palm Fatty Acid Distillate
PFR	Plug-Flow Reactor
PAHs	Polycyclic Aromatic Hydrocarbons
PG	Propyl allate
POP	Oleo-dipalmitin
POS	Oleo-palmitin – stearin
POSt	Oleo-palmitin – stearin
PStP	Stearo-dipalmitin
SSHEs	Scraped-Surface Heat Exchangers
Silver-ion HPLC	Silver ion High Performance Liquid Chromatography
SFC	Solid Fat Content
SBDD	Soybean Deodoriser Distillate
SOS	Oleo-distearin
SUS	Saturated Unsaturated Saturated triglyceride
StOSt	Oleo-distearin
UUS	Unsaturated Unsaturated Saturated triglyceride
USU	Unsaturated Saturated Unsaturated triglyceride
USS	Unsaturated Saturated Saturated triglyceride
TBHQ	Tertiary Butyl hydroquinone

Introduction

In the years since the first edition of *Edible Oil Processing* was published (in 2000), there have been many changes in the processing of oils. Two major factors have been involved: first, the need to reduce the hydrogenated fats in food products, and second, the move to use enzymes. These two issues both originate from an overall increased awareness of the possible impact of processing on consumers' health and on the environment. This edition tries to bring this awareness, and the way in which it has altered the nature of edible oil processing, to the forefront of the discussion.

In Chapter 1, Gunstone outlines the makeup of fats and oils, from the major components such as triacylglycerols (TAGs) to minor constituents such as squalene. He illustrates the changes in oils that have been obtained by seed breeding procedures, such as Nu Sun oil. He also deals with the physical properties on which much of the processing of oils is based.

In Chapter 2, Hamm explains how multi-compartmented parcel tankers play a major role in the transport of oils and fats. He highlights the systems and regulations pertaining to oil shipments, and he deals with the role of FOSFA and NIOP in greater detail than in the first edition.

In Chapter 3, van Doosselaere describes how important seed handling and storage are to the overall production of good-quality oils. In sampling incoming seeds, moisture, foreign material, damaged or broken seeds, protein content and oil content must all be controlled. He explains the methods of storing seeds used to maintain their high quality. Preparation and extraction of seeds are covered in a general way before the special care that must be taken for soybean, rapeseed, cottonseed, corn germ, copra, peanut, rice bran, olive and of course palm oils is discussed.

In Chapter 4, Kemper describes how hexane became the industry's solvent of choice for the extraction of oils, and considers the effects of various plant and processing parameters on solvent extraction plant performance. He also records how important solvent recovery and heat recovery are to the overall economy of the process. The chapter provides a comprehensive overview of solvent extraction as used in edible oil production.

In Chapter 5, De Greyt deals with the refining of food oils in a sustainable manner. He explains how new technologies have become available and how some have been employed commercially, such as hydrodynamic Nano Reactors and enzymatic degumming. Some processes are still at the pilot plant stage, such as the use of chlorophyllases. He finishes with a look at the future for short-path distillation and supercritical processing, and what this might bring to this field of oil processing.

In Chapter 6, Kellens and Calliauw describe how hydrogenation, interesterification and fractional crystallisation are still used to modify oils and fats. Health concerns have led to a large reduction (6–30 million tonnes) in the amount of oil being hydrogenated, and the authors touch on the proposed newer methods of cutting down on *trans* fatty acid composition. They elaborate on the discussion of fractional crystallisation given in the first edition, examining everything from intersolubility to industrial practice, and noting that multistage processing and continuous operation hold the most promise for oil modification technology.

In Chapter 7, Cowan shows the considerable change that has occurred in the use of enzymes since the first edition of this book. By using gene transfer between microorganisms and low-cost immobilisation techniques, it has been possible to move the technology from one restricted to high-value products to one with much wider applications. He covers the use of cellulases, proteolytic enzymes, phospholipases, esterase and lipases, and considers their environmental impacts.

Chapter 8 deals with the applications of edible oils and the considerable reformulation resulting from the reduction of the use of hydrogenated oils. Bot and Flöter also explain fat crystal networks, the polymorphic changes in spreads, the lower-fat versions of mayonnaises and the use of tropical fats in nondairy creams.

Verhoeff and van Duijn concisely describe in Chapter 9 the methods used to measure the natural components of edible oils, including free fatty acids (FFA), peroxides, phosphorus, moisture, dirt, colour, metals and tocopherols, as well as contaminants such as polycyclic aromatic hydrocarbons, pesticides, hydrocarbons and mycotoxins. The authors go on to describe the crude oil risk matrix and finish with a consideration of hazard analysis and critical control points (HACCP).

In Chapter 10, van Duijn and den Dekker explain the steps needed to decide whether the building of a new refinery can be justified. They outline the process routes to a fully refined oil based on lowest costs. Batch and continuous processes and chemical and physical refining are contrasted, and the design parameters for storage tanks and piping are fully covered. The authors provide estimates based on best-practice data, which can be used for first-design purposes. They then explain that occupational safety hazards must be considered from an early stage in the planning.

Wolf Hamm
Richard J. Hamilton
Gijs Calliauw

1

Composition and Properties of Edible Oils

Frank D. Gunstone
Professor emeritus, St Andrews University, St Andrews, UK

1.1 Introduction

According to US Department of Agriculture (USDA) statistics, the production of nine vegetable oils from seven seeds and from palm fruit and olive was 153 million tonnes worldwide in 2010/11 (Table 1.1). In addition, production of four animal fats (butter, lard, tallow and fish oil) amounted to about 25 million tonnes. Over time, animal fats have fallen in market share, and they now make up only 15% of total annual production. Among vegetable oils, palm, soya, rape and sun oils have become increasingly important, with palm and soya dominant (Table 1.1). It is interesting that these four vegetable oils are produced in different parts of the world (Table 1.2). It should also be noted that crops grown in the southern and northern hemispheres are harvested at different times of the year, with the exception that palm oil is produced in all months of the year. This is particularly significant for soybeans, grown predominately in North and South America. Palm oil and olive oil are obtained by pressing the fruits in the countries where they grow, and trade is confined to the oil or to downstream products. Exports/imports of vegetable oils represent 41% of total production, but there is also considerable trade in unprocessed seeds (24%), especially in soybeans, with extraction occurring in the importing country.

Oils and fats are used mainly for food purposes, but both oilseeds and extracted oil are also used in some part as animal feed. Oils also have industrial uses. Traditionally, these have been mainly in the production of soap and other surface-active molecules, but increasingly they are for energy-producing purposes, such as transport use by automobiles, trains, aeroplanes

Edible Oil Processing, Second Edition. Edited by Wolf Hamm, Richard J. Hamilton and Gijs Calliauw.

Table 1.1 Annual production of major vegetable oils (million tonnes) between 2007/08 and 2010/11, 2011/12 (estimate) and 2012/13 (forecast).

	07/08	08/09	09/10	10/11	11/12 (e)	12/13 (f)
Palm	41.08	44.02	45.87	47.95	50.67	52.77
Soya bean	37.83	35.90	38.88	41.24	41.85	43.62
Rapeseed	18.43	20.56	22.44	23.58	23.76	23.52
Sunflower	10.03	11.95	12.11	12.21	14.14	14.52
Cottonseed	5.21	4.78	4.62	4.99	5.32	5.24
Groundnut	4.86	5.08	4.74	5.10	5.24	5.37
Palm kernel	4.88	5.17	5.50	5.56	5.84	6.09
Coconut	3.53	3.54	3.63	3.83	3.56	3.52
Olive	2.78	2.78	3.05	3.04	3.10	3.09
Total	**128.62**	**133.78**	**140.84**	**147.50**	**153.48**	**157.74**

Source: USDA figures (June 2012).

Table 1.2 Major geographical regions for the production of oilseeds and vegetable oils in 2011/12.

Product	Weight (million tonnes)	Major producing countries/regions (percentage of total)
		Seven oilseeds
Total	437.0	
Soya	236.4	USA (35), Brazil (28), Argentina (18), China (6), India (5)
Rape	60.7	EU-27 (31), Canada (23), China (21), India (11)
Sunflower	39.1	Russia (25), Ukraine (24), EU-27 (21), Argentina (9)
Cottonseed	46.6	China, India, USA, Pakistan
Groundnut	35.5	China, India
Palm kernel	13.3	Indonesia, Malaysia
Copra[a]	5.5	Philippines, Indonesia, India
		Nine vegetable oils[b]
Total	153.48	
Palm	50.67	Indonesia (50), Malaysia (37), Thailand (3)
Soya	41.85	China (25), USA (21), Argentina (17), Brazil (17), EU-27 (5), India (4)
Rape	23.76	EU-27 (37), China (23), Canada (12), India (10), Japan (4)
Sunflower	14.14	Ukraine (26), Russia (23), EU-27 (21), Argentina (10)
Cottonseed	5.32	China (28), India (23), USA (6)
Groundnut	5.24	China (48), India (26)
Palm kernel	5.84	Indonesia, Malaysia
Coconut	3.56	Philippines, Indonesia, India
Olive	3.10	EU-27 (77)

[a]Copra is the source of coconut oil.
[b]Vegetable oils may be extracted from indigenous and/or imported seeds.
Source: USDA figures (June 2012).

Table 1.3 Consumption of vegetable oils in 2011/12 in China, EU-27, India and the USA.

	Population (millions)	Million tonnes	Percentage of world total	kg/person/year
China	1345	29.05	19.2	21.6
EU-27	502	23.99	15.9	47.8
India	1198	16.93	11.2	14.1
USA	315	12.94	8.6	41.1
World total	7022	151.16	–	21.5

Source: USDA figures (June 2012).

or boats, or the direct production of energy. These new uses underlie the food versus fuel debate (Gunstone, 2011).

Total consumption covers all these differing uses and is not to be equated with food consumption. It should also be remembered that dietary intake of fat goes beyond these commodity oils and includes sources such as nuts, meat products and dairy products other than butter (milk and cheese). The major consuming countries/regions of vegetable oils are China, EU-27, USA and India, as shown in Table 1.3. It is sometimes convenient to express consumption (for all purposes) on a *per capita* basis by dividing it by population. In 2011/12, the world average was 21.5 kg for vegetable oils, but the figure shows great variation for individual countries/regions. The world figure has grown steadily over the last 60 years and production of vegetable oils has grown more quickly than population. The figure for China has increased recently and is now close to the world average. The Indian figure has changed less and remains well below average. Higher figures are apparent for the USA and Europe, with the European figure inflated by the significant production of biodiesel, made mainly from rapeseed oil. The very large kg/person figure of 159 for Malaysia reflects the presence of a large oleochemical industry in a country with modest population (27.5 million).

The lower section of Table 1.2 shows the major producing countries/regions for nine vegetable oils. Since these oils can be produced, in some part, from imported seeds, the upper part of the table is a better indication of their geographical origin.

1.2 Components of natural fats

The oils and fats of commerce are mixtures of organic molecules. They are mainly triacylglycerols (commonly referred to as triglycerides), accompanied by lower levels of diacylglycerols (diglycerides), monoacylglycerols (monoglycerides) and free fatty acids, and by other minor components, some of which are important materials in their own right. Materials (1–3%) that are not soluble in aqueous alkali after hydrolysis are sometimes referred to as nonsaponifiable or unsaponifiable material. Although oils and fats are the source of dietary lipids, they are also an important source of other essential dietary requirements. These minor components include phospholipids, phytosterols, tocols (tocopherols and tocotrienols, including vitamin E) and hydrocarbons. Phospholipids are recovered during degumming and sterols and tocols are enriched in deodoriser distillate. Thus soybeans are not only the source of soybean oil and soybean meal (protein) but are also the major source of lecithin (a crude mixture containing phospholipids), sterols and sterol esters, and of natural vitamin E (Clark, 1996; Ghosh and Bhattacharyya, 1996; Gunstone, 2011; Walsh *et al.*, 1998).

1.2.1 Fatty acids and glycerol esters

Over 1000 natural fatty acids have been identified. These vary in chain length (commonly C_{12}–C_{22}), degree of unsaturation (usually in the range 0–6 *cis* olefinic centres) and the presence or absence of other functional groups such as hydroxy or epoxy. However, only a limited number – perhaps 25–50 – are likely to be important to most lipid scientists and technologists. The most common members of this group are detailed in Table 1.4. They are divided into four categories: saturated acids, monounsaturated acids

Table 1.4 Structures of the most common fatty acids.

Common name	Systematic name[a]	Shorthand[b]
Saturated		
Lauric	Dodecanoic	12:0
Myristic	Tetradecanoic	14:0
Palmitic	Hexadecanoic	16:0
Stearic	Octadecanoic	18:0
Monounsaturated		
Oleic	9-octadecenoic	18:1
Erucic	13-dodecenoic	22:1
Polyunsaturated (n-6)		
Linoleic	9,12-octadecadienoic	18:2
γ-linolenic	6,9,12-octadecatrienoic	18:3
Arachidonic	5,8,11,14-eicosatetraenoic	20:4
Polyunsaturated (n-3)		
α-linolenic	9,12,15-octadecatrienoic	18:3
EPA	5,8,11,14,17-eicosapentaenoic acid	20:5
DHA	4,7,10,13,16,19-docosahexaenoic acid	22:6

[a]The unsaturated centres in these acids have *cis* configuration.
[b]The shorthand designation indicates the number of carbon atoms and of *cis* unsaturated centres in the molecule. It is not necessary to prefix the numbers with the letter 'C'.

and polyunsaturated acids of the n-6 and n-3 families (also referred to as omega-6 and omega-3 acids). The terms 'n-6' and 'n-3' refer to the positions of the first double bond with respect to the end methyl group. For the most part, unsaturation is confined to olefinic systems with *cis* configuration, and the polyunsaturated fatty acids (PUFAs) have methylene-interrupted patterns of unsaturation. They will thus contain one or more pentadiene group (—CH=CHCH$_2$CH=CH—) with a doubly activated CH$_2$ function, which has an important influence on their properties. The (largely unnatural) *trans* acids differ from their *cis* isomers in their physical properties (especially higher melting points) and in their nutritional properties. There has been wide recognition of the undesirable nutritional properties of most *trans* acids in the past 10 years, which has had important consequences for food processors. In some countries, the content of *trans* acids above a certain level has to be reported on the packaging; even where this is not required by law, processors have sought to keep levels to a minimum. This has had important consequences for the blends of fats used in spreads and in the production of baking fats, as processors have struggled to maintain desirable physical properties while achieving higher nutritional status. Another nutritional factor that has become more significant in the last 10 years is the recognition of the importance of omega-3 (n-3) acids, particularly those with more than 18 carbon atoms.

These common fatty acids are easily recognised and separated by gas chromatography of their methyl esters, and this technique is a standard analytical procedure in quality-control laboratories (see Chapter 9). Other analytical procedures used in research laboratories, including mass spectrometry (MS) and nuclear magnetic resonance (NMR), are also starting to be used in some quality-control centres.

An oil or fat will usually contain at least 95% triacylglycerols before refining. After refining, this number will generally be in the range 97–99%, depending on the level of unsaponifiable material the oil or fat still contains. Triacylglycerols are fatty acid esters of the trihydric alcohol glycerol (1,2,3-trihydroxypropane) and contain three acyl chains in each molecule, usually from two or three different fatty acids (Figure 1.1). In the biosynthesis of a vegetable oil, acylation of a glycerol phosphate is enzyme-promoted, and the fatty acids are not distributed in a random manner. If the natural mixture is randomised, the resulting material has the same total amount of fatty acids but different triacylglycerols and, consequently, different melting behaviour (see Chapter 6). In vegetable oils, the *sn*-2 position is esterified almost entirely by unsaturated fatty acids, while saturated acids and the remaining unsaturated acids are in the *sn*-1(3) positions.

An oil with n different fatty acids could contain $(n^3 + 3n^2 + 2n) \div 6$ triacylglycerols if all possibilities of isomerism were included. This corresponds to values of 10, 20 and 35 for 3, 4 and 5 fatty acids, respectively. In reality,

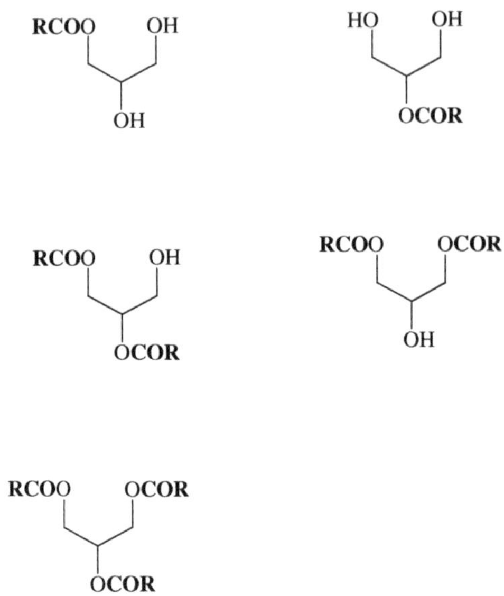

Figure 1.1 Glycerol esters (1- and 2-monoacylglycerols, 1,2- and 1,3-diacylglycerols and triacylglycerols). **RCO** represents the acyl group from the fatty acid **RCOOH**. All other letters relate to atoms derived from the glycerol molecule.

these values are too low, since all the minor acids have been ignored. The number becomes very much greater in fats, such as dairy fats, fish oils and partially hydrogenated oils, with very complex fatty acid compositions. There are methods of triacylglycerol analysis, but these are not trivial, and the results can be complex. This level of analysis is therefore not routine. There are, however, standardised procedures for distinguishing fatty acids in the *sn*-2 position from those in the *sn*-1(3) positions.

Accompanying the triacylglycerols are low levels of diacylglycerols, monoacylglycerols and free acids. These can result from incomplete biosynthesis in immature seeds or from post-harvest lipolysis. Almost all of the free acids and most of the monoacylglycerols will be removed by refining, but diacylglycerols tend to remain in the product. These are usually in the range 0–2%, but refined palm oil contains 3–8% diacylglycerols (Wai-Lin & Wee-Lam, 1995).

After conventional refining, some oils, such as rape/canola, corn, rice bran and sunflower, contain high-melting material that slowly crystallises during storage at ambient temperature. This causes a haze, which – though harmless from a nutritional standpoint – does not find favour with users of salad oil and frying oil. This haze is caused mainly by wax esters and can be removed by holding the oil at ~5 °C for several hours and then filtering (at a slightly higher temperature, to reduce viscosity) with the assistance of a filter aid. Undesirable solids present in some biodiesel samples have been identified as monoacylglycerols and sterol glucosides (Tang *et al.*, 2008).

1.2.2 Phospholipids

Crude oils generally contain phospholipids, which are removed during refining at the degumming stage (Chapter 4). The valuable crude product containing phospholipids and other lipid molecules is termed 'lecithin'. It is the basis of the phospholipid industry, and phospholipids are used extensively in food products, animal feed and industrial products; their uses are based mainly on their amphiphilic properties (i.e. different parts of the molecule show lipophilic and hydrophilic properties). The major components (phosphatidylcholines, phosphatidylethanolamines and phosphatidylinositols) are accompanied by smaller proportions of other phospholipids (Figure 1.2). Soybean oil, rapeseed oil and sunflower seed oil contain 1.5–2.5%, $\leq$2.5% and $\sim$1% phospholipids, respectively. Soybean oil is the major source of commercial lecithin, and this raises a problem in that most soybean oil now comes from genetically modified sources. Those who want to avoid GM products must either find identity-preserved soybean lecithin or use sunflower lecithin from non-GM seeds. The typical composition of a commercial deoiled soybean lecithin is 81% phospholipids (mainly PCs, PEs and PIs), 10% glycolipids and 6% carbohydrates (Gunstone, 2008). Palm oil contains little or no phospholipid.

1.2.3 Sterols

Most vegetable oils contain 1000–5000 ppm (1–5 g/kg) of sterols, partly as free sterols and partly as esterified sterols. Higher levels are present in rapeseed oil (5–11 g/kg, mean $\sim$7.5 g/kg) and in corn oil (8–22 g/kg, mean 14 g/kg). β-sitosterol (Figure 1.3) is generally the major phytosterol (50–80% of total sterol), with campesterol, stigmasterol and Δ^5-avenasterol frequently attaining significant levels (Tables 1.5 and 1.6). Brassicasterol is virtually absent from the major seed oils, apart from rapeseed oil, in which it makes up 10% of total sterol. Kochhar (1983) reviewed sterol composition and sterol content in edible vegetable oils and the changes that take place in these as a result of processing (Section 1.6). Verleyen *et al.* (2002a, 2002b) have described an analytical procedure by which to measure free sterols and sterol esters and have examined the changes that occur during refining. Cholesterol (Figure 1.3) is considered to be a zoosterol and is not present in plant systems at a significant level. The normal value of 20–50 ppm in vegetable oils is much lower than the levels reported for animal fats (up to 1000 ppm), fish oils (up to 7000 ppm), dairy fats (2000–3000 ppm) and egg yolk (12 500 ppm).

Phytosterol (plant sterol) esters are now being added to spreads at significant levels up to 10% because they are considered to reduce cholesterol levels (Sato *et al.*, 2003). These phytosterols are recovered during wood

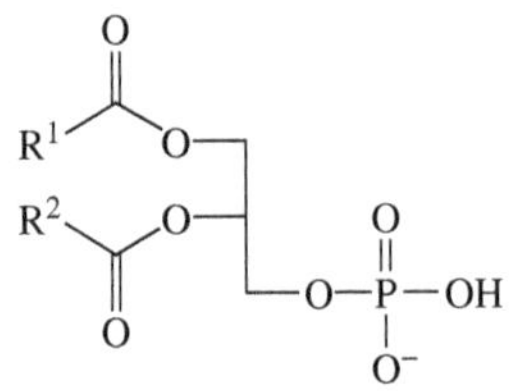

phosphatidic acids (PAs)

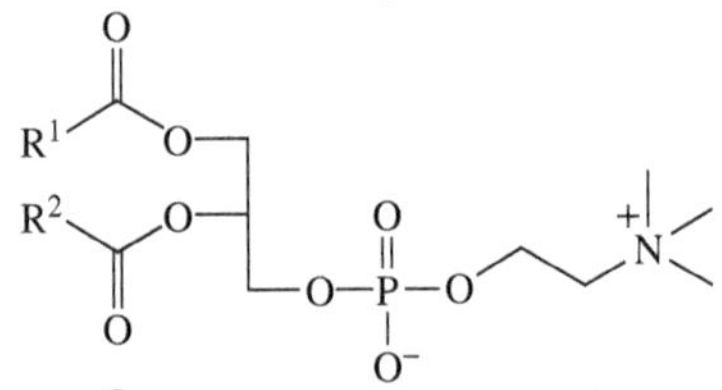

phosphatidylcholines (PCs)

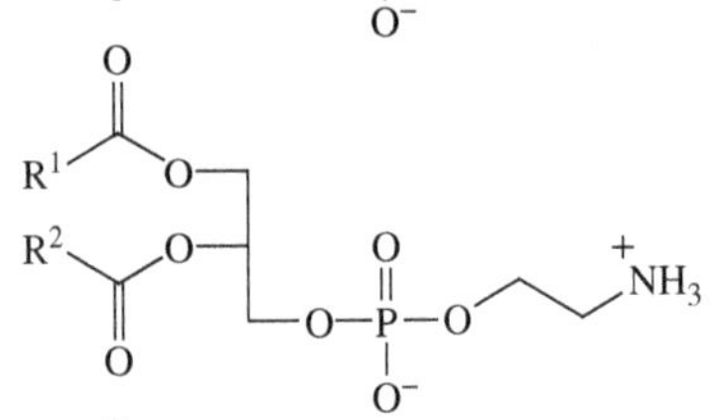

phosphatidylethanolamines (PEs)

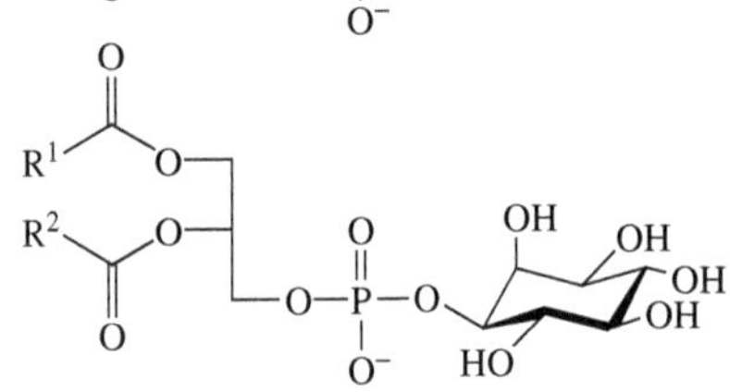

phosphatidylinosititols

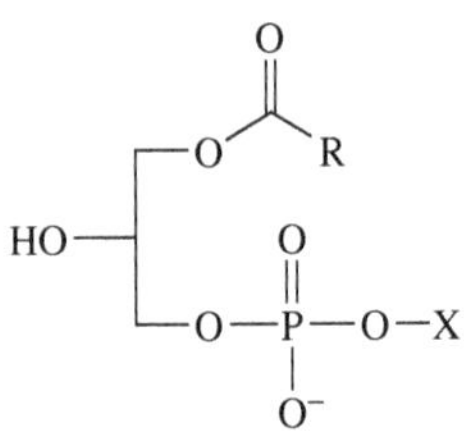

2-lysophospholipids:
LPC, X = choline;
LPE, X = ethanolamine

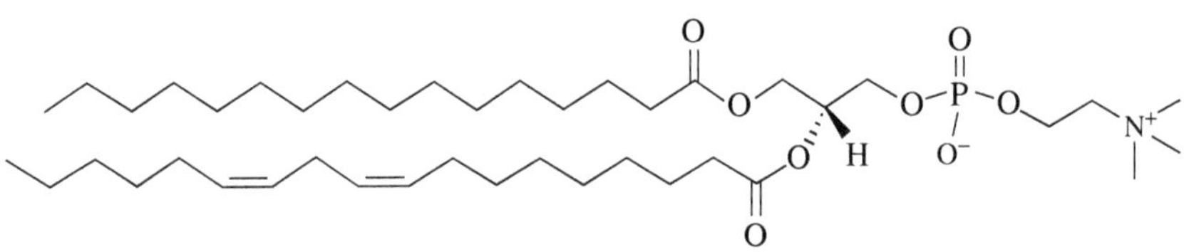

Figure 1.2 Structures of selected phospholipids (PAs, PCs, PEs, PIs, lysoPLs). These are correctly named in the plural because natural products are mixtures of compounds which vary in the nature of the acyl groups R^1CO and R^2CO. The final structure is an alternative representation of a PC containing palmitic acid and linoleic acid. These molecules (apart from phosphatidic acid) contain four ester bonds. On complete hydrolysis they furnish fatty acids, glycerol, phosphoric acid and a hydroxy compound (choline etc.). A series of phospholipases which catalyse selective hydrolysis (lipolysis) of these ester groups exists.
Source: Most of these structures have been taken from "Lipid Glossary 2" (The Oily Press, 2000) which can be downloaded free via The Oily Press website by permission of the authors and the publisher.

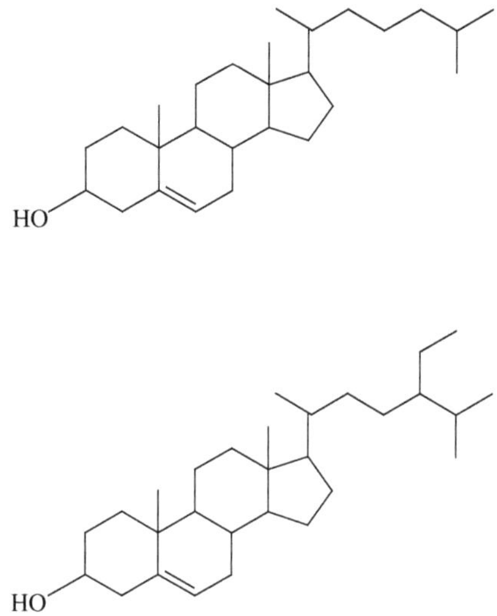

Figure 1.3 Cholesterol (top) and sitosterol (bottom).

Table 1.5 Major sterols (campesterol, stigmasterol and β-sitosterol) in vegetable oils as percentage of total sterols.

	Total sterols (mg/kg)	Camp	Stig	β-sito
Palm	300–700	19–27	8–14	50–62
Rape[a]	4500–11300	25–39	0–1	45–58
Soybean	1800–4500	16–24	15–19	47–60
Sunflower	2400–5000	6–13	6–13	50–70

[a]Rape also contains brassicasterol 5–13% (see Table 1.6).
Source: Codex Standard for Named Vegetable Oils, Codex-Stan 210–1999 (adopted 1999, revised 2001, amendments 2003, 2005), Table 3 (available from www.codexalimentarius.org).

processing or are obtained from soybean deodoriser distillate. During high-temperature deodorisation (see Chapter 5), the following are removed in the distillate: aldehydes, ketones and other short-chain compounds resulting from oxidation, tocopherols (vitamin E), sterols, carotene degradation products, nitrosamines, residual extraction solvent, organochlorine pesticides and volatile sulfur compounds (Kao *et al.*, 1998; Torres *et al.*, 2009).

1.2.4 Tocols and other phenolic compounds

Tocol extracts are mixtures of up to eight compounds. There are four tocopherols with a saturated, branched, polyisoprenoid C_{16} side chain and

Table 1.6 Content (mg/100 g) of major esterified and free sterols in crude and refined vegetable oils.

	Esterified sterols					Free sterols					Total sterols				
	sum	Cp	Sg	Si	Av	sum	Cp	Sg	Si	Av	sum	Cp	Sg	Si	Av
Crude oils															
Palm	25	5	2	17	-	51	10	5	36	-	79	20	7	52	-
Soya	59	6	5	40	9	255	63	55	137	-	327	71	61	184	10
Rape	475	193	-	257	26[a]	336	97	-	171	68[a]	824	293	-	420	111[a]
Sun															
Refined oils															
Palm	28	6	3	17	2	29	6	4	18	-	60	14	7	36	3
Soya	88	11	9	58	10	193	39	40	113	-	267	47	48	159	12
Rape	485	191	-	255	39[a]	278	93	-	158	26[a]	767	300	-	390	77[a]
Sun	124	13	4	81	26	192	19	22	138	12	330	36	27	225	42

[a]These numbers in rapeseed oil relate to the content of brassicasterol.
Cp, campesterol; Sg, stigmasterol; Si, β-sitosterol; Av, Δ^5-avenasterol.
Source: Adapted from Verleyen 2002a and 2002b.

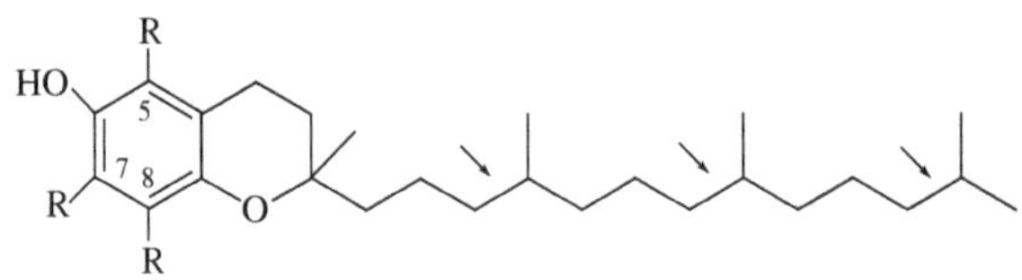

Figure 1.4 Tocopherols and tocotrienols. Tocopherols have a saturated C_{16} side chain, tocotrienols have double bonds at the three positions indicated by the arrows. R=H or CH_3. α = 5,7,8-trimethyltocol, β = 5,8-dimethyltocol, γ = 7,8-dimethyltocol, δ = 8-methyltocol.

four tocotrienols with three double bonds in the side chain (Figure 1.4). The tocotrienols, though significant in palm oil and in rice bran oil, are less common than the tocopherols, and less is known about their biological properties. The four tocopherols differ in the number of methyl groups attached to the heterocyclic moiety (chroman). They are designated α (5,7,8-trimethyl), β (5,7-dimethyl), γ (7,8-dimethyl) and δ (8-methyl). These compounds are sometimes incorrectly described as isomers, but this is true only for the β and γ compounds.

The tocols have two valuable properties: they show vitamin E activity and they are powerful antioxidants (Elmadfa & Wagner, 1997). These two properties are not identical. For vitamin E activity, the order is α (1.0), β (0.5), γ (0.1), δ (0.03), with total activity expressed in α-tocopherol units. For antioxidants, this order is reversed. Some typical levels are presented in Table 1.7. Among the readily available oils, palm and sunflower (as well as walnut and wheatgerm) are good sources of vitamin E because of the

Table 1.7 Tocols in the major vegetable oils (mg/kg equivalent ppm).

Vegetable oil	Total (mg/kg)	Tocopherols				Tocotrienols		
		α	β	γ	δ	α	γ	δ
Palm	150–1500	4–193	0–234	0–526	0–123	4–336	14–710	0–377
Rape	430–2680	100–386	0–140	189–753	0–22	ND	ND	ND
Soybean	600–3370	9–352	0–36	89–2307	154–932	0-69	0-103	ND
Sunflower	440–1520	409–935	0–45	0–34	0–7	ND	ND	ND
Wheatgerm	2540	1210	650	240	250	20	170	
PFAD	744–8192	(21%)	–	–	–	(16%)	(39%)	(24%)

Deodoriser distillates are enriched in tocols but have variable composition. SBDD is reported to contain 19 and 11% tocopherols (mainly gamma and delta) in two reports, and PFAD typically has the composition shown in the table.
Further information on the four major oils is available in appropriate chapters of Gunstone (2011) and in Yang (2003).
Source: Codex Standard for Named Vegetable Oils, Codex-Stan 210–1999 (adopted 1999, revised 2001, amendments 2003, 2005), Table 4 (available from www.codexalimentarius.org, last accessed 8 January 2013).

Table 1.8 Levels (ppm, equivalent to mg/kg) of the four tocopherols in crude rapeseed, palm, soybean and sunflower oils.

Oil	α	β	γ	δ
Rapeseed	175	0	415	10
Palm	190	0	0	0
Soybean	120	10	610	190
Sunflower	610	10	30	10

Source: Adapted from Warner (2007).

high levels of the α compound, whereas soybean tocopherols are effective antioxidants due to their high levels of γ and δ compounds (Evans *et al.*, 2002; Wagner & Isnardy, 2006; Warner, 2007; Warner *et al.*, 2008). The tocopherols are recovered from refinery byproducts such as palm fatty acid distillate (PFAD) and soybean deodoriser distillate (SBDD) (Table 1.8). The compositions of PFAD and SBDD are somewhat variable depending on the refining conditions employed.

Netscher reported in 1999 that production of vitamin E was about 20 000 tonnes. This included synthetic vitamin E (90%) – a mixture of eight racemic forms – made from trimethylhydroquinone and (all-*rac*-)-phytol and natural vitamin E (10%) principally from soybean. The latter product is an excellent antioxidant but its vitamin E activity is limited because of the low proportion of the α compound. This can be raised by a per-methylation reaction, which converts the mono- and dimethyl compounds to the trimethyl derivative. These products, whether natural or synthetic, are used in the animal feed, food and pharmaceutical industries.

Crude palm oil contains up to 800 ppm of tocols, of which α-tocopherol represents 22% and β-, γ- and δ-tocotrienol represent 20, 46 and 12%, respectively. About 70% of this mixture remains in the oil after refining, with the remainder present in PFAD at a level 5–10 times higher than in the original oil. This is used as a source of Palm Vitee^{TM}, which is 95% tocols rich in tocotrienol (>60%) (Basiron, 2005). Tocols in other oils have been discussed by Clark (1996), Ghosh & Bhattacharyya (1996) and Walsh *et al.* (1998).

Natural tocopherol mixtures are usually used as antioxidants at levels up to 500 ppm (along with ascorbyl palmitate, which extends antioxidant activity). At higher levels (>1000 ppm), α-tocopherol acts as a prooxidant. Since vegetable oils generally contain tocols at 200–800 ppm, further additions show only a limited effect. Evans *et al.* (2002) have discussed the optimal tocopherol blend for inhibiting soybean oil oxidation. The tocols are themselves very sensitive to oxidation and are more stable in an esterified form when the all-important hydroxyl (phenolic) group is not free. However, such compounds do not show antioxidant activity until they have been hydrolysed *in vivo* to the free phenolic form.

Many plant sources of lipids contain phenolic compounds other than the tocols. Some of these are water-soluble and are not extracted with the nonpolar lipids. However, they may be present in oils that are obtained by pressing rather than by hexane extraction. This holds for olive oil, which contains a wide range of phenolic compounds (Boskou, 2011), and for the growing range of cold-pressed oils. Sesame and rice bran oils are known for their high oxidative stability. They contain phenolic compounds which act as powerful antioxidants, including the sesamin lignans in sesame oil and the oryzanols (esters of ferulic acid – 3-methoxy 4-hydroxy cinnamic acid – $MeO(HO)C_6H_3CH{=}CHCOOH$) in ricebran oil (Kochhar, 2011).

1.2.5 Chlorophyll

Chlorophyll and its magnesium-free derivative (phaeophytin) are not wanted in refined oils because they produce an undesirable green hue and act as sensitisers for photooxidation (Section 1.5.2). No general listing of chlorophyll/phaeophytin levels has been discovered, but the following information has been gleaned from a range of sources (the levels cited for chlorophyll include phaeophytin):

- *Olive oil*: chlorophyll pigment levels vary with the maturity of the olive and the method of extraction. Unrefined oil contains 10–30 ppm chlorophyll.

- *Canola oil*: levels of chlorophyll in crude oil (5–35 ppm) are much reduced (<50 ppb) by alkali refining and bleaching (Przybylski, 2011).
- *Soybean oil*: low levels of chlorophyll in crude oil (1.0–1.5 ppm) are reduced to about 15 ppb after refining.
- *Sunflower oil*: crude oil contains 200–500 ppb chlorophyll, but in refined oil this is reduced to <30 ppb.
- *Palm oil*: crude palm oil contains 250–1800 ppb chlorophyll (mean 900 ppb, SD 100). The level falls with increasing maturity of the palm fruit.

1.2.6 Hydrocarbons

Though hydrocarbons are minor components of oils and fats, they are of dietary and legislative interest. They include alkanes, alkenes (such as squalene and carotenes) and polycyclic aromatic hydrocarbons (PAHs).

1.2.6.1 Alkanes

Many studies of alkanes ignore the more volatile compounds (up to C_{12} and including C_6, used as an extraction solvent) because of analytical difficulties arising from their volatility. They are not likely to be significant in refined oils that have been submitted to high-temperature deodorisation. Levels of C_{13}–C_{33} alkanes in crude oils are usually between 40 and 100 mg/kg (ppm), with lower levels for refined oils. Typical values, in ppb, reported by McGill *et al.* (1993) are 30–100 for olive, 100–170 for sunflower, 25–35 for corn and 25–35 for groundnut oil in samples purchased from retail outlets. There is a preference for odd-chain molecules, as illustrated in Table 1.9. The variation between different oils can be used to fingerprint them, and the consistency in the proportion of different alkanes – if not of the total levels present – suggests that they may be endogenous and not exogenous artefacts. Kao *et al.* (1998) have described some C_8–C_{18} unsaturated hydrocarbons present in deodoriser distillate, but these are probably thermal-decomposition products of glycerol esters.

1.2.6.2 Squalene

Squalene ($C_{30}H_{50}$, Figure 1.5) is a highly unsaturated open-chain triterpene used in the cosmetics industry after hydrogenation to squalane ($C_{30}H_{62}$). The most abundant source of squalene is the liver oil of the deep-sea dogfish (*Squalus acanthus*, hence the name 'squalene') and of some other marine species. Vegetable sources of potential interest include olive oil and amaranthus oil. Squalene levels of 100–1200 mg/100 ml of oil have been reported in olive oil, with most samples containing 200–500 g/100 ml (de Leonardis *et al.*, 1998). This rises to 200–500 mg/100 ml in the deodoriser distillate (Bondioli *et al.*, 1993). Amaranthus contains 6–8% squalene and this

Table 1.9 Odd-chain alkanes in selected seed oils (mg/kg, ppm).

Alkane (carbon atoms)	Sunflower	Olive (extra virgin)	Sesame
23	<1	19	<1
25	2	18	1
27	11	16	6
29	50	12	18
31	48	9	14
33	4	6	7
23–33	115	80	46
Total alkanes[a]	105–166 (5)	28–99 (6)	22–82 (4)

[a]Number in brackets = number of samples examined.

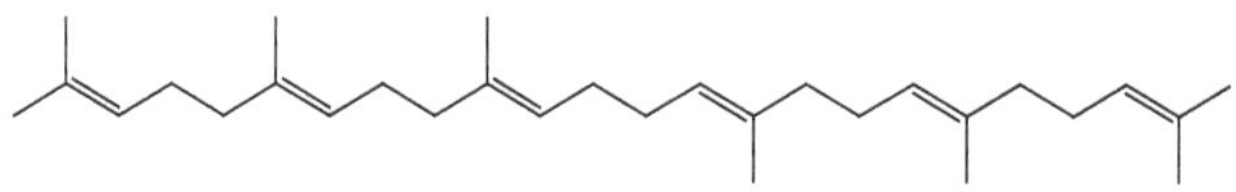

Figure 1.5 Squalene ($C_{30}H_{50}$).

concentration can be raised 10-fold after short-path high-vacuum distillation (Sun *et al.*, 1997).

1.2.6.3 Carotenes

Carotenes are minor components of most vegetable oils but occur to a greater degree in palm oil. These molecules contain a long chain of conjugated unsaturation and are yellow/orange in colour (Figure 1.6). Crude palm oil normally contains 500–700 ppm carotenes. These are mainly α-carotene (24–42% of total carotenes) and β-carotene (50–60%), with low levels of several other carotenes. Carotenes are also present in palm leaves and in the pressed fibre that remains when oil has been expressed from palm fruits. This fibre still contains 5–6% of oil that is very rich in carotenes (4000–6000 ppm). When palm oil is refined, bleached and deodorised in the normal way, the carotenes are completely destroyed. Carotenes are a biological source of vitamin A, act as powerful antioxidants against both autoxidation and photooxygenation (Section 1.5.2) and show anticancer activity. Attempts have therefore been made to retain these valuable materials in refined palm oil or to recover them in concentrated form.

Products such as red palm oil and Nutrolein™ are palm oils or palm oleins that retain most of the original carotene obtained by carrying out deodorisation at temperatures below 150 °C. Carotenes can be recovered from palm methyl esters, prepared by methanolysis of palm oil and produced in large quantities for biodiesel and other purposes. This is achieved by

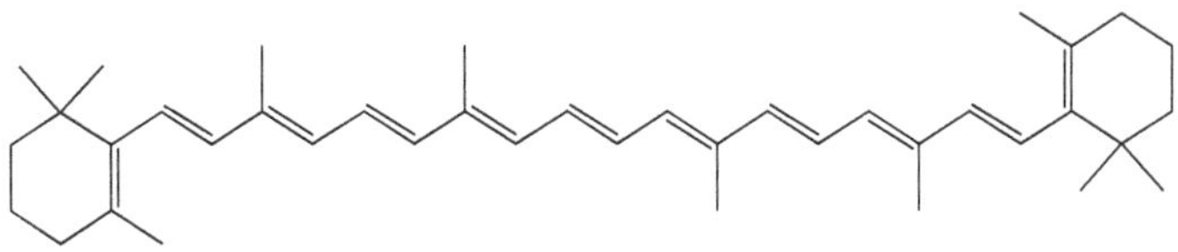

Figure 1.6 β-carotene ($C_{40}H_{56}$). Other carotenes vary in the nature of the cyclic end groups.

chromatography in an open column or by molecular distillation. The latter option gives a carotene concentrate (8%) that can be further purified (>90%) by chromatography (Baharin *et al.*, 1998; Ooi *et al.*, 1994; Yanishlieva *et al.*, 1998). The various methods for obtaining carotene from palm oil have been reviewed by Thyrion (1999).

Muller (1995) has reported the daily intake of individual carotenes and Yanishlieva and coworkers (1998) have reviewed the role of β-carotene as an antioxidant. Stanley (1999) has described some of the conflicting results concerning the biopotency of carotene supplied as a concentrate rather than as part of a food.

1.2.6.4 Polycyclic aromatic hydrocarbons

PAHs are present at levels up to about 150 μg/kg (ppb) in a number of crude vegetable oils, but less after refining (<80 ppb). They are removed to a small extent during bleaching and somewhat more during deodorisation. This is particularly the case for the more volatile tri- and tetracyclic compounds. The pentacyclic and other less volatile compounds are best removed with activated charcoal, which can be added to earth during bleaching. These values do not hold for crude coconut oil when the copra is dried with combustion gases, where values around 3000 ppb are normally recorded. Normal values are obtained after charcoal treatment (Larsson *et al.*, 1987). In Finland, Hopia and coworkers (1986) examined margarines, butters and vegetable oils for their levels of 38 different PAHs. Apart from a sample of crude coconut oil (4600 ppb), they gave values between 1 and 90 ppb. These compounds probably result from PAHs present in the atmosphere as a result of human-induced combustion of gas, coal or oil. Gertz & Kogelheide (1994) reported on PAHs in 40 native and refined vegetable oils. Extracted oils may contain pesticides resulting from agricultural processes, but these are usually removed during deodorisation.

Gossypol is a toxic hexaphenolic C_{40} compound present in cotton boll cavities. When the seed is extracted, the gossypol adheres to the protein meal and only a small proportion remains in the crude oil. Residual gossypol gives a red-brown colour to crude cottonseed oil but is largely removed during chemical refining and is present only at safe levels of 1–5 ppm in the final product. Kenar (2006) has reviewed the reaction chemistry of gossypol and its derivatives.

1.2.6.5 Contaminants and specifications

A typical specification includes the following impurities and limits for refined oils based on customer requirements, industry standards and EU legislation: taste and colour (bland), moisture (max. 0.05%), phosphorus (max. 5 ppm), insolubles (not visible), free fatty acids (max. 0.1%), peroxides (max. 1 meq/kg), iron (max. 0.5 ppm), copper (0.05 ppm), lead (max. 0.01 ppm), hexane (max. 5 ppm), benz(a)pyrene (max. 2 ppb), pesticides (maximum residue level in seeds, limit of detection in oils), dioxins (0.75 pg), aflatoxins (2 ppb aflatoxin B1, 4 ppb aflatoxin B1, B2, G1, G2), mineral oil (LOD) and residues of previous cargoes (complete removal).

1.3 Fatty acid composition

The food uses of lipids depend on their physical, chemical and nutritional properties, which are linked to their fatty acid and triacylglycerol composition. The latter is important but can be quite complex and for most practical purposes lipids are discussed in terms of their fatty acid composition. Typical values for the fatty acid composition of a range of oils and fats are presented in Table 1.10. These will not be considered in detail, but a few general points will be made. Figures cited in these tables must be considered merely as typical values. Debruyne (2007), Wilkes (2008) and Watkins (2009) have described some of the new varieties being investigated.

Coconut and palm kernel oils (Table 1.11) are typical lauric oils and differ from most of the other vegetable oils. They are important in both the food and the oleochemical industries and are characterised by high levels of lauric acid (12:0), significant levels of myristic acid (14:0) and useful levels of octanoic (8:0) and decanoic acids (10:0). The lauric oils are rich in saturated acids (80–90%) and contain very little unsaturated acid. Palm kernel oil is one of two products from the oil palm and must not be confused with the very different palm oil, which is the major product from this tree.

Most vegetable oils contain palmitic, oleic and linoleic (Table 1.10). The writer has calculated that the world's commodity oils in 2004/05 – both vegetable and animal fats – contained 83% of these three acids (Gunstone, 2005). Calculations were based on the fatty acid composition of each oil and on the level of production in that year. Palmitic as the major saturated acid reaches significant levels in palm oil (46%) and in cottonseed oil (27%). Some oils are rich in oleic acid (olive, canola), some in linoleic acid (corn, cottonseed, linola, soybean and sunflower) and some in both acids (groundnut). Seed breeders have produced oleic-rich varieties of many of

Table 1.10 Typical fatty acid compositions (%wt) of selected oils and fats.

	14:0	16:0	16:1	18:0	18:1	18:2	18:3
Cocoa butter	–	26	–	34	35	–	–
Corn	–	13	–	3	31	52	1
Cottonseed	–	27	–	2	18	51	tr
Groundnut	–	13	–	3	38	41	tr
Linola	–	6	–	3	16	72	2
Linseed	–	6	–	3	17	14	60
Olive	–	10	–	2	78	7	1
Palm	–	46	–	4	40	10	tr
Palm olein	–	40	–	4	43	11	tr
Rape[a]	–	3	–	1	16	14	10
Rape[b]	–	4	–	2	56	26	10
Soybean	–	11	–	4	22	53	8
Sunflower	–	6	–	5	20	60	tr
Sunola[c]	–	4	–	5	81	8	tr
Nusun	–	4	–	5	65	26	–
Butter[d]	12	26	3	11	28	2	1
Lard	2	27	4	11	44	11	–
Beef tallow	3	27	11	7	48	2	–
Mutton tallow	6	27	2	32	31	2	–

[a]High erucic (also 20 : 1 6% and 22 : 1 55%).
[b]Low erucic.
[c]High oleic sunflower.
[d]Also 4 : 0 (3%), 6 : 0 (2%), 8 : 0 (1%), 10 : 0 (3%) and 12 : 0 (4%).
tr, trace (<1%).

Table 1.11 Typical fatty acid compositions (%wt) of lauric oils.

	8:0	10:0	12:0	14:0	16:0	18:0	18:1	18:2
Coconut	8	7	48	16	9	2	7	2
Palm kernel	3	4	45	18	9	2	15	3

these oils. For example, commodity sunflower oil normally contains about 20% oleic acid and 60% linoleic, but two other varieties are now commercially available with higher levels of oleic acid and lower levels of linoleic acid. NuSun contains about 65% oleic acid and high-oleic sunflower is at least 80% oleic acid. These have been produced by conventional seed breeding procedures and are not genetically modified products (Anon, 1998; Watkins, 2009; Wilkes, 2008).

Linolenic acid (18 : 3) is the major acid in linseed oil (60%) and is the basis for most of the industrial uses of this oil. 'Linola' is the name given to a chemically induced mutant with low levels of linolenic acid and high levels of linoleic acid (Table 1.10). Linolenic acid is also present in soybean oil (8%) and in rapeseed oil (10%). There is some ambivalence about this acid. Its presence promotes undesirable oxidation and foods containing it have reduced shelf life. This problem has been overcome traditionally by a very light hydrogenation (brush hydrogenation), which halves the level of linolenic acid. More recently, new varieties of these oils have been developed with lower levels of linolenic acid – some of them by genetic modification.

Cocoa butter, the lipid component in chocolate, is an unusual vegetable fat with saturated (~60%) and monoene (~35%) acids in such proportions that its triacylglycerols are mainly of the SOS type (S = saturated, O = oleic). These are responsible for the characteristic melting behaviour of this fat, which is so important in chocolate (Timms, 2003). Other vegetable fats with similar compositions and similar melting characteristics are designated cocoa butter equivalents (CBEs).

These comments hold for the major vegetable oils and fats and also for most of the minor seed oils, but some other oilseeds illustrate the rich diversity of plants in their ability to generate unusual fatty acids, sometimes at very high levels. Examples include castor oil (90% ricinoleic acid – 12-hydroxyoleic acid), coriander oil (80% petroselinic acid – 6-octadecenoic acid), *Vernonia galamensis* seed oil (75% vernolic acid – *cis*-12,13-epoxyoleic acid) and the seed oil of *Picramnia sow* (95% tariric acid – 6-octadecynoic acid).

The major animal fats are more saturated than vegetable oils and contain only low levels of PUFAs. They generally consist of 40–60% saturated acids and 30–60% monounsaturated acids. Butter has acids with a wide range of chain lengths (4–18 carbon atoms) but, like the animal depot fats, it is rich in saturated and monoene acids and low in polyunsaturates. Because of the large number of fatty acids in milk fat, differing in chain length and unsaturation, the triacylglycerol composition is much more complex than that of most vegetable oils. This makes fractionation of anhydrous milk fat (AMF), based on only slightly different properties among the many triacylglycerols, very difficult. Some indication of triacylglycerol complexity was given in a paper by Robinson & MacGibbon (1998). Using silver ion thin-layer chromatography (TLC) and reversed-phase high-performance liquid chromatography (RP-HPLC) they isolated 61 fractions, each of which contained one to four major triacylglycerol components. Some of the difficulties of fractionation were discussed by Bhaskar and coworkers (1998) in a paper comparing the physical and chemical properties of milk fat fractions obtained by commercial melt crystallisation and supercritical carbon dioxide extraction.

Fish oils are characterised by the wide range of acids present and, particularly, by the highly unsaturated members. Saturated (14:0 and 16:0), monoenoic (16:1, 18:1, 20:1 and 22:1) and omega-3 polyenoic acids (eicosapentaenoic acid, 20:5 and docosahexaenoic acid, 22:6) are frequently major components, and fish oils are valued for the latter.

1.4 Physical properties

1.4.1 Polymorphism, crystal structure and melting point

Important physical properties relevant to this book are polymorphism, crystal structure and melting point, which combine in the melting behaviour of lipid mixtures.

In the solid state, long-chain compounds frequently exist in more than one crystalline form and may consequently have more than one melting point. This property of polymorphism is of both scientific and technical interest. Understanding this phenomenon is essential for the satisfactory blending and tempering of fat-containing materials, such as baking and confectionery fats, which must attain a particular physical appearance during preparation and maintain it during storage. Problems of graininess in spreads and of bloom in chocolate, for example, are both related to polymorphic changes. The experimental methods used most extensively to examine melting and crystallisation involve dilatometry, low-resolution pulsed ^{1}H NMR spectroscopy, differential scanning calorimetry (DSC), infrared spectroscopy and X-ray diffraction (Larsson *et al.*, 2006; Timms, 2003).

Alkanoic acids exist in three polymorphic forms, designated A, B and C for acids with an even number of carbon atoms. Form C has the highest melting point and is physically the most stable. It is obtained by crystallisation from the melt or from polar solvents. Crystallisation from nonpolar solvents gives form A or forms B and C.

For the purpose of this book, the melting point of triacylglycerols is more important. It has long been known that fats show multiple melting points. As far back as 1853, glycerol tristearate was known to have three melting points (52, 64 and 70 °C). When the melt of a simple triacylglycerol is cooled quickly, it solidifies in the form with lowest melting point (α) with perpendicular alkyl chains in its unit cell (angle of tilt is 90°). When heated slowly, this melts, and held just above the melting point, it will resolidify in the β' crystalline form. In the same way, a more stable β form can be obtained from the β' form. The β form has the highest melting point and is obtained directly from solvent by crystallisation. The β' and β forms have tilted alkyl chains, which permit more efficient packing of the triacylglycerol molecule in the crystal lattice. Glycerol esters with only one type of acyl chain have been thoroughly studied. The results have provided useful guidance, but such molecules are not generally significant components of natural fats (except perhaps after complete hydrogenation). With mixed saturated triacylglycerols such as PStP (P = palmitic, St = stearic), the β form is only obtained with difficulty, and such compounds usually exist in their β' form. Among unsaturated triacylglycerols, symmetrical compounds (SUS and USU, where S = saturated and U = unsaturated acyl chains) have higher melting β forms (more stable) but the unsymmetrical compounds (USS and UUS) have stable β′ forms.

Crystallisation occurs in two stages: nucleation and growth. A crystal nucleus is the smallest crystal that can exist in solution and is dependent on concentration and temperature. Spontaneous (homogeneous) nucleation rarely occurs in fats. Instead, heterogeneous nucleation occurs on solid particles (dust etc.) or on the walls of the container. Once crystals are formed, fragments may drop off and either redissolve or form nuclei for further crystals. The latter is not desirable in fat crystallisation, so agitation during fractionation should be kept to the minimum required to facilitate heat transfer (see Chapter 6).

Nucleation rates for the different polymorphs are in the order $\alpha > \beta' > \beta$ so that α and β′ are more readily formed in the first instance, even though the β polymorph is the most stable and favoured thermodynamically. Crystal nuclei grow by incorporation of other molecules from the adjacent liquid layer, at a rate depending on the amount of supercooling and the viscosity of the melt (Gibon, 2006; Lawler & Dimock, 2002; Mori, 1988; Timms, 2005).

In the production of spreads and shortenings, the β′ crystalline form is preferred over the β form. β′ crystals are relatively small and can incorporate a large amount of liquid. This gives the product a glossy surface and a smooth texture. β crystals, on the other hand, though initially small, grow into needle-like agglomerates. These are less able to incorporate liquid and produce a grainy texture. Spreads and shortenings made from rape/canola, sunflower or partially hydrogenated soybean oil generally develop crystals. This can be inhibited or prevented by incorporation of some palm oil or palm olein, which stabilises the crystals in the β′ form. These changes in crystallisation pattern are linked with the larger amount of palmitic acid in the palm products. Glycerol esters with C_{16} and C_{18} acyl chains are more likely to be stable in the β′ form than glycerol esters with three C_{18} chains.

Cocoa butter is particularly rich in three 2-oleo-1,3-disaturated glycerol esters, namely POP, POSt and StOSt. The solid fat has been identified in six crystalline forms, designated I–VI (the melting points and the nature of the double/triple chain lengths are indicated in Table 1.12). Of these, form V (β_2) is preferred for chocolate. This crystalline form gives good demoulding characteristics and has a stable gloss and a favourable snap at room temperature. Two procedures have been employed to promote the formation of this particular crystalline form. The most widely used is tempering; that is, putting molten chocolate through a series of cooling and heating processes. This optimises the production of the appropriate polymorph. An alternative procedure requires seeding of the molten chocolate with cocoa butter already prepared in form V (β_2) or VI (β_1), but this method is restricted by the difficulty of obtaining adequate supplies of these crystalline forms. The synthetic glycerol ester, 2-oleo-1.3-dibehenin (BOB, O = 18 : 1, B = 22 : 0), may

Table 1.12 Polymorphism in cocoa butter.

	I	II	III	IV	V	VI
MP (°C)	17.3	23.3	25.5	27.3	33.8	36.3
Chain length	D	D	D	D	T	T

D, double chain length; T, triple chain length; MP, melting point.

be added to cocoa butter to prevent bloom formation by keeping it in its form V at temperatures above 30 °C (Longchampt & Hartel, 2004; Norberg, 2006; Timms, 2003; other relevant references are Gibon, 2006; Martini *et al.*, 2006; and Smith, 2009).

Oils rich in saturated acids may contain high-melting triacylglycerols that crystallise from the oil when stored. When this is considered to be undesirable, the oil is subjected to winterisation (see Chapter 6). This process is applied to cottonseed oil and to partially hydrogenated soybean oil.

1.4.2 Density

Density is very important in the oil trade since fatty oil shipments are sold on a weight basis but measured on a volume basis. Since these two values are related by density, it is important to have correct and agreed values for this unit. Density is not the same for all oils but depends on fatty acid composition and on minor components, as well as on temperature. An equation taking these variables into account is based on iodine value (IV), saponification value (SV) and temperature (Pantzaris, 1985):

$$d = 0.8543 + 0.000308\,(SV) + 0.000157\,(IV) - 0.000681t \qquad (1.1)$$

where d is apparent density (g/ml or kg/l) and t is temperature.

Density can be defined in various ways and the correct form must be used when relating volume to weight:

- Density (absolute density or density in vacuum) is the 'mass in vacuum of a volume of oil at t °C ÷ volume of the oil at the same temperature', expressed in g/ml or kg/l.
- Apparent density (density in air, weight-by-volume or litre-mass) is the 'mass in air of a volume of oil at t °C ÷ volume of the oil at the same temperature', expressed in g/ml or kg/l.
- Relative density (specific gravity, density in relation to water) is the 'mass in air of a given volume of oil at t_1 °C ÷ mass in air of same volume of water at t_2 °C'. This is a ratio without units. It is important to note that two temperatures are involved and the value is meaningless unless both

figures are cited. Relative density is the value most commonly employed and equations exist to connect these three expressions.

Halvorsen and coworkers (1993) have described a method for estimating the density of fatty acids and vegetable oils based on critical volume, critical temperature, critical pressure and a modified Racket equation. Some data have been published by Coupland & McClements (1997), and Topallar and coworkers (1995) have reported the effect of hydrogenation on the density and viscosity of sunflower seed oil.

1.4.3 Viscosity

Viscosity can be reported as kinematic viscosity or dynamic viscosity, with the two values related through density. The viscosity of a vegetable oil depends on its chemical composition (summarised in its IV and SV) and the temperature of measurement. Equations have been derived which permit the calculation of viscosity from a knowledge of these three parameters. They have been developed empirically from observation of a range of oils at different temperatures and have been reported by Duff & Prasad (1989), Toro-Vazquez & Infante-Guerrero (1993), Rabelo *et al.* (2000), Azian *et al.* (2001) and Fasina *et al.* (2006). Coupland & McClements (1997) and Fisher (1998) have related viscosity with density, refraction, surface tension and other physical properties. The relation between temperature and viscosity has been described for coconut oil, palm kernel oil, palm oil and mixtures (Timms, 1985), and for several vegetable oils (Noureddini *et al.*, 1992). Changes in viscosity have been used to monitor interesterification (De Filippis *et al.*, 1995) and hydrogenation (Topallar *et al.*, 1995).

1.4.4 Refractive index

The refractive index is easily measured using small amounts of material. The refractive index increases with chain length (though not in a linear fashion) and with increasing unsaturation. Geometric isomers differ from one another and methylene-interrupted polyenes differ from those with conjugated unsaturation. Triacylglycerols have higher values than free acids. Values for commercial oils are given in Table 1.13.

1.4.5 Solubility of gases in oils

A recent discussion (Hilder, 1997) of the solubility of gases in vegetable oils included the data for oxygen, nitrogen and air presented in Tables 1.14 and 1.15. When an oil is in contact with air, the dissolved gases will depend on

Table 1.13 Physicochemical properties of selected commodity oils and fats.

	Specific gravity (temperature °C)	Refractive index (40 °C)	Refractive index (25 °C)	Iodine value	Saponification value	Titre (°C)	Unsaponifiable (%)	Mp (°C)
Cocoa butter	0.973–0.980 (25/25)	1.456–1.458	–	32–40	192–200	45–50	0.2–1.0	31–35
Coconut	0.908–0.921 (40/20)	1.448–1.450	–	6–11	248–265	–	<1.5	23–26
Corn	0.917–0.925 (20/20)	1.465–1.468	1.470–1.473	107–128	187–195	–	1–3	–
Cottonseed	0.918–0.926 (20/20)	1.458–1.466	–	100–115	189–198	–	<2	–
Linseed	0.930–0.936 (15.5/15.5)[a]	1.472–1.475	1.477–1.482	170–203	188–196	19–21	0.1–2.0	–
Olive	0.910–0.916 (20/20)	–	1.468–1.471	75–94	184–196	–	1.5	−3–0
Palm kernel	0.899–0.914 (40/20)	1.452–1.488	–	14–21	230–254	–	<1.1	24–26
Palm	0.891–0.899 (50/20)	1.449–1.455[b]	–	50–55	190–209	–	<1.4	33–40
Palm olein	0.899–0.920 (40/20)	1.459–1.459	–	>55	194–202	–	<1.4	–
Palm stearin	0.881–0.891 (60/20)	1.447–1.451	–	<49	193–205	–	<1.0	–
Peanut	0.914–0.917 (20/20)	1.460–1.465	–	86–107	187–196	–	<1.1	–
Rape[c]	0.910–0.920 (20/20)	1.465–1.469	–	94–120	168–181	–	<0.21[d]	–
Rape[e]	0.914–0.920 (20/20)	1.465–1.467	–	110–126	182–193	–	<0.21[d]	–
Sesame	0.915–0.923 (20/20)	1.465–1.469	–	104–120	187–195	–	<2.1	–
Soybean	0.919–0.925 (20/20)	1.466–1.470	–	124–139	189–195	–	<1.6	–
Sunflower	0.918–0.923 (20/20)	1.467–1.469	1.472–1.476	118–145	188–194	–	<1.6 (max. 2.0)	–
Sunflower[f]	0.915–0.920 (20/20)	–	1.467–1.469	75–90			0.8–1.0 (max. 2.0)	–

[a] Also 0.924–0.930 (25/25).
[b] 50 °C.
[c] High-erucic rapeseed oil.
[d] These values are correctly copied from the source but they are in error. Better values are 0.5–1.2%.
[e] Low-erucic rapeseed oil.
[f] High-oleic sunflower seed oil.
Source: Adapted from Firestone (1999).

Table 1.14 Solubility of gases in oils.

Temp. (°C)	Oxygen (ppm, 1 bar)	Nitrogen (ppm, 1 bar)
0	170	80
25	180	85
50	165	90
75	190	95
100	200	105
125	[a]	110
150	[a]	115

[a]Oxygen solubilities at higher temperatures are not reliable because oxidation occurs.
Source: Adapted from Hilder (1997).

Table 1.15 Gas content of oil saturated with air.

	Solubility (ppm)	Air dissolved in oil (ppm)
Oxygen	180	38
Nitrogen	85	66
Argon	270	S

Source: Adapted from Hilder (1997).

their individual solubility as well as on their concentration in air. The high solubility of monatomic argon enhances its concentration, so that 1% in air becomes 3% in oil.

Koetsier (1997) has summarised data on the solubility of hydrogen in vegetable oil. This information is important for hydrogenation. He cites solubility values (maximum concentration in oil at a given temperature and pressure) from two sources at 1 bar and 100–200 °C of 2.60–3.36 and 2.76–3.40 mol/m^3. The concentration of hydrogen is therefore much lower than the concentration of unsaturated centres; for a fish oil hydrogenated at 5 bar and 180 °C, Koetsier gives concentrations of ~7000 and 16 mol/m^3 for the olefinic groups and the hydrogen, respectively.

1.4.6 Other physical properties

Gross heats of combustion (HGs) for saturated and unsaturated triacylglycerols can be related to the number of valence electrons (ENs). Freedman & Bagby (1989) have given equations for saturated (Equation 1.2) and unsaturated (Equation 1.3) triacylglycerols, while Krisnangkura (1991) has expressed this relationship in terms of SV and IV (Equation 1.4).

$$HG = -109.20 + 26.38\ EN \tag{1.2}$$

$$HG = 115.87 + 25.88\ EN \tag{1.3}$$

$$HG = 1\,896\,000/SV - 0.6\ IV - 1600 \tag{1.4}$$

In a useful paper, Coupland & McClements (1997) reported several physical properties (density, viscosity, adiabatic expansion coefficient, thermal conductivity, specific heat, ultrasonic velocity and ultrasonic attenuation coefficient) for a number of liquid oils (coconut, corn, cottonseed, crambe, grapeseed, groundnut, olive, palm, palm-olein, palm kernel, rape, rice bran, safflower, sesame, soybean and sunflower). Timms (1978) reviewed and significantly extended information on the heats of fusion of glycerol esters. He derived an equation for the heat of fusion of mono-acid triacylglycerols in the β polymorph form and showed how this could be adapted to calculate the heat of fusion of most glycerol esters of commercial interest. Chumpitaz *et al.* (1999) have recently reported the surface tensions of four fatty acids (lauric, myristic, palmitic and oleic) and two triacylglycerols (tricaprylin and tripalmitin) over a range of temperatures. These data are important for processes involving gas–liquid contact, such as distillation and stripping columns, deodorisers, reactors and equipment for physical refining. Fisher (2000) has presented equations correlating several properties of *n*-fatty acids and derivatives with chain length.

Some useful data in this section (Table 1.13) have been taken from the AOCS publication *Physical and Chemical Characteristics of Oils, Fats, and Waxes* (Firestone, 1999).

1.5 Chemical properties

1.5.1 Hydrogenation

Hydrogenation and, more importantly, partial hydrogenation of some of the unsaturated centres in a liquid oil to convert it into a solid or semisolid fat is an important procedure in making them usable as spreads. However, with current concern about *trans* acids, this process has become less useful. The topic is discussed in Chapter 6 and will not be pursued further in this chapter.

1.5.2 Oxidation

Part of the refining process (see Chapter 5) involves removal of oxidation products, with their undesirable flavour and aroma, after which further oxidation must be inhibited as efficiently as possible during processing of oils and fats, food processing and storage up to the moment of consumption. The word 'inhibited' is used because it is virtually impossible to prevent oxidation. It is therefore important to understand this reaction in order that the lipid is always handled under appropriate conditions. It is only possible to give a brief account of this topic in the present volume. The best, fullest and most relevant accounts are to be found in two recent books by Frankel (2005, 2007); the topic is also fully reviewed in *Food Lipids*, edited by Akoh & Min (2008).

Non-enzymatic oxidation occurs by two routes, and it is necessary to protect against both. Lipid oxidation is accelerated by metals, light, heat and several initiators (prooxidants) and can be inhibited by avoiding prooxidants and including antioxidants. The primary products are allylic hydroperoxides. Double bonds remain, though they may have changed configurations and position in the fatty acid chain. These compounds are not directly responsible for the undesirable flavour and aroma associated with rancid fat, but they are unstable molecules which readily undergo a series of secondary reactions, including the formation of short-chain aldehydes:

$$RCH = CHCH_2CHR' \rightarrow RCH(OOH)CH = CHR' \rightarrow RCHO \qquad (1.5)$$

and other compounds.

1.5.3 Autoxidation

This is a radical chain reaction; that is, the intermediates are radicals (odd electron species) and the reaction involves an initiation step, a propagation sequence and one or more termination steps:

Initiation	$RH \rightarrow R\bullet$ resonance-stabilised alkyl radical
Propagation	$R\bullet + O_2 \rightarrow RO_2\bullet$ fast reaction to give a peroxy radical
	$RO_2\bullet + RH \rightarrow RO_2H + R\bullet$ rate-determining step
Termination	$RO_2\bullet + RO_2\bullet \rightarrow$ stable products
	$RO_2\bullet + R\bullet \rightarrow$ stable products
	$R\bullet + R\bullet \rightarrow$ stable products

where RH represents an olefinic compound in which H is attached to an allylic carbon atom and RO_2H is a hydroperoxide.

There is usually an induction period, during which oxidation occurs only slowly, followed by a more rapid reaction. It is desirable to extend the induction period (and hence the shelf life of the product) as long as possible.

The detailed nature of the initiation step is not fully understood, but any or all of three reactions may be involved: (1) metal-catalysed decomposition of existing hydroperoxides produces initiating radicals (it is very difficult to obtain olefinic compounds entirely free of oxidation products); (2) photooxygenation (a very rapid reaction – see below) may be responsible for the first-formed hydroperoxides; and (3) thermal initiation is possible in a heated sample. In the propagation sequence, given an adequate supply of oxygen, the reaction between alkyl radical (R•) and molecular oxygen is fast, and the subsequent reaction of peroxy radical (ROO•) with another olefinic molecule is rate-determining. Autoxidation can be inhibited by minimising the initiation step and/or promoting a termination step so that the propagation cycle goes through as few cycles as possible. The methods of achieving these ends are discussed later.

There is some evidence that PUFAs are more stable to oxidation when located in the *sn*-2 position of triacylglycerols than when in the *sn*-1(3) (Wijesundera *et al.*, 2008).

1.5.4 Photooxidation

Photooxidation mainly involves interaction between a double bond and a highly activated singlet oxygen molecule produced from ordinary triplet oxygen. Energy from light is transferred to oxygen via a sensitiser such as chlorophyll, erythrosine, rose bengal or methylene blue. This reactive oxygen species (ROS) reacts with olefins to give allylic hydroperoxides. Photooxidation differs from autoxidation in several important respects:

- It involves reaction with singlet oxygen produced from triplet oxygen by light and a sensitiser.
- It is an ene reaction and not a radical chain process.
- It displays no induction period.
- It is unaffected by many of the antioxidants used to inhibit autoxidation but is inhibited by singlet oxygen quenchers such as carotene.
- Oxygen addition is confined to olefinic carbon atoms but the double bond moves and changes configuration from *cis* to *trans*.
- It gives allylic hydroperoxides that are similar in type but not identical in composition to those obtained by autoxidation.
- It is a quicker reaction than autoxidation, especially for monounsaturated acids. The rate is related to the number of olefinic centres and not to the number of doubly allylic allylic groups (photooxidation of oleate is ~30 000 times quicker than autoxidation).

1.5.5 Decomposition of hydroperoxides to short-chain compounds

Hydroperoxides are unstable compounds which readily undergo further change, giving, among other products, a range of short-chain compounds. The volatile compounds include aldehydes, ketones, alcohols, hydrocarbons, acids, esters, lactones and ethers, of which the aldehydes are of most concern for odour and flavour. They are produced from the hydroperoxides mainly by homolytic fission and also, in a minor way, by heterolytic breakdown. Each hydroperoxide (there are many) can produce two aldehydes, of which the short-chain volatile member is the more significant. In a glycerol ester, the other aldehyde remains as a glycerol derivative – sometimes called the core aldehyde – and may not be removed during refining.

Most of the short-chain aldehydes have a very low threshold value, so they need only be present at minute levels in order to exert their olfactory effect. For example, the 9-hydroperoxide from linoleate gives 2,4-decadienal with a deep-fried flavour at a concentration equivalent to 0.5 ppb.

1.5.6 Antioxidants

The antioxidants which can be added to fats and to fat-containing foods are rigorously controlled. Only permitted substances can be used, and then only below agreed maximum levels. The matter is further complicated by the fact that not all of these substances are universally accepted. For example, tertiary-butyl hydroquinone is allowed in the USA but not in the EU. Antioxidants permitted in Europe have E numbers (the European Community designation for permitted food additives) assigned to them. Antioxidants can be classified according to their mode of action and, in addition, can be described as natural or synthetic. There is an increasing demand for the former, even though the latter are cheaper and there are not enough natural antioxidants to meet total demand (Section 1.2.4). Much of the large processed-food industry would be impossible to run without antioxidants of some kind. They are essential to inhibiting the development of rancidity and thereby extending shelf life.

Important synthetic antioxidants include butylated hydroxyanisole (BHA), butylated hydroxytoluene (BHT), propyl gallate (PG) and tertiarybutyl hydroquinone (TBHQ) (Figure 1.7). Natural antioxidants include vitamin E (tocopherols), ascorbyl palmitate, β-carotene and compounds present in a range of spices and herbs.

1.5.6.1 Primary antioxidants

Primary antioxidants promote the termination process (Figure 1.7) and thereby shorten the propagation sequence. They are mainly phenols or

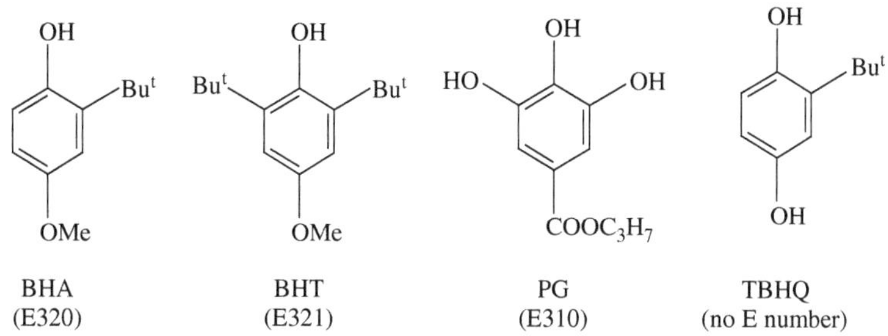

Figure 1.7 The structures and E numbers of synthetic antioxidants. TBHQ has no E number because it is not a permitted antioxidant in the EU.

amines, though the latter are not accepted as food antioxidants. Some are compounds with extensive conjugated unsaturation, such as carotene. The intermediate radicals (A• or ROOB•) are stabilised by extensive delocalisation of the odd electron and do not support the propagation sequence. They are usually converted to dimers or to substitution products and it is important that, in addition to the antioxidants themselves, their oxidation products are also acceptable in food products.

$$ROO\bullet + AH \rightarrow ROOH + A\bullet - - - \rightarrow \text{products} \quad (1.6)$$

$$ROO + B \rightarrow ROOB\bullet - - - \rightarrow \text{products} \quad (1.7)$$

where AH = amines or phenols and B = β-carotene and so on.

β-carotene and similar substances containing extensive conjugated unsaturation inhibit photooxygenation due to their ability to quench singlet oxygen, but they also inhibit autoxidation through their ability to react with and remove peroxy radicals. When this happens, the odd electron is delocalised over the conjugated polyene system. Under other conditions, β-carotene can act as a prooxidant.

Antioxidants act in a sacrificial manner and the induction period ends when they have been expended. However, some compounds (Section 1.5.6.2) are able to regenerate spent antioxidants and to extend their useful life. Moreover, some antioxidants react twice with peroxy radicals, and sometimes the oxidised antioxidants themselves have further antioxidant activity.

Natural phenolic antioxidants are present in a wide range of plant sources, such as rosemary, thyme, sage, myrtle, tea and oats. Sesame and rice bran oils are very rich in antioxidants and may be added to other oils to stabilise them.

1.5.6.2 Secondary antioxidants

Secondary antioxidants operate by inhibiting the initiation process, acting mainly through chelation of the metal ions that promote initiation, particularly copper and iron. The concentration of metals required to reduce the keeping time of lard at 98 °C is 0.05 ppm for copper and 0.6 ppm for iron. In addition to avoiding these metals in equipment used to handle oils and fats, it is common to add a metal chelator such as ethylenediamine tetra-acetic acid (EDTA), citric acid, phosphoric acid or certain amino acids. These are often added along with chain-breaking primary antioxidants. Some phospholipids are also able to chelate metals.

Citric acid can be used in refining processes in various ways: to assist degumming, in the bleaching step, to convert soaps to the more easily removed free acids and, of greatest importance, to act as a metal chelating agent. It may also be added during storage of crude oils to inhibit oxidation. Citric acid decomposes rapidly above 150 °C and so should be added after deodorisation, even if it has been used earlier in the refining process. It is usually added at the rate of 50–100 ppm in the form of a 30–50% aqueous solution (Law & Berger, 1984).

Other compounds can also be used to enhance antioxidant activity. Vitamin C, for example, is useful because it reacts with spent tocopherols (vitamin E), causing regeneration. However, vitamin C is a water-soluble compound with low lipid solubility and is more commonly employed as ascorbyl palmitate, which is more lipophilic. Ascorbyl palmitate is also reported to act as an oxygen scavenger. Phospholipids promote antioxidant activity through chelation of metal ions and/or by acting as an emulsifying agent, bringing antioxidant and fat together.

These materials are concerned with the inhibition of autoxidation and have no effect on photooxygenation, which proceeds along a different reaction pathway. This process is inhibited by singlet oxygen quenchers, of which the best known are the carotenes. These may be present in natural fats, and if not can be added at levels of 5–10 ppm.

It is important to recognise that antioxidants do not prevent oxidation. They serve only to extend the induction period during which oxidation is very slow and of no great consequence. It follows that appropriate antioxidants should be added before oxidation has started. No amount of antioxidant can regenerate a fat that is already oxidised. The best antioxidant mixtures combine primary and secondary antioxidants and an emulsifying agent. Conditions which promote oxidation must be scrupulously avoided during all handling and storage. This involves avoiding unnecessarily elevated temperatures, unnecessary contact with air (by nitrogen-blanketing when possible, avoiding splashing, which increases access to air, and avoiding half-full containers) and exposure to light. Storage should always be under the best possible conditions and exposure to iron and copper in storage vessels and pipelines should be avoided (see Chapter 9). Changes which can occur during storage have been discussed by Patterson (1989).

1.5.7 Stereomutation

Natural unsaturated acids are almost entirely *cis* isomers. These acids can also exist in the *trans* form, which is thermodynamically more stable and is therefore the dominant form in an equilibrium mixture of the two. For example, oleic (*cis*) and elaidic (*trans*) acids form a 1 : 4 equilibrium mixture. The *trans* isomers are usually higher-melting and have different nutritional properties to the *cis* compounds. The change of configuration from *cis* to *trans* is described as stereomutation. It can be promoted by chemical reagents (not discussed here) or by exposure to high temperatures during processing (see Section 1.6.2).

1.5.8 Double-bond migration and cyclisation

Double-bond migration (accompanied by stereomutation, particularly in polyene acids) is promoted by acidic and basic reagents, but the conditions required are generally vigorous and migration does not present a serious problem during processing. Double-bond migration occurs during partial hydrogenation (see Chapter 5). At higher temperatures, the migration process may continue to give cyclised products. Monocyclic derivatives may contain a five- or six-membered carbocyclic ring. Such compounds have been recognised in overheated frying oils (Dobson, 1998; Le Querre & Sebedio, 1996).

1.5.9 Hydrolysis

Fats can be hydrolysed to free acids by water, in what is probably a homogeneous reaction between fat and water dissolved in the fat phase. Loncin (1952), in a study covering the hydrolysis of various vegetable oils, suggested that the reaction is autocatalytic, accelerating once a certain level of free fatty acid has been reached. His report indicates the risk of hydrolysis occurring when oils are stored for extensive periods at temperatures above ambient. Crespo (1973) studied the hydrolysis of beef tallow and showed that an increase in partial glyceride content accompanies the formation of free acid during this process.

As a result of lipolysis, crude oils frequently contain some free fatty acid, which is removed at appropriate stages during refining (see Chapter 5). The presence, at low levels, of hydrolysis during deodorisation or physical refining has sometimes been given as the reason for the difficulty experienced in removing free fatty acids completely in this part of the refining process, but this has proved difficult to confirm.

Complete hydrolysis of fats is applied on a large scale for the production of fatty acids for the soap and oleochemical industries, using high pressures (20–60 bar) and a temperature of approximately 250 °C. This hydrolytic reaction can be carried out under milder conditions using biocatalysts (lipases), but such reactions have not yet achieved industrial status.

1.5.10 Ester formation

The formation of esters is important in lipid science and technology. Esters can be made by the catalysed reaction of fatty acids with an appropriate alcohol. They can also be produced from existing triacylglycerols (or other esters) by reaction with an alcohol, leading to an exchange of alcohol moieties (alcoholysis), with an acid, leading to an exchange of acyl functions (acidolysis), or with another ester, leading to randomisation of all the possible esters (interesterification). All these processes require a catalyst, which may be acidic, basic or an enzyme (lipase). The latter provides opportunities for specificity that are not possible with wholly chemical operations. Examples of all of these are significant as industrial procedures.

1.5.11 Methanolysis

Fatty acid methyl esters are important in gas chromatographic analysis (mg or less) and also on an industrial scale. The methyl esters rank among the basic oleochemicals and are used as solvents, as biodiesel and as intermediates in the preparation of fatty alcohols. Large-scale methanolysis involves appropriate oils and fats, an excess of methanol and preferably a basic catalyst. Many recipes have been reported. One paper suggests that at a molar ratio of 27 : 1 at 23 °C, methanol converts soybean and other oils into methyl esters in a yield >99% in only 7 minutes. This molar ratio represents about equal weights of the two substrates (Boocock *et al.*, 1998). Glycerol is also formed in this reaction, and it can be recovered as a secondary product and then employed as a 'platform chemical' to produce other valuable molecules, such as 1,2-propanediol, 1,3-propanediol, epichlorohydrin (2,3-epoxypropylchloride), acrolein (propenal), glycerol carbonate, polyglycerols and others (Kenar, 2007).

1.5.12 Glycerolysis

When a triacylglycerol is heated with glycerol and a basic catalyst such as sodium hydroxide or sodium methoxide, the following equilibrium is established:

$$\text{triacylglycerol} + \text{glycerol} \rightleftharpoons \text{monoacylglycerol} + \text{diacylglycerol} \qquad (1.8)$$

The composition of the equilibrium mixture depends on the amount of glycerol dissolved in the lipid phase. This is an important route to mixtures of monoacylglycerols and diacylglycerols, and 90–95% concentrates of the former are obtained by molecular distillation. Monoacylglycerols and some derivatives of these are important food emulsifiers (Krog, 1997).

1.5.13 Interesterification

The redistribution of acyl chains among glycerol molecules in an oil or mixture of oils is an important way of modifying physical properties and is of greater importance now that partial hydrogenation is less commonly employed for this purpose. This process is known as interesterification and requires either an alkaline or an enzymatic catalyst. Details can be found in Chapters 6 and 7.

1.6 Effect of processing on food oil components

Some changes take place in oils and fats during bleaching in the presence of an earth at 80–160 °C, but more extensive alterations are associated with the deodorisation process conducted at 200–260 °C. Ferrari and coworkers (1996) have charted the decline of sterols and tocols and the rise of *trans* acids and polymers during processing. In an early study of cottonseed oil (Gumuskesen & Cakaloz, 1992), it was reported that in order to keep changes to a minimum deodorisation should be carried out at temperatures not exceeding 220 °C for 3 hours at most. Changes in the levels of tocopherols and carotenes were discussed in Sections 1.2.4 and 1.2.5. There have been additional reports by Willner and coworkers (1997) and Schone and coworkers (1998).

In a study of 70 fat samples (vegetable fats, olive oils, animal fats and margarines), sterol dienes were observed at levels between 1 and 200 mg/kg. These are formed by dehydration of 3-hydroxy sterols (Figure 1.8). Since these compounds are probably absent from the native oils, their presence provides some evidence of the history of the oil. The detection of these compounds is indicative that certain refining processes have been carried out, and Grob and coworkers (Grob *et al.*, 1992; Grob & Bronz, 1994) showed that claims that many oils were cold-pressed and therefore nonrefined were false. Similar changes have been reported by Kochhar (1983), Schulte (1994) and Amelia and coworkers (1998).

Vegetable oils, particularly soybean, rape/canola and olive, are significant dietary sources of vitamin K_1. However, this compound is converted to

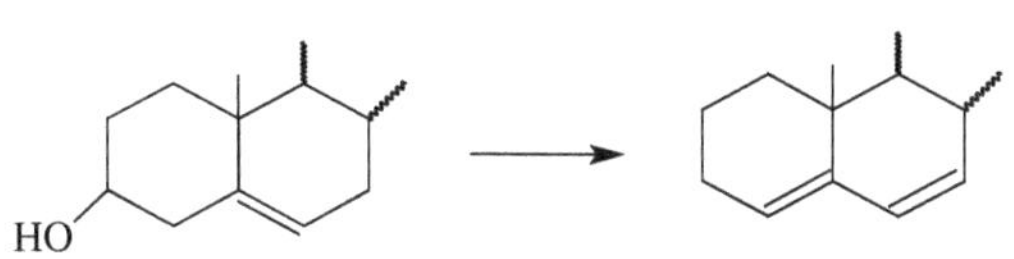

Figure 1.8 Formation of steradienes by thermal dehydration.

its 2′3′-dihydro derivative during partial hydrogenation of an oil and it is estimated that ~30% of total vitamin K_1 intake is in this dihydro form in the US diet. Further investigation of the biological activity of this artefact is clearly desirable (Booth *et al.*, 1996a, 1996b).

Heat-induced stereomutation of linolenic acid and its esters is very slow at 190 °C but quicker under the normal conditions for semicontinuous or continuous deodorisation at 210–270 °C. At these temperatures, about 35% of natural (all-*cis*) linolenic acid is converted to four of the seven possible *trans* isomers (9*c*12*c*15*t*, 48–50% of total *trans* isomers; 9*t*12*c*15*c*, 41%; 9*c*12*t*15*c*, 6–7%; and 9*t*12*c*15*t*, 4–5%). The similar reaction with linoleate esters is 12–15 times slower (Wolff, 1992).

Recently there has been some concern about the presence of traces of glycidol (epoxypropanol) in vegetable oils and of 3-chloropropane-1,2-diol (monochloropropanediol, MCPD, $ClCH_2CH(OH)CH_2OH$) in prepared foods. There is some evidence that glycidol may be carcinogenic. Both compounds may be present as acyl esters. Appropriate methods of analysis are being devised. It is believed that glycidol is a product of high-temperature deodorisation. MCPD is produced from the epoxide in a reaction involving a chloride ion, probably from salt (MCPD Web site: http://www.aocs.org/tech/3-mcpd.cfm, last accessed 8 January 2013).

When exposed to high temperatures (especially under frying conditions), PUFAs undergo cyclisation to give monocyclic compounds with five- or six-membered rings (Dobson, 1998; Le Querre & Sebedio, 1996). The formation of polycyclic compound is less well understood (Gertz, 2006).

A study of fish oil deodorisation indicates that these oils should not be heated at temperatures above 180 °C, in order to avoid stereomutation of eicosapentaenoic acid and docosahexaenoic acid. At higher temperatures, polymers, cyclic monomers and geometrical isomers are formed (Fournier *et al.*, 2006).

Cottonseed oil is unusual in that it contains malvalic acid (C_{18}) and sterculic acid (C_{19}) at combined levels up to 1%. These acids contain a cyclopropene unit and are toxic, but they are removed or modified to less toxic materials during processing – especially deodorisation and hydrogenation – and the product is entirely safe.

References

Akoh, C.C. & Min, D.B. (2008) *Food Lipids – Chemistry, Nutrition and Biochemistry*. Boca Raton, FL, USA: CRC Press, pp. 245–266.

Amelia, M., Rizzo, R. & Varazini, F. (1998) Separation of stigmasta-3,5-diene, squalene isomers and wax esters from olive oils by a single high-performance liquid chromatography run. *J. Am. Oil Chem. Soc.*, **75**, 527–530.

Anon (1998) US growers plant mid-oleic sunflower. *INFORM*, **9**, 715.

Azian, M.N., Kamal, A.A.M., Panau, F. & Ten, W.K. (2001) Viscosity estimation of triacylglycerols and of some vegetable oils based on their triacylglycerol composition. *J. Am. Oil Chem. Soc.*, **78**, 1001–1005.

Baharin, B.S., Rahman, K.A., Karim, M.I.A., Oyaizu, T., Tanaka, K. & Takagi, S. (1998) Separation of palm carotene from crude palm oil by adsorption chromatography with a synthetic polymer adsorbent. *J. Am. Oil Chem. Soc.*, **75**, 399–404.

Basiron, Y. (2005) Palm oil. In F. Shahidi (ed.) *Bailey's Industrial Oil and Fat Products*, Volume II, 6th edn. New York, NY, USA: Wiley Interscience, pp. 333–429.

Bhaskar, A.R., Rizvi, S.S.H., Bertoli, C., Fay, L. & Hug. B. (1998) A comparison of physical and chemical properties of milk fractions obtained by two processing technologies. *J. Am. Oil Chem. Soc.*, **75**, 1249–1264.

Bondioli, P., Mariani, C., Lanzani, A., Fedeli, E. & Muller, E. (1993) Squalene recovery from olive oil deodorizer distillate. *J. Am. Oil Chem. Soc.*, **70**, 763–766.

Boocock, D.G.B., Konar, S.K., Mao, V., Lee, C. & Biligan, S. (1998) Fast formation of high-purity methyl esters from vegetable oils. *J. Am. Oil Chem. Soc.*, **75**, 1167–1172.

Booth, S.L., Davidson, K.W., Lichtenstein, A.H. & Sadowsi, J.A. (1996a) Plasma concentrations of dihydro-vitamin K_1 following dietary intake of a hydrogenated vitamin K_1-rich vegetable oil. *Lipids*, **31**, 709–713.

Booth, S.L., Pennington, J.A.L. & Sadowski, J.A. (1996b) Dihydrovitamin K_1: primary food sources and estimated dietary intakes in the American diet. *Lipids*, **31**, 715–720.

Boskou, D. (2011) Olive oil. In: F.D. Gunstone (ed.) *Vegetable Oils in Food Technology*, 2nd edn. Oxford, UK: Blackwell, pp. 243–272.

Chumpitaz, L.D.A., Coutinho, L.F. & Meirelles, A.J.A. (1999) Surface tension of fatty acids and triglycerides. *J. Am. Oil Chem. Soc.*, **76**, 379–382.

Clark, J.P. (1996) Tocopherols and sterols from soybeans. *Lipid Technology*, **8**, 111–114.

Coupland, J.N. & McClements, D.J. (1997) Physical properties of liquid edible oils. *J. Am. Oil Chem. Soc.*, **74**, 1559–1564.

Crespo, F. (1973) Die Autokatalystische Spaltung der Fette. *Seifen-Oele-Fette-Wachse*, **99**, 55–57.

Debruyne, I. (2007) Novel soybean products for a healthier nutrition – recent developments, market introduction and targetted commercialization. *Lipid Technology*, **19**, 128–131.

De Filippis, P., Giavarini, C., Searsella, M. & Sorrentino, M. (1995) Transesterification process for vegetable oils: a simple control method of methyl ester content. *J. Am. Oil Chem. Soc.*, **72**, 1399–1404.

De Leonardis, A., Macciola, V. & De Felice, M. (1998) Rapid determination of squalene in virgin olive oil using gas–liquid chromatography. *Ital. J. Food Sci.*, **10**, 75–80.

Dobson, G. (1998) Cyclic fatty acids: qualitative and quantitative analysis. In: R.J. Hamilton (ed.) *Lipid Analysis in Oils and Fats*. London, UK: Blackie, pp. 136–180.

Duff, N.V.K. & Prasad, D.H.L. (1989) Interrelationships among the properties of fatty oils. *J. Am. Oil Chem. Soc.*, **66**, 701–703.

Elmadfa, I. & Wagner, K-H. (1997) Vitamin E and stability of vegetable oils. *Fett/Lipid*, **99**, 234–238.

Evans, J.C., Kodali, D.R. & Addis, P.B. (2002) Optimal tocopherol concentrations to inhibit soybean oil oxidation. *J. Am. Oil Chem. Soc.*, **79**, 47–51.

Fasina, O.O., Hallman, H., Craig-Schmidt, M. & Clements, C. (2006) Predicting temperature-dependence viscosity of vegetable oils from fatty acid composition. *J. Am. Oil Chem. Soc.*, **83**, 899–903.

Ferrari, R.A.P., Schulte, E., Esteves, W., Bruhl, L. & Mukherjee, K.D. (1996) Minor constituents of vegetable oils during industrial processing. *J. Am. Oil Chem. Soc.*, **73**, 587–592.

Firestone, D. (1999) *Physical and Chemical Characteristics of Oils, Fats, and Waxes*. Champaign, IL, USA: AOCS Press.

Fisher, C.H. (1998) Correlating viscosity with temperature and other properties. *J. Am. Oil Chem. Soc.*, **75**, 1229–1232.

Fisher, C.H. (2000) Equations correlating properties of *n*-fatty acids and derivatives with chain length. *J. Am. Oil Chem. Soc.*, **77**, 569–571.

Fournier, V., Destaillats, F., Juaneda, P., Dionisi, F., Lambelet, P., Sebedio, J.-L. & Berdaux, O. (2006) Thermal degradation of long-chain polyunsaturated fatty acids during deodorisation of fish oil. *Eur. J. Lipid Sci. Technol.*, **108**, 33–42.

Frankel, E.N. (2005) *Lipid Oxidation*, 2nd edn. Bridgwater, UK: The Oily Press.

Frankel, E.N. (2007) *Antioxidants in Food and Biology – Facts and Fiction*. Bridgwater, UK: The Oily Press.

Freedman, B. & Bagby, M.O. (1989) Heats of combustion of fatty esters and triglycerides. *J. Am. Oil Chem. Soc.*, **66**, 1601–1605.

Gertz, C. (2006) Developments in frying oils. In: F.D. Gunstone (ed.) *Modifying Lipids for Use in Food*. Cambrdige, UK: Woodhead Publishing, pp. 517–538.

Gertz, C. & Kogelheide, H. (1994) Investigation and legal evaluation of polycyclic aromatic hydrocarbons in vegetable fats and oils. *Fat Sci. Technol.*, **96**, 175–180.

Ghosh, S. & Bhattacharyya, D.K. (1996) Isolation of tocopherol and sterol concentrate from soybean deodoriser distillate. *J. Am. Oil Chem. Soc.*, **73**, 1271–1274.

Gibon, V. (2006) Fractionation of lipids for use in food. In: F.D. Gunstone (ed.) *Modifying Lipids for Use in Food*. Cambrdige, UK: Woodhead Publishing, pp. 201–233.

Grob, K. & Bronz. M. (1994) Tolerable concentration of olefinic sterol degradation products in edible oils and fats sold as non-refined. *Fat Sci. Technol.*, **96**, 252–255.

Grob, K., Artho, A. & Mariana, C. (1992) Determination of raffination of edible oils and fats by olefinic degradation products of sterols and squalene, using coupled LG-GC. *Fat Sci. Technol.*, **94**, 394–400.

Gumuskesen, A.S. & Cakaloz, T. (1992) Chemical and physical changes in cottonseed oil during deodorisation. *J. Am. Oil Chem. Soc.*, **69**, 392–393.

Gunstone, F.D. (2005) Fatty acid production for human consumption. *INFORM*, **16**, 736–737.

Gunstone, F.D. (2008) *Phospholipid Technology and Applications*. Bridgwater, UK: The Oily Press.

Gunstone, F.D. (2011) The food-fuel debate. In: F.D. Gunstone (ed.) *Vegetable Oils In Food Technology*, 2nd edn. Oxford, UK: Blackwell, pp. 19–21.

Halvorsen, J.D., Mammel, W.C.J. & Clements, L.D. (1993) Density estimation for fatty acids and vegetable oils based on their fatty acid composition. *J. Am. Oil Chem. Soc.*, **70**, 875–880.

Hilder, M.H. (1997) Oil storage, transport and handling. In: F.D. Gunstone & F.B. Padley (eds) *Lipid Technologies and Applications*. Dekker, pp. 169–198.

Hopia, A., Pyysalo, H. & Wickstrom, K. (1986) Margarines, butter and vegetable oils as sources of polycyclic aromatic hydrocarbons. *J. Am. Oil Chem. Soc.*, **63**, 889–893.

Kao, J-W., Hammond, E.G. & White, P.J. (1998) Volatile compounds produced during deodorisation of soybean oil and their flavouring significance. *J. Am. Oil Chem. Soc.*, **75**, 1103–1107.

Kenar, J.A. (2006) Reaction chemistry of gossypol and its derivatives. *J. Am. Oil Chem. Soc.*, **83**, 269–302.

Kenar, J.A. (2007) Glycerol as a platform chemical. *Lipid Technology*, **19**, 249–253.

Kochhar, S.P. (1983) Influence of processing on sterols of edible vegetable oils. *Prog. Lipid Res.*, **22**, 161–188.

Kochhar, S.P. (2011) Minor and specialty oils. In: F.D. Gunstone (ed.) *Vegetable Oils in Food Technology*, 2nd edn. Oxford, UK: Blackwell, pp. 291–342.

Koetsier, W.T. (1997) Hydrogenation of edible oils: technology and applications. In: F.D. Gunstone & F.B. Padley (eds) *Lipid Technologies and Applications*. Dekker, pp. 169–198.

Krisnangkura, K. (1991) Estimation of heat of combustion of triglycerides and fatty acid methyl esters. *J. Am. Oil Chem. Soc.*, **68**, 56–58.

Krog, N. (1997) Food emulsifiers. In: F.D. Gunstone & F.B. Padley (eds) *Lipid Technologies and Applications*. Dekker, pp. 521–534.

Larsson, B.K., Eriksson, A.T. & Cervanka, M. (1987) Polycyclic aromatic hydrocarbons in crude and deodorised vegetable oils. *J. Am. Oil Chem. Soc.*, **64**, 365–370.

Larsson, B.K., Quinn, P., Sato, K. & Tiberg, F. (2006) *Lipids: Structure, Physical Properties and Functionality*. Bridgwater, UK: The Oily Press.

Law, K.S. & Berger, K.G. (1984) Citric acid in the processing of oils and fats. *PORIM Technology*, **11**.

Lawler, P.J. & Dimock, P.S. (2002) Crystallisation and polymorphism of fats. In: C.C. Akoh & D.B. Min (eds) *Food Lipids – Chemistry, Nutrition and Biochemistry*. Boca Raton, FL, USA: CRC Press, pp. 245–266.

Le Querre, J.-L. & Sebedio, J.-L. (1996) Cyclic monomers of fatty acids. In: E.G. Perkins & M.D. Erikson (eds) *Deep Frying: Chemistry, Nutrition and Practical Applications*. Champaign, IL, USA: AOCS Press, pp. 49–88.

Loncin, M. (1952) L'hydrolise spontanee autocatalytique des triglycerides. *Oleagineux*, **7**, 695–698.

Longchampt, P. & Hartel, R.W. (2004) Fat bloom in chocolate and compound coatings. *Eur. J. Lipid Sci. Technol.*, **106**, 241–274.

Martini, S., Awad, T. & Marigini, A.G. (2006) Structure and properties of fat crystal networks. In F.D. Gunstone (ed.) *Modifying Lipids for Use in Food*. Cambridge, UK: Woodhead Publishing, pp. 142–169.

McGill, A.S., Moffat, C.F., Mackie, P.R. & Cruickshank, P. (1993) The composition and concentration of *n*-alkanes in retail samples of edible oils. *J. Sci. Food Agric.*, **61**, 357–362.

Mori, H. (1988) Solidification problems in the preparation of fats. In: N. Garti & K. Sato (eds) *Crystallisation and Polymorphism of Fats and Fatty Acids*. New York: Dekker, pp. 423–442.

Muller, H. (1995) Carotenoids in the total diet: determination of the daily intake of individual derivatives. *Fat Sci. Technol.*, **97**, 397–402.

Netscher, T. (1999) Synthesis and production of vitamin E. In: F.D. Gunstone (ed.) *Lipid Synthesis and Manufacture*. Sheffield, UK: Sheffield Academic Press, pp. 250–267.

Norberg, S. (2006) Chocolate and confectionery fats. In: F.D. Gunstone (ed.) *Modifying Lipids for Use in Food*. Cambridge, UK: Woodhead Publishing, pp. 488–516.

Noureddini, H., Teoh, B.C. & Clements, L.D. (1992) Viscosities of oils and fatty acids. *J. Am. Oil Chem. Soc.*, **69**, 1189–1191.

Ooi, C.K., Choo, Y.M., Yap, S.C., Basiron, Y. & Ong, A.S.H. (1994) Recovery of carotenoids from palm oil. *J. Am. Oil Chem. Soc.*, **71**, 423–426.

Pantzaris, T.P. (1985) The density of oils in the liquid state. *PORIM Technology*, **12**.

Patterson, H.B.W. (1989) *Handling and Storage of Oilseeds, Oils, Fats and Meal*. London, UK: Elsevier Applied Science, pp. 181–264.

Przybylski, R. (2011) Canola/rapeseed oil. In: F.D. Gunstone (ed.) *Vegetable Oils in Food Technology*, 2nd edn. Oxford, UK: Blackwell, pp. 107–136.

Rabelo, J., Batista, E., Cavaleri, F.W. & Meirelles, J.A. (2000) Viscosity prediction for fatty systems. *J. Am. Oil Chem. Soc.*, **77**, 1255–1261.

Robinson, N.P. & MacGibbon, A.K.H. (1998) The composition of New Zealand milk fat triacylglycerols by reversed-phase high-performance liquid chromatography. *J. Am. Oil Chem. Soc.*, **75**, 993–999.

Sato, P., Wester, I. & Hopia, A. (2003) Phytosterols. In: F.D. Gunstone (ed.) *Lipids for Functional Foods and Nutraceuticals*. Bridgwater, UK: The Oily Press, pp. 183–224.

Schone, F., Fritsche, J., Bargholz, J., Leiterer, M., Jahreis, G. & Matthaus, B. (1998) Changes of rapeseed and linseed during processing. *Fett/Lipid*, **100**, 539–545.

Schulte, E. (1994) Determination of edible fat refining by HPLC and Δ3,5-steradienes. *Fat Sci. Technol.*, **96**, 124–128.

Smith, K.W. (2009) Crystallisation and physical properties of fats and subsequent papers. *Eur. J. Lipid Sci. Technol.*, **111**, 217–302.

Stanley, J. (1999) Beta-carotene – friend or foe. *Lipid Technology*, **11**, 14–16.

Sun, H., Wiesenborn, D., Tostenson, K., Gillespie, J. & Rayas-Duarte, P. (1997) Fractionation of squalene from amaranth seed oil. *J. Am. Oil Chem. Soc.*, **74**, 413–418.

Tang, H., DeGuzman, R.C., Esteves, W. & Ng, K.Y.S. (2008) Formation of insolubles in palm oil-, yellow grease-, and soybean oil-based biodiesel blends after cold soaking at 4 °C. *J. Am. Oil Chem. Soc.*, **85**, 1173–1182.

Thyrion, F.C. (1999) The production of natural antioxidants (other than vitamin E). In: F.D. Gunstone (ed.) *Lipid Synthesis and Manufacture*. Sheffield, UK: Sheffield Academic Press, pp. 268–287.

Timms, R.E. (1978) Heats of fusion of glycerides. *Chem. Phys. Lipids*, **21**, 113–129.

Timms, R.E. (1985) Physical properties of oils and mixtures of oils. *J. Am. Oil Chem. Soc.*, **62**, 241–248.

Timms, R.E. (2003) *Confectionery Fats Handbook – Properties, Production and Application*. Bridgwater, UK: The Oily Press.

Timms, R.E. (2005) Fractional crystallization – the fat modification process for the 21st century. *Eur. J. Lipid Sci. Technol.*, **107**, 48–57.

Topallar, H., Bayrak, H. & Iscan, M. (1995) Effect of hydrogenation on density and viscosity of sunflower oil. *J. Am. Oil Chem. Soc.*, **72**, 1519–1522.

Toro-Vazquez, J.F. & Infante-Guerrero, R. (1993) Regressional models that describe absolute viscosity. *J. Am. Oil Chem. Soc.*, **70**, 1115–1119.

Torres, C.F., Fornari, T., Torrelo, G., Senorans, F.J. & Regiero, G. (2009) Production of phytosterol esters from soybean oil deodoriser distillates. *Eur. J. Lipid Sci. Technol.*, **111**, 459–463.

Verleyen, T., Forcades, M., Verhe, R., Dewettinck, K.M, Huyghebaert, A. & De Greyt, W. (2002a) Analysis of free and esterified sterols in vegetable oils. *J. Am. Oil Chem. Soc.*, **79**, 117–122.

Verleyen, T., Sosinska, U., Ioannidou, S., Verhe, R., Dewettinck, K.M., Huyghebaert, A. & De Greyt, W. (2002b) Influence of the vegetable oil refining process on free and esterified sterols. *J. Am. Oil Chem. Soc.*, **79**, 947–953.

Wagner, K-H. & Isnardy, B. (2006) New insights into tocopherol research with emphasis on gamma-tocopherol and delta-tocopherol. *Lipid Technology*, **18**, 36–40.

Wai-Lin, S. & Wee-Lam, N. (1995) Partition coefficients of diglycerides in crystallisation of palm oil. *J. Am. Oil Chem. Soc.*, **72**, 591–595.

Walsh, L., Winters, R.L. & Gonzalez, R.G. (1998) Optimising deodoriser distillate tocopherol yields. *INFORM*, **9**, 78–83.

Warner, K. (2007) Increasing gamma- and delta- tocopherols in oils improves oxidative stability. *Lipid Technology*, **19**, 229–231.

Warner, K., Miller, J. & Demurin. Y. (2008) Oxidative stability of crude mid-oleic sunflower oils from seeds with high γ- and δ- tocopherol levels. *J. Am. Oil Chem. Soc.*, **85**, 529–533.

Watkins, C. (2009) Oilseeds of the future. *INFORM*, **20**, 276–279, 342–344, 408–410.

Wijesundera, C., Ceccato, C., Watkins, P., Fagan, P., Fraser, B., Thienthong, N. & Perlmutter, P. (2008) Docosahexaenoic acid is more stable to oxidation when located in the *sn*-2 position of triacylglycerol compared to *sn*-1(3). *J. Am. Oil Chem. Soc.*, **85**, 543–548.

Wilkes, R.S. (2008) Low linolenic soybeans and beyond. *Lipid Technology*, **20**, 277–279.

Willner, T., Jess, U. & Weber, K. (1997) Effect of process parameters on the balance of tocopherols in the production of vegetable oils. *Fett/Lipid*, **99**, 138–147.

Wolff, R.L. (1992) *Trans*-polyunsaturated fatty acids in French edible rapeseed and soybean oils. *J. Am. Oil Chem. Soc.*, **69**, 106–110.

Yang, B. (2003) Natural vitamin E: activities and sources. *Lipid Technology*, **15**, 125–130.

Yanishlieva, N.V., Aitzetmuller, K. & Raneva V.G. (1998) β-carotene and lipid oxidation. *Fett/Lipid*, **100**, 444–462.

2 Bulk Movement of Edible Oils

Wolf Hamm
Retired, Harpenden, UK

2.1 Oil production and exports

Developments in oilseed production in recent years have resulted in a very active international trade in vegetable oils. This trade encompasses a comprehensive range of crude and refined oils, extending from soft oils such as soybean, rapeseed and sunflower oils to palm oil, its fractions and the lauric oils, which have a higher content of the more solid triacylglycerols than the soft oils. The global production of vegetable oils is dominated by two oils, soybean oil and palm oil, and the output of these makes up a significant proportion of oil consumed in the producer country and that exported, often over large distances. The international long-distance movement of the many other vegetable oils is on a much smaller scale but is nevertheless subject to similar considerations as far as oil quality is concerned.

Argentina and Brazil, as major producers of soybeans and soybean oil, are now responsible for a large share of the exports of this oil. A growing proportion of the soybean oil shipped by South American producers is used for the production of fatty acid methyl esters (FAMEs) for blending to produce biodiesel. Production of soybean oil in the USA is also substantial, but exports in this case are more limited, due to the dominant use of the oil produced for domestic consumption.

Argentina is also a major producer and substantial exporter of sunflower oil, and the Ukraine has in recent years become a key producer of this oil and in consequence an important exporter of it. In the last decade (2001–2011), its share of sunflower exports has grown from 20% of total oil exports to 50% of a significant growth (~60%) in global sunflower oil production.

Edible Oil Processing, Second Edition. Edited by Wolf Hamm, Richard J. Hamilton and Gijs Calliauw.

Palm oil production and export are at present located principally in South East Asia, with Indonesia and Malaysia responsible for the bulk of both. The pressure to minimise the ecological damage attributed to oil palm plantations has led producers of these oils in South East Asia to look outside this region for further oil palm cultivation opportunities. It is therefore possible that in future, palm oil exports will be sourced from a wider geographical area. Palm oil production in Indonesia and Malaysia has led to the large-scale production of its fractions – palm olein and palm stearin – which have important applications as components of fat-based foods and have found important markets in various parts of the world. As is the case with soybean oil shipments by South American producers, the use of palm oil as a feedstock for FAME production for use in biodiesel has contributed significantly to the overall quantity of the oil shipped.

Palm oil and its fractions shipped from Indonesia and Malaysia may be in crude or refined form. Palm oil exports (including fractions) from the region are now close to 40 million tonnes per annum. Much of the oil produced is exported to various parts of mainland Asia, with the industrial giants China and India taking a large share of the total. Imports by the EU countries are rather lower: a significant reflection of the growing importance of Asia in shaping the future of the uses of this oil. Palm oil exports have accounted for over 60% of global oil exports, and palm oil shipments therefore account for more than 50% of the quantity of vegetable oil shipped worldwide. This growth in the trade in palm oil is paralleled by a proportionately massive growth in exports of palm kernel oil.

The destination of vegetable oil cargoes exported by the various producer countries obviously reflects the demand for the major oils in various parts of the world. As already indicated, soybean oil and palm oil and its fractions account for the bulk of the oil shipped. A large proportion of the soybean oil exported from South America is shipped to the principal receiver countries in Asia – China and India. In the case of palm oil, both crude and processed, more than 12 million tonnes of the oil and its fractions were shipped to the major user countries in Asia, particularly China and India, and more than 5 million tonnes were shipped to countries belonging to the EU in 2011. Lesser quantities were shipped to destinations in West Asia and in Africa. Comparing import data for 2011 and for a decade earlier shows that Asian countries have become major consumers of these oils and that therefore this region has become a major destination for the long-distance shipment of oils.

This trade nowadays makes use of multicompartmented parcel tankers and in most cases entails long-distance journeys over lengthy periods and traversing zones with different climatic conditions. Consequently, cargoes may experience fluctuating weather, and in particular temperature conditions, which can have an adverse effect on their conditions.

Table 2.1 Major oil production and export data, 2001–2011, thousand tonnes.

Oil		2001	2006	2011
Palm oil	Production	23 920	37 415	50 518
	Export	17 688	29 971	39 184
	% exported	74%	80%	78%
	% exported to India/China	32%	29%	33%
	% exported to EU	17%	15%	14%
Soybean oil	Production	27 788	35 196	41 562
	Export	8078	10 435	9314
	% exported	29%	30%	22%
	% exported to India/China	20%	30%	22%
Sunflower oil	Production	8145	11 217	13 098
	Export	2383	4470	5262
	% exported	29%	40%	40%
Rapeseed oil	Production	13 691	18 446	23 657
	Export	1239	2102	3707
	% exported	9%	11%	16%

Source: Oil World Annual, 2002, 2006 Part 1, 2012 Part 1, copyright ISTA Mielke GmbH (available from www.oilworld.biz/annual).

The key trends in vegetable oil exports (see Table 2.1) are:

(1) The strong growth in the shipment of palm oil and its fractions.
(2) The growth in the Indonesian contribution to the total palm oil availability for export.
(3) The growing demand for vegetable oils generally and for palm oil in particular by the Asian giants, China and India.

It follows that the movement of vegetable oils in large quantities and over considerable distances requires that rigorous attention be paid to the maintenance of quality throughout the period from production to delivery to the final processor.

The diversity of the vegetable oil cargoes carried has encouraged growth in the number of parcel tankers designed for efficient loading and discharge of numerous parcels of various oils, requiring the control of cargo temperature at loading and on discharge. The temperature range considered necessary for different oils at all stages of shipment has been agreed by the Federation of Oils, Seeds and Fats Associations (FOSFA) and is given in Table 2.2. In the decade beginning in 2000, the seaborne trade in vegetable oils doubled from 30 million to more than 60 m tonnes per annum. This growth reflects to a considerable extent the use of soybean and palm oils for biodiesel production in the USA and Europe.

Table 2.2 FOSFA heating recommendations.

	Temperature during voyage[a]		Temperature at discharge[a]	
Oil type	**Min. (°C)**	**Max. (°C)**	**Min. (°C)**	**Max. (°C)**
Castor oil	20	25	30	35
Coconut acid oil	27	32	40	45
Coconut oil	27	32	40	45
Cottonseed oil[b]	Ambient		20	25
Fish acid oil	20	25	35	40
Fish oil	20	25	25	30
Grapeseed oil[b]	Ambient		15	20
Grease	37	42	50	55
Groundnut oil[b]	Ambient		20	25
Hydrogenated oils[c]	Various		Various	
Illipe butter	37	42	50	55
Fatty acid methyl esters (FAMEs) from maize/rapeseed/soybean/sunflower[b]	Ambient		Ambient	
Fatty acid methyl esters (FAMEs) from coconut/palm/palm kernel/tallow	25	30	30	40
Lard	38	45	50	55
Linseed oil[b]	Ambient		15	20
Maize (corn) oil[b]	Ambient		15	20
Maize/soya/sun acid oil	30	35	45	55
Mixed soft rape acid oil	20	25	30	35
Oiticica oil	24	32	35	40
Olive oil[b]	Ambient		15	20
Palm acid oil	45	50	55	72
Palm fatty acid distillate	45	50	55	72
Palm kernel acid oil	27	32	40	45
Palm kernel fatty acid distillate	27	32	35	45
Palm kernel oil	27	32	40	45
Palm kernel olein	25	30	30	35
Palm kernel stearin	32	38	40	45
Palm oil	32	40	50	55
Palm olein	25	30	30	35
Palm stearin[d]	40	45	55	70
Rapeseed oil (HEAR type)[b]	Ambient		15	20
Rapeseed oil (LEAR type or canola)[b]	Ambient		15	20
Safflower oil[b]	Ambient		15	20
Sesame oil[b]	Ambient		15	20
Sheanut butter	37	42	50	55
Soybean oil[b]	Ambient		20	25
Sunflowerseed oil[b]	Ambient		15	20
Tallow (for voyages of 10 days or less)[b]	Ambient		55	65
Tallow	44	49	55	65

[a]The maximum temperature specified during the voyage is lower than the minimum required for discharge, in some cases by as much as 15 °C. It should be recognised that in some cases ships' officers will need to apply heat a few days prior to arrival in order to reach the appropriate discharge temperature.

[b]It is recognised that in some cases the ambient temperatures may exceed the recommended maximum figures shown in the heating recommendations.

[c]Hydrogenated oils can vary considerably in their slip melting points, which should always be declared. It is recommended that during a voyage, the temperature should be maintained at around the declared melting point and that this should be increased prior to discharge to give a temperature of between 10 and 15 °C above that point to effect a clean discharge.

[d]Different grades of palm stearin may have wide variations in their slip melting points, and the temperatures quoted may need to be adjusted to suit specific circumstances.

2.2 Cargo damage

An important aspect of the trading rules set by the organisations involved concerns the problem of contamination of a cargo by other cargoes. The use of tankers for the movement of oil cargoes has in the past led to occasional contamination problems, in most cases at the level of parts per million (ppm) of an undesirable component traced to an earlier cargo carried by the vessel. Contamination by a foreign substance may be regarded as making a carried oil unsuitable for human consumption. The contaminants most widely encountered have been organic intermediates, in some cases having documented toxicological and/or carcinogenic characteristics. Hydrocarbons in the diesel range have also been detected. As the clear distinction between naturally present hydrocarbons and external mineral oils cannot be made for every individual oil, the full range of $C_{10}-C_{56}$ will be determined. Shorter-chain hydrocarbons than C_{10} are volatile and will be detected by the flash point analysis. Longer-chain hydrocarbons than C_{56} are solid and can be detected as solid impurities in the standard analysis. The current action limit for all imported vegetable oils is 300 mg/kg (ppm) presence of total hydrocarbons ($C_{10}-C_{56}$). All shipments with hydrocarbon levels of 300 mg/kg or below are considered to be free from contamination by mineral oil.

Extraneous hydrocarbons may also be introduced during oil production and transport from the production site to the port of cargo loading (Tan & Kuntom, 1994). Evidence of the presence of polycyclic aromatic hydrocarbons in food oils, due to contamination of a shipment by an oil of petrogenic origin, will generally lead to rejection of the cargo by the receivers (although virtually complete removal of many volatile contaminants is possible), and in some cases to costly legal disputes concerning responsibility for the damage caused and the cost of reconditioning the cargo.

More recently, the ingress of seawater into a vegetable oil cargo has been the most frequently encountered form of cargo contamination. This is a more serious problem in the case of refined oils, since re-refining then becomes unavoidable, whereas in the case of crude oils damaged in this manner a simpler form of remediation or a financial adjustment of the cargo value is the most likely consequence. Loncin (1952) studied the rate of formation of free fatty acid (FFA) in various oils by hydrolysis at various temperatures and demonstrated that the rate of formation is autocatalytic, with the rate rising rapidly after an initial slow rise when hydrolysis takes place at elevated temperature. The rate of formation of FFA in a refined oil is far slower than in the corresponding crude oil, and minor components, such as phosphatides and partial glycerides, do not significantly affect it. For hydrolysis at 60 °C, Loncin found that the rate of formation is given by the equation:

$$da/dt = k.a \qquad (2.1)$$

where a is FFA, expressed as palmitic acid, t is the time in days and k is a constant dependant on the oil being studied.

For palm oil, the rate constant k was found to be 0.125. This study was subsequently extended (Crespo, 1973) to quantify the rate of formation of partial glycerides and glycerine in the process of hydrolysis.

Even when shipped under optimal conditions in a parcel tanker, an oil subjected to regular temperature fluctuations and vessel-induced agitation is likely to show some deterioration in quality. Berger (1985) reported comprehensively on various aspects of quality control in the storage and shipment of palm oil, both as crude oil and in refined, bleached, deodorised (RBD) form. The results of measurements on pipeline contamination and a range of oil quality parameters on both crude and refined oil demonstrate the importance of tight control of oil handling at all stages of storage and transportation. Recommendations for improved conditions of storage and transport were also included. Comparison of oils shipped normally – that is, without replacement of the air above the oil surface in the ship's tank by an inert gas – with those shipped under nitrogen showed that the latter arrived with significantly lower levels of free fatty acid (FFA). It is important to operate within the range of temperature conditions laid down by FOSFA, as given in Table 2.2.

The loading and discharge of oils with a relatively high melting point, such as palm oil and its fractions and the lauric oils, obviously requires pumping at a slightly elevated temperature. It is then important to prevent thermal damage to the oil, which can result in both colour deterioration and the formation of oxygenated by-products. Overheating of an oil can lead to reduced bleachability. The pumping arrangements normally installed on parcel tankers can, if not managed efficiently, lead to unacceptable admixture of one parcel of oil with another (of different composition) during loading or discharge. This can cause rejection of the cargo by the receiver. Cases of deliberate adulteration with oils of inferior value are now only rarely encountered, but a number of very specific analytical tests can be used to verify the authenticity of an oil (Jee, 2002).

The growing importance of the shipment of refined oils in recent years has led to consideration of improved storage conditions on ocean-going vessels, in order to permit the use of the oils after discharge without the need for re-refining. Nitrogen blanketing of cargoes is possible in some cases, but the use of International Standards Organization (ISO)-tanks is considered to offer a better guarantee of quality preservation. The use of this method of oil movement is, however, considerably more costly than conventional parcel tanker carriage. The quality aspects of the long-distance transport of oils and fats have been considered in detail by Rossell (1998). A comprehensive review of the effect of storage and shipment conditions on oil quality can be found in List *et al.* (2005).

2.3 Quality of oils shipped

2.3.1 Palm oil

The long-distance movement of palm oil has been the dominant feature of oil exports since the 1970s and has, for this reason, attracted special attention. Palm oil and its fractions are shipped in both crude and refined form, though crude oil shipments are mainly of the oil rather than its fractions. Exports from Malaysia are predominantly in the form of RBD oil or fractions, whereas Indonesia exports considerable quantities of crude palm oil. The importance of being able to produce a light-coloured refined oil has led to the development of a test to assess the bleachability of palm oil, this being the Deterioration of Bleachability Index (DOBI), which is now an international standard (ISO 17932:2005). Siew & Mohamad (1992) have used this ratio to develop a discriminant function that can be used to characterise the quality of crude palm oil, with particular reference to long-term stability. They have suggested that for good bleachability, palm oil should have a DOBI value of at least 2.3, with a DOBI grade of 3–4 giving the best bleaching results. Palm oil loaded with FFA below 0.05% also shows lower hydrolysis after shipment than oil shipped containing 0.05–0.10% FFA (Berger, personal communication 2010).

Conditions for handling of the oil at the receiving port were discussed in Section 2.2.

2.3.2 Soybean oil and other seed oils

Soybean oil is traded primarily in the form of crude degummed soybean oil, degumming taking the form of water degumming, which leaves a residual phosphatide content not exceeding 200 ppm (expressed as phosphorus). As its phosphatide content is considerably lower than that of phosphatide-rich oils, sunflower oil may be shipped as either crude or RBD oil.

Hydrocarbon contamination of various seed oils can be caused by at least two factors: the use of badly cleaned road tankers to transport oil from the production site to the loading port and inadequate cleaning of ships' tanks that have carried hydrocarbon oils prior to loading seed oils. Moffat and coworkers (1995) documented the range of n-alkanes of biogenic origin present in various seed oils, including soybean, rapeseed and sunflower oils. By adding controlled quantities of hydrocarbons of petrogenic origin to oils of known alkane concentration profile, they were able to show that it is possible to distinguish between n-alkanes of biogenic and of petrogenic origin. However, to some extent this ability to detect a contaminant of petrogenic origin requires knowledge of the composition of the contaminant. Oils such as soybean and sunflower which have been transported in parcel tankers have sometimes been found to contain unusual patterns of hydrocarbons on discharge, and this has led to the suspicion of contamination with hydrocarbons derived from petrogenic sources, such as diesel or other fuel oils.

As a result of FOSFA/National Institute of Oilseed Products (NIOP) actions concerning the carriage of previous cargoes, the problem of petrogenic hydrocarbon contamination appears to have been virtually eliminated from the intercontinental trade in oils. Ingress of moisture into parcel tanker tanks is now the most prevalent source of receiver complaints, as it gives rise to an increase in the FFA content of the oil, possibly leading to a need for additional refining as well as to the removal of free water.

2.3.3 Shipment of oils intended for production of FAMEs

As already indicated, a growing proportion of the soybean and palm oils shipped is used for the production of FAMEs, which are then blended to form biodiesel. No difference is made in the regulations governing the shipment of oils between those intended for the production of FAMEs and those intended for food use. In the case of FAME production prior to shipment, the producer may or may not take it out of the food chain; an important consideration is its possible use in the oleochemical industry for personal care products, which must use food-grade materials. FOSFA Contracts 60 and 61 (Contract for Fatty Acid Methyl Esters in Bulk, FOB and CIF) are for food-grade products and have the same conditions with respect to previous cargoes as the standard commodity contracts.

2.4 Codex Alimentarius

The Codex Alimentarius is a series of food standards and related texts which aims to provide a high level of consumer protection and fair practice in the international trade of food and agricultural products. The organisation charged with the development of the Codex standards and related texts is the Codex Alimentarius Commission (CAC), which is an intergovernmental body jointly sponsored by the Food and Agriculture Organization (FAO) and the World Health Organization (WHO). Membership is open to all Member Nations and Associate Members of the FAO/WHO, and currently comprises over 160 countries. International nongovernmental organisations, such as consumer, academic and industry bodies, may attend Codex meetings as observers.

The Codex standards are freely available from their Web site (www.codexalimentarius.net). They are updated as necessary during the biennial meetings of the Codex Committee on Fats and Oils (CCFO). They are based on a general standard for fats and oils, with additional standards for some specific oils such as olive oil. The standards are broad in their definitions of fatty acid profiles as they include material sourced from worldwide growing regions. They are not yet relevant to the oils and fats industry, as no companies currently trade on the basis of the Codex. The CCFO is currently developing a standard for fish oils, primarily for their nutritional fatty acid content.

In addition to compositional standards, the CCFO has published a Code of Practice for the storage and transport of edible oils in bulk (CAC/RCP 36 – 1987, Rev.3-2005), which includes information on aspects such as the design and construction of tanks so as to reduce the effects of oxidation and hydrolysis and to prevent contamination. It also stresses the need to avoid overheating of oils in storage, and the desirability of protecting oil by inert gas blanketing or sparging is noted. There are useful notes on the loading and discharge of ships and on the sequence in which oils should be pumped through pipelines in order to reduce the amount of contamination that can occur.

The Codex Code of Practice also includes a List of Banned Previous Cargoes. These are chemicals that are frequently shipped by parcel tankers which because of their toxicity, persistent taste or smell properties, or the difficulty of cleaning out their residues from the tanks, must not be followed by edible oil in the same tank. The Committee is in the process of developing a List of Acceptable Previous Cargo, which is effectively a list of materials which when used as prior cargoes for vegetable oils, reduce the risk to consumers of any possible contamination. The development of these lists means that the Codex Code of Practice will mirror the contract rules devised over many years of experience by the oils and fats trade.

2.5 Oil shipments: systems and regulations

2.5.1 The parcel tanker

This type of vessel is capable of carrying numerous consignments of different triglyceride oils and has become the carrier most widely used for intercontinental oil trade, as well as for some regional trade. The typical larger tanker of this type has 35–45 tanks with appropriate pumps for loading and discharging a variety of cargoes. Whereas vessels built in the past may have had a majority of their tanks made of stainless steel, the remainder being coated, vessels built in recent decades are more likely to be fully equipped with stainless-steel tankage. The total capacity of such a vessel can be of the order of 38 000 m^3 of oil. Vessels with a smaller capacity are also used for edible oil transport.

In the case of higher-melting oils, such as palm oil or palm stearin, it is important to raise the temperature of the cargo slowly before discharge in order to avoid scorching it, which may require gentle heating for several days before arrival at the vessel's destination. Table 2.2 provides guidance on this, as well as on the recommended temperature conditions at loading. In the case of oils and fats with a significant solids content at ambient temperature, it is particularly important to ensure that the cargo has reached a temperature at discharge at which the solids content is sufficiently low to ensure that cargo discharge is essentially complete, as any residual quantity may not be readily recovered as suitable for edible purposes. Where possible, such residues are transferred to slops tanks for subsequent recovery.

2.5.2 Parcel tanker categories: IMO classification

The International Maritime Organization (IMO) has classified parcel tankers into three categories, of which types 2 and 3 apply particularly to the carriage of vegetable oils and fats. A type 2 tanker is a bulk chemical tanker that has significant preventative measures (in order to avoid environmental damage) and a double hull of at least 0.75 m between the inner and outer hulls, with a 6 m double bottom. A type 3 vessel is a bulk chemical carrier that has a single hull only. IMO 2 ships are not allowed to load over 3000 m^3 of product into any single tank, but IMO 3 ships are allowed to carry larger volumes. A subsequent derogation to the type of oil carried by parcel tankers has had the effect that vegetable oils can be carried in certain type 3 ships, but fatty acid distillates and acid oils must be carried in IMO type 2 ships.

The International Convention for the Prevention of Pollution from Ships (MARPOL) has important implications for the oils and fats trade, and the International Code for the Construction and Equipment of Ships carrying Dangerous Chemicals in Bulk (IBC Code) defines types of vessel and their operating requirements. Under the generic heading ‘Vegetable Oils’, some 31 oils and fats are listed. These have a category Y rating, and have been assigned a hazard profile by GESAMP, a technical committee sponsored by eight UN agencies, including the IMO. Vegetable oil products can only be carried in bulk by sea if their name is included in the MEPC.2/Circ document.

The MARPOL regulations have had the effect of imposing a limit placed by the IMO on the amount of residue from a ship’s tanks that can be discharged to sea (Hancock, 2011). If the oil is viscous at the time of discharge, which would lead to significant quantities remaining in the ship’s tank at the end of the discharge operation, the first tank washings must be pumped to shore and not to sea. On shore, these washings can be disposed of in an environmentally acceptable way.

2.5.3 Trade regulation: the role of the FOSFA and NIOP

The international trade in edible oils is catered for and regulated by a number of trade associations, of which the FOSFA and NIOP are responsible for most contracts. An estimated 85% of the current world trade in oils and fats uses FOSFA contracts. The main advantage of this is that the use of standard-form contracts reduces the risk of trading parties misunderstanding the procedures they need to follow in order to enable trade to go smoothly. The contracts also reduce the risk in trade, as their clauses are well known by all parties and reflect longstanding trade practices. This allows the parties to discuss and agree on the important features, such as quality, quantity, price and shipment/delivery dates. Their confirmation letters include these details and usually a statement saying, 'All other terms as per FOSFA 80' (for crude palm oil, by way of an example).

The contracts also reduce risk as they include rules for the hygienic carriage of oils and fats in bulk by sea. These rules are tried and tested and have been developed for over 2 decades now, and with much experience. For an FOSFA contract, these rules are contained in the publication generally referred to as 'The Carriage of Oils and Fats'. Of particular importance in these rules are the two lists of banned previous cargoes and acceptable previous cargoes.

The standard FOSFA trading contracts are based on 'banned list terms'. The banned list includes cargoes that led to problems in the 1980s. These materials have persistent properties, are difficult to remove and clean from tanks and are generally toxic. The risk of contamination from previous cargoes or from poor cleaning of the tanks is reduced if these substances are not allowed to be carried prior to vegetable oils. Therefore, the basic terms mean that a receiver will accept a parcel of oil only if the previous cargo is not on the banned list. In the 1990s, FOSFA developed an 'acceptable' list of cargoes, which then became an optional clause added to the standard FOSFA contract. An 'acceptable list' trade is now increasingly used internationally. The term 'good merchantable quality' is often used in describing an oil quality requirement and the concept is inherent in a number of FOSFA contracts. The meaning and significance of the term are discussed by Backlog (1990).

In 2008, FOSFA International issued a revised List of Acceptable Previous Cargoes which includes the entry 'Fatty acid esters – mono-alkyl esters of fatty acids produced by the reaction of oils and fats and fatty acids with an alcohol'. This served the purpose of clarifying that FAME products derived from any vegetable oil or animal fat are acceptable as immediately previous cargoes to the carriage of oils and fats. The NIOP Acceptable List 2, comprising acceptable prior cargoes for edible oils which will undergo further processing, includes entries for the FAMEs of palm and coconut oils, as well as a generic entry for FAMEs with a list of examples. The EU list of Acceptable Previous Cargoes is more specific with regard to this category of product, and at present is confined to the methyl esters of four fatty acids: lauric, palmitic, stearic and oleic acids (Strode, 2009).

One code that has been adopted by several countries is the use of hazard analysis and critical control points (HACCP). This control scheme for safe food manufacture has been included in the legislation of many countries, including those of the EU and the USA. The HACCP scheme and its seven principles can readily be applied to the transport of oils and fats by sea. The international body that is concerned with worldwide food safety and fair world trade is the Codex Alimentarius Commission.

The movement of high-quality oils or fractions, such as refined oils, means that it has become more important to take all possible steps to ensure that quality is maintained from loading to discharge of cargo. It is therefore vital to prevent ingress of potentially harmful compounds. The ingress of moisture should always be avoided, even in the case of unrefined cargoes, due to the danger of hydrolysis (see Section 2.3), but refined cargoes also need protection from oxygen, as oxidative damage to a cargo may lead to the need to re-refine it upon arrival at its destination. Nitrogen blanketing is resorted to in some cases, where possible by prior displacement of dissolved oxygen by bubbling nitrogen through the cargo.

The use of ISO tanks for the purpose of quality preservation and authentication has also been recommended in certain cases, such as with virgin olive oils, though this obviously adds to the cost of transport and is only appropriate for the movement of smaller quantities.

2.6 Shore storage

The growth in the bulk movement of edible oils on an intercontinental basis has necessitated a substantial increase in the capacity of shore-based oil storage capacity. This growth has taken place particularly in the principal ports where these oils are loaded and/or discharged. Hamburg and Rotterdam are prime examples of such installations, and there are similar installations in other parts of the world capable of handling parcel tankers. In addition, many smaller so-called tank farms provide storage capacity for more limited quantities of vegetable oil supplies. Tanks of varying capacities can be found in the larger installations, ranging from 300 to 40 000 m^3 capacity. These tanks are available in mild steel, stainless steel and coated steel, but the vegetable oil industry is increasingly requiring stainless-steel tankage. In some cases, tank heating is also available. ISO 9001:2008 (an updated version of the earlier ISO 9002:1987) has now been widely adopted within the tank storage sector.

ISO 9001:2008 specifies requirements for a quality management system where an organisation:

- Needs to demonstrate its ability to consistently provide product that meets customer and applicable statutory and regulatory requirements.
- Aims to enhance customer satisfaction through the effective application of the system, including processes for continual improvement of the system

and the assurance of conformity to customer and applicable statutory and regulatory requirements.

All requirements of ISO 9001:2008 are generic and are intended to be applicable to all organisations, regardless of type, size and product provided. Apart from the need for the management of shore-based storage installations to handle cargoes efficiently and speedily, they must meet the various environmental and safety regulations now in place.

2.7 Movement and storage costs

The cost of oil movement between the major international terminals, such as the main Malaysian ports and the principal oil cargo-handling ports in Western Europe, is regularly recorded in publications, such as *Oil World*, which report on developments in oil production and trade, and in publications specialising in aspects of the trade in edible oils.

The cost of shipping oil from Malaysia to Europe has fluctuated over a wide range in recent years, reaching a high point of US$80/tonne in 2008. The cost of shipment of oil from South America to North West Europe in corresponding periods was slightly lower. Cost of oil movement can rise appreciably where transshipment is involved or where delivery at a destination not on a major shipping lane is required. The cost of oil storage in port tank farms will obviously depend on the conditions (material of construction of storage tank, heating requirements) and duration of storage.

2.8 Refinery location

Edible oil processing in Europe and North America has witnessed a growing concentration of ownership – oil production and oil processing being increasingly integrated both physically and financially. This has beneficial effects on the quality of refined oil and will also reduce production costs by facilitating direct transfer of oil produced to the refinery, thus eliminating the cost of double handling. Refineries processing significant quantities of imported oil will obviously benefit from being located close to a port that has oil storage facilities.

Acknowledgement

The author wishes to acknowledge his indebtedness to Dr J.N.S. Hancock of FOSFA International, London, for his expert guidance on those aspects of this chapter relating to the Codex Alimentarius as well as to FOSFA procedures.

References

Backlog, P.B. (1990) Storage, handling and shipping practices buyer/recipient viewpoint. In: D.R. Erickson (ed.) *Edible Fats and Oils Processing: Principles and Modern Practice*. Champaign, IL, USA: AOCS Press.

Berger, K.G. (1985) *J. Amer. Oil Chem. Soc.*, **62**, 438.

Crespo, F. (1973) The autocatalytic hydrolysis of fats [in German]. *Seifen, Öle, Fette, Wachse*, **99**, 55.

Hancock, J.N.S. (2011) Regulatory issues associated with the international oils & fats trade. *INFORM*, **22**(1), 47.

Jee, M.J. (2002) *Oils and Fats Authentication*. Oxford, UK: Blackwell.

List, G.R., Wang, T. & Shukla, V.K.S. (2005) Storage, handling and transport of oils and fats. In: F. Shahida (ed.) *Bailey's Industrial Oil and Fat Products*, 6th edn. New York, NY, USA: Wiley Interscience.

Loncin, M. (1952) Spontaneous autocatalytic hydrolysis of triglycerides [in French]. *Oleagineux*, **7**, 695.

Moffat, C.F., Cruickshank, P., Brown, N.A., Mennie, D., Anderson, D.A. & McGill, A.S. (1995) The concentration and composition of *n*-alkanes in edible oils. Report submitted to Food Science Division, MAFF.

Rossell, J.B. (1998) Transport of oils and fats. In: R. Heap, M. Kierstan & G. Ford (eds) *Food Transportation*. London, UK: Blackie A&P.

Siew, W.L. & Mohamad, N. (1992) The effect of fruit storage on palm oil bleachability. *J. Am. Oil Chem. Soc.*, **69**(12), 1266–1268.

Strode, J. (2009) Work progresses on biofuels shipping. *Oils and Fats International*, **25**(6), 22.

Tan, Y.A. & Kuntom A. (1994) Hydrocarbons in crude palm kernel oil. *J. AOAC Internat.*, **77**, 67–73.

3

Production of Oils

Philippe van Doosselaere

Retired (formerly Product Manager, Crushing, Desmet Ballestra Oils and Fats), Brussels, Belgium

3.1 Introduction

The demand for edible oil is increasing with increasing world population and the improvement in buying power. The yearly per capita consumption of fatty matter ranges from over 40 kg/tonne in the USA and Western Europe to less than 10 kg/tonne in the world's poorest countries. The world average is around 15 kg/year, while the intake recommended by the World Health Organization (WHO) is 20–25 kg/year. Countries with huge populations and growing buying power like China and India are well below the recommended figures. Their consumptions keep growing and put pressure on the demand for oilseeds. With improving prosperity, the increasing world population both increases consumption and becomes more discriminating. Oil producers must not only meet the increasing demand for edible oil but also improve product quality and variety, as these become major factors in the marketplace.

In the first decade of the 21st century, the rapid increase in demand for biodiesel placed an additional pressure on the demand for crude oil. Though the biodiesel market has calmed down recently, EU directives may put renewed pressure on the edible oils and fats markets in the future.

Oilseeds are also the main source of protein meal for animal feed. The demand for protein meal increases with world population, but mainly with buying power, as the demand for meat grows with increasing standard of living. In addition, the recent appearance of mad cow disease and the presence of dioxin in fish meal have pushed a switch from animal to vegetal meal as a source of protein for animal feed.

Over the last decade, the growth in world demand for edible oil has mostly been satisfied by huge increases in cultivated areas of soybeans in

Edible Oil Processing, Second Edition. Edited by Wolf Hamm, Richard J. Hamilton and Gijs Calliauw.

South America and of oil palm trees in Malaysia and Indonesia. Agriculture has also responded to the increase in demand by selective seed breeding. The emphasis has been on improving the physical characteristics of oils and meals and improving the yield per hectare or tolerance to climatic variations. The ability to modify seed genetically promises a revolution in agriculture. However, whereas genetically modified organisms are relatively well accepted in the Americas, their cultivation and their incorporation in food are fiercely opposed to, in most of Europe, as well as in many other countries.

A prerequisite for the production of top-quality oils and meals is a high-quality raw material that is not damaged during transportation, storage and processing.

The industry uses a number of process steps to produce crude oil, the application of which depends on the type of seed and the scale of the operation. The main steps and their importance are discussed in this chapter.

3.2 Seed handling and storage

3.2.1 Seed arrival

Seeds can reach a plant in lots ranging from just a few bags to 60 000-tonne shiploads. Clearly, the handling requirements for these extremes vary enormously. It is essential for the plant to control the weight and the quality of each load of seed that is reaching the plant. Weighing and sampling protect the interests of both the buyer and the seller and provide the basis for any necessary price adjustment.

For locally harvested seed, the control at the plant may be the first it is subjected to. Imported seeds have been controlled at departure, but the receiving plant still controls the lot for any variation in quality. A slight loss in moisture, for instance, is not uncommon during overseas shipping.

Trade rules and classifications based on quality exist for all major seeds and serve as a basis for settling any dispute between seller and buyer regarding seed quality and for defining price penalties, if required.

3.2.1.1 Seed weighing

Weighbridges are usually used for trucks or rail cars, and various types of inline scale are used in other cases.

The reliability and precision of the weighing equipment is of utmost importance and must be controlled, preferably by a specialised body, at regular intervals. The precision sought must be less than one in a thousand.

Equally important is the use of well-proven, well-defined and fixed procedures to weigh the incoming seeds, in order to prevent any fraud. It is for instance important to weigh trucks both as they enter and as they leave the plant, and to make sure that no dead weight is left on them as they came in and taken out when they leave: 'Has the driver left his truck?' 'Are you sure there are no buckets full of water somewhere at the back of the truck?' and so on. Such questioning might seem trivial but is essential in daily practice.

3.2.1.2 Sampling

As the seed arrives at the plant, it must be sampled.

The quality of the raw material determines how it will have to be processed and hence at what cost, how it will be stored and for how long, what kind of handling it will require and so on. These parameters, which are implied in the specifications of the seed purchase contract, must be checked carefully and unambiguously. Deteriorations during transport cannot be excluded, nor can frauds. The latter are more common when seeds arrive on trucks, and if left uncorrected may mean a serious loss at the end of the day.

Parameters that are controlled include, as a minimum:

- moisture;
- foreign material;
- damaged and broken seeds or kernels;
- protein content;
- oil content.

3.2.2 Seed reception and precleaning

This operation – and all others throughout the process – requires gentle handling of the seed.

Seeds received by trucks or rail cars are generally discharged in underground receiving pits. Seeds received by ships or barges are unloaded by means of specialised equipment.

The plant receiving system, until storage, must be of high capacity in order to avoid unnecessary backlog or even penalty (demurrage). All plant conveying systems must be designed to avoid breakage of seeds but also to allow full and free flow of the products (no dead pockets where product accumulates), and to enable internal inspection and easy cleaning where some product accumulation cannot be avoided. All transition devices between pieces of equipment or conveyors must be carefully designed and sized. Dust emission must be minimised and dust must be captured. Finally, conveying speeds must be low enough to permit continuous operation without excessive wear.

Belt conveyors are the preferred type for transporting products at high capacity over long distances. Chain conveyors and bucket elevators have the advantage of sealing easily and preventing dust emission. Pneumatic conveyors and screw conveyors cause more product breakage. Pneumatic conveyors are excellent for the transportation of light products, like hulls. Screw conveyors are adequate for small capacities and short conveying distances.

Before entering the storage facility, whether a silo or a flat store, the bulk material should be precleaned. This is usually done using a coarse screen, to remove stalks, stones and other foreign material that might damage the mechanical handling or processing equipment.

3.2.3 Storage

Oilseeds arrive at the plant on an intermittent basis, by shiploads, by trainloads or by truckloads. The plant must be able to store the grain for a while, between reception and process. The optimum capacity will depend on the frequency of arrivals and on their tonnage. A storage capacity equivalent to 1 month's processing is often considered a good rule of thumb. However, plants situated in countries or areas where it is advantageous to buy as much seed as possible at harvest, or where seed supply is unsecure, sometimes have a much larger storage capacity. Plants receiving seeds by ship must be able to store the equivalent of 1.2 shiploads, at least.

Secure storage is vital to preserving the quality and therefore the value of a seed. As a natural product, seed deteriorates over time, and this is accelerated at higher temperatures. The critical factor in determining the storage life of seed material is its moisture content: the higher the moisture and the hotter the climate, the shorter the secure storage time. Moisture induces the appearance and growth of moulds or fungi, as well as biological respiration of the grain itself. Respiration or 'sweating' induces exothermic reactions that cause the temperature to rise. Once the temperature reaches 70 °C, the seed deteriorates in a matter of hours, which may lead to total destruction in a few days after localised wet zones have been produced and colonised by fungi. Respiration and biological activity are particularly active in newly harvested grains.

Soybeans can safely be stored for 1 year or more at 11% moisture or less. However, soybeans are traded at up to 14% moisture and can only be stored for a limited time. Seeds are therefore usually segregated by moisture at reception, and the wetter grain is processed first.

Plants receiving high-moisture seeds on a regular basis must consider whether to equip themselves with a grain dryer. This is generally encouraged for plants receiving locally harvested grain. The need for a dryer can also be linked to a process requirement, such as dehulling. Seed lots whose temperature rises above 40 °C should also be processed at once.

Safe conditions for storing rapeseed and sunflower seed are in the range of 8% and below. The difference with soybean is due to the higher oil content in these seeds. Indeed, moisture is present in the nonfat part of the seeds. The critical moisture of the nonfat part is around 15%. This corresponds to overall moisture of 9% for a seed with 40% oil and to overall moisture of 8% for a seed with 47% oil. For storage of longer than 5 months, rapeseed or canola should be stored at a maximum of 8% moisture.

Oilseeds and grains can be stored in vertical silos made of concrete or steel, generally galvanised, or in horizontal warehouse-type silos. Considerations are the cost of construction, available space, ease of filling and emptying, and segregation by type or quality. Manual intervention in the handling of a grain should be limited to the strict minimum. Apart from concrete silos, large storage systems generally have a flat bottom, and the system for fully emptying such silos must be carefully considered.

Provided the storage system is well designed, concrete silos, steel silos and flat storage houses are all adequate for storing oilseeds.

Plants susceptible to receiving high-moisture seeds and storing them for more than 2 weeks should have at least some silos equipped with temperature detectors and aeration systems. Aeration is used primarily for cooling, not for drying. An aeration system must be used with care, particularly in areas where the climate is hot and humid. If used wrongly, aeration may cause more harm than good. Fans should be operated only when the air temperature is at least 3 °C colder than the grain, and at times when the air relative humidity is low.

Table 3.1 shows the equilibrium moisture of soybean at 20 and 30 °C, for various relative humidities of air. These are the moistures that beans will reach if left under aeration with such airs for a long time.

Soybeans (and other oilseeds) must not be aerated when the air relative humidity is greater than 70%.

Insects and rodents are another potential problem deserving of particular attention. Fumigation of oilseeds is generally not permitted, so it is important to have silos that are built to prevent infestation.

A final problem with oilseeds is acidification of the oil inside the seed during storage. This generally occurs due to the presence of an excessive percentage

Table 3.1 Equilibrium fraction of moisture in soybeans.

Relative humidity of air (%)	Air temperature	
	30 °C	20 °C
20	5.0	5.4
30	5.72	6.45
40	6.4	7.1
50	7.17	8.0
60	8.86	9.5
70	10.63	11.6
80	14.5	15.29
90	20.15	20.86

of broken or damaged seeds, as a result of receiving low-quality seed, or of careless or inadequate handling, perhaps as a result of poor equipment or hurried unloading from ships or railway wagons.

Deteriorated raw materials cannot be made good again and are the cause of very severe losses in capacity at the refining level – specifically, degumming or neutralising. They also require expensive equipment and more reagents at the refining level.

3.3 Preparation of oilseeds

3.3.1 Reason for and purpose of preparation

Oilseeds have a cellular structure. They are made of a large number of small cells, each containing oil, protein, carbohydrates and so on. A typical soybean cell is approximately 0.02 mm in diameter.

The oil inside a cell consists of hundreds of very small oil bodies each clinging to the inside surface of the cell wall and to the outside surface of the protein bodies. Oil bodies in rapeseed cells are 0.5–3.0 μm in diameter, and there are over 300 oil bodies per cell.

The oil is well protected inside the cells. It cannot be taken out in an efficient way without changing the shape of the seed and affecting its internal structure. This is why preparation is a prerequisite for oil extraction (more details of this process are given in Chapter 4).

The objectives of oilseed preparation are:

- To weaken or break the walls of the oil-containing cells, in order to enable the oil bodies to move from inside to outside the cells.
- To shape the material so as to give the solvent a short path access to the oil.
- To shape the material in such way that the solvent can percolate through a bed of prepared material in the extractor.
- To mechanically press some of the oil out of seeds with high oil content before solvent extraction (otherwise the solid structure crumbles in the extractor); seeds must also be prepared prior to mechanical pressing.

Achieving a high extraction rate requires the following:

- The capillary paths must be short, so that the distance over which diffusion occurs is as small as possible. Short capillary paths are achieved by fine grinding of the feed material.
- Since the extraction is carried out on the percolation principle, the feed material must be prepared in such a way that the solvent can percolate freely.

- For this reason, most oilseeds are rolled into thin flakes in order to produce a feed material with short capillary paths in one direction and good percolation properties.
- Even materials which are mechanically pressed before solvent extraction are generally flaked first.
- Some feed materials, such as rice bran and fish meal, cannot be formed in resistant flakes. Pelletising is the choice method of preparation for these materials.

The structure of oil cells must be weakened to the point where the oil can flow out, partially during mechanical pressing, and more fully during solvent extraction.

After preparation, the oilseed flakes, cake or pellets must be strong enough to resist the impact of liquid washing in the percolation extractor, but permeable enough to allow the solvent to penetrate into their structure.

3.3.2 Milling defect

The milling defect is an analytical method for measuring the lack of preparation. Its basic principle is:

- extract the prepared material (3 hours' Soxhlet);
- grind finely;
- extract the ground material (3 hours' Soxhlet).

The quantity of oil extracted the second time is a measure of the milling defect. The higher the milling defect, the higher the percentage of oil that cannot be extracted in the extractor and is left in the meal.

To summarise, it can be said that the purpose of the preparation is to form the material into a shape adequate for extraction, with the objective of obtaining the desired capacity, a high extraction yield (low residual oil after extraction) and high-quality products (crude oil and meal), while maintaining a low production cost. It is important to keep in mind that the extraction result is dictated in large part by the preparation efficiency.

3.4 Preparation of soybean

Soya is a leguminous annual plant with an aspect similar to that of the black bean.

The plant has dense foliage and can reach a height of 0.5–1.0 m according to its variety. The bean is a pod, green before maturity and yellow to black at maturity. The pods are self-opening at maturity and have a length of 3–11 cm;

they contain one to four beans. The beans are spherical, with a diameter of 5–10 mm and a colour from yellowish to dark brown depending on variety. The yellow beans are the richest in oil.

Soya may be cultivated in every tropical or subtropical country and even in most of the temperate climates. It has been cultivated since ancient times in China and in Manchuria, where it constitutes the basis of human food, as it does in Japan. Since early in the 20th century, soya cultivation has grown considerably in the USA, and later in South America.

The major factors behind its expansion are:

- Cultivation can be fully mechanised.
- Seeds can be stored easily and without much damage.
- Soybean is a very rich source of proteins.
- Many varieties exist, which can accommodate very different soils and climates.

Soybean is today the most widely cultivated oilseed in the world. World production is currently approximately 250 million tonnes per year. The USA is the largest producer, but its share has been decreasing steadily as Brazil and Argentina have increased their production in impressive ways.

For commercial purposes, US soybeans are divided into four grades (1–4). Most soybean lots exported from the USA are classified as grade 2 and meet the following specifications:

- bulk density: min. 0.695 tonnes/m^3;
- foreign material: max. 2%, stones <0.1%;
- splits: max. 20%;
- damaged kernels: max. 3.0% (max. 0.5% heat damaged);
- soybeans of other colour: max. 2.0%.

Generally, soybean processed as oilseed has the following composition:

- 9–14% moisture;
- 18–22% oil;
- 33–39% protein;
- 15–25% carbohydrate;
- ≤7% fibre;
- ≤6% ash.

3.4.1 Cleaning and weighing

While the usefulness of these operations is sometimes overlooked, it is highly recommended that the process be begun by weighing and thoroughly cleaning the incoming seed. Weighing at the initiation of the process allows the plant to be fed at a constant and controlled rate. Uniform feed will improve yields, reduce energy consumption and reduce hexane losses in the extraction plant. Cleaning eliminates components from the stream that might damage the equipment or negatively impact upon the quality of the finished product.

When entering the preparation plant, the seeds pass over a magnetic separator, which removes any ferrous particles that might be present. They then go on to the weighing system, where a buffer bin receives them. Its function is to absorb incoming feed variations, allowing a constant and regular feed to the rest of the plant. The size of the bin depends of course on the plant capacity, but also on the manner in which the seeds are fed from storage to process.

The buffer feeds the scale itself. The scale is preferably a high-precision hopper scale with digital control. It is equipped with gates at its inlet and its outlet that open and close sequentially. The purpose of the scale is not only to measure the feed process but also to maintain the desired capacity, simply by regulating the intervals between dumping of the hopper load.

From weighing, the seeds are conveyed to cleaning, in order to remove impurities.

Cleaning is important to reducing wear on the cracking and flaking rolls and in conveyors in general, and for obtaining high-quality products. Fines, pods, sticks, stones and other impurities must be removed. This is particularly true when producing high-protein meal.

Cleaning is carried out on a multideck screener. The cleaner is composed of screens which has opening sizes that are usually designed to segregate three fractions:

- The oversized impurities or pods are retained by the upper screen. Foreign materials, such as leaves, sticks, most stones and so on, are separated here. These impurities are evacuated to a trash container, or else are recovered (after stone separation) and sent to a hammer mill, where they are ground. The ground pods are added back to the hulls and sold as byproduct, for their fibre content.
- The clean fraction is the intermediate fraction. It passes through the first screen, and stays on top of the bottom one. This fraction is also subject to aspiration through a current of air. The aspiration is used to remove light particles, mainly loose hulls, and dust.
- The small fraction – fines and sand – passes through the bottom screen. This fraction is also evacuated to a trash container, or depending on its composition, might be added back to the product before extraction, or to the extracted meal. The composition of this stream will depend on the origin of the seeds and their harvesting method, and should be checked periodically.

Small stones of the same size as the seed are not separated by screening. If present in excessive number, it is prudent to employ destoners. A destoner might be a gravity table that uses the difference in density to separate heavy stones from lighter seeds. A simpler way of separating stones is to use a current of air, which carries seeds but not stones.

Equipment, conveyors and bins must be dust-tight and connected to a dust-control system. Air is aspirated from the pieces of equipment through ducts that are connected to a bag filter. The purpose of this is to contain the dust within the equipment and ducts and conduct it to the filter, where it is recovered. Clean air is expelled to the atmosphere and the dust is discharged from the filter by means of a rotary lock. The dust consists mainly of loose hulls and is thus sent to the hulls grinding section.

In order to reduce the overall emission load from the plant, it is recommended that air be recirculated and reused where possible. In many parts of the world, environmental permitting requires identification of all emissions points and definition of all air flows according to their quantity and quality.

3.4.2 Cracking

Clean soybeans are cracked into pieces in cracking mills. These are generally equipped with two pairs of corrugated rolls (see Figure 3.1): the top pair breaks the beans into halves or quarters, the bottom pair into quarters or eighths. Each mill is equipped with a feeder, which distributes the beans uniformly over the lengths of the rolls, at a constant and adjustable rate.

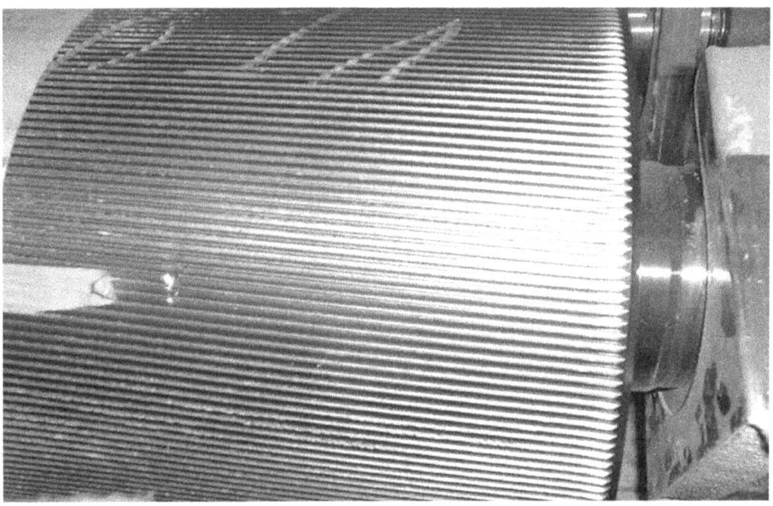

Figure 3.1 Detail of a corrugated roll in a cracking mill. Courtesy of Allocco. The full colour version can be found on the plates.

A permanent magnetic plate is located in the feed chute of each cracking mill, in order to protect the corrugated rolls from damage by stray iron or steel.

3.4.3 Cooking–conditioning

The cracked soybeans or grits proceed to the cooker–conditioner.

The purpose of cooking–conditioning is to heat the grits to 60–70 °C and soften them. A slight moisture adjustment, to 10–11%, can also be made. These are the best conditions for good flaking.

Two types of cooker are used by the industry.

The vertical stack cooker consists of a succession of horizontal steam-heated cooking pans stacked in a vertical shell. Each successive compartment is equipped with agitating arms, connected to a central rotating shaft. The grits enter at the top and are swept successively over each cooking pan, from top to bottom. Each pan has an opening which allows the cracks to flow from one stage to the next. These openings are equipped with actuated gates that allow maintenance of the product level at a preset height, according to the required residence time.

Large-capacity soybean plants tend to use horizontal rotary cookers instead of a vertical stack. These are made of a cylindrical shell housing parallel steam-heated tubes over their entire length. The grits are continuously introduced into the cylinder at one end of the machine and gradually progress to the discharge at the opposite end. The advancement of the material is due to the action of the rotating tubes. In some models, the tubes rotate inside the shell; in others the entire shell rotates along with the tubes. The rotation of the tubes continually lifts the material and drops it after about a third to a half-revolution. The machine is slightly inclined, with the discharge end being lower than the feed end, so that the material advances every time it drops.

3.4.4 Flaking

The hot, soft cracks leaving the conditioner are finally sent to flaking. The flaking mills (Figure 3.2) are equipped with one pair of large-diameter smooth rolls, which laminate the grits into flakes. Oil cells are weakened in the process, and the oil becomes accessible to the solvent in the extractor. In addition, the shape of the flakes, with their large surface-to-volume ratio, plus the short distance from the oil cells to the flake surface, facilitates liquid–solid contact and the migration of the oil to the liquid phase. Soybean flakes of the right moisture, temperature and thickness also form a solid bed of adequate resilience and permeability in the extractor. Optimum flake thickness varies between 0.25 and 0.38 mm, depending mostly on the type of extractor.

Figure 3.2 Industrial flaking mill. Courtesy of Allocco. The full colour version can be found on the plates.

High roll pressure is necessary to form flakes of uniform thickness. This pressure is applied by means of a hydraulic system.

Each mill is equipped with a feeder that distributes the material uniformly over the length of the rolls, at a constant and adjustable rate. Uniform feed is even more critical than in the cracking mills.

A permanent magnetic plate is located in the feed chute of each flaking mill, to protect the rolls from damage by stray iron or steel.

Each feed hopper is also equipped with a level switch that stops the motor of the corresponding feeder and opens the main rolls when a low level is reached. An alarm warns the operator.

Flaking releases moisture, which migrates to the surface of the flakes and might be a hindrance to solvent penetration in the extractor. The surface moisture is removed by passing a strong current of air through the flaker's discharge hoppers and conveyor.

The flakes are collected in a slow-moving chain conveyor and conveyed through the safety zone to the extraction plant. An air break must be provided in the succession of conveyors transporting the flakes from the preparation plant to the solvent extractor.

3.4.5 Expander

The expander (Figure 3.3) is an optional additional step in soybean preparation.

Figure 3.3 Expander with feeder and steam injection nozzles. Courtesy of Allocco. The full colour version can be found on the plates.

It is an extrusion cooking technique which consists in heating the flakes in a few seconds by mixing with live steam, then pushing them through a restriction. The equipment itself consists of a horizontal barrel through which the material is pushed forward by means of a rotating worm assembly. Live steam is introduced inside the barrel by means of nozzles and mixed with the product, raising its temperature and moisture, as well as the pressure inside the barrel. Then the material is pushed through the outlet section, which can be either a die plate with several orifices or a hydraulic cone.

Due to the high pressure reached inside the equipment, an expansion phenomenon of the product and a flash evaporation of the water in the product take place at the outlet. As a result, the product has a 'sponge'-like texture.

The expander is an excellent tool for increasing the capacity of an existing extraction plant, because it:

- Increases the bulk density of the material in the extractor.
- Produces a higher percolation in the extractor.
- Reduces the solvent retention of the extracted material entering the meal desolventiser.
- Enables the same residual oil content to be achieved after extraction, with higher full miscella concentration.

3.4.6 Soybean dehulling

3.4.6.1 Traditional process

The soybean is composed of a kernel protected by an external shell or hull. The hull represents 7–8% of the weight of the bean.

The kernel contains approximately 20% oil, 38% protein and less than 3% fibre. The hull contains about 40% fibre, 10% protein and 1% oil.

Dehulling consist in separating the hull from the kernel, in order to produce high-protein, low-fibre meal. To be considered hi-pro, soya meal must have a protein content of 48% or more, and less than 3.5% fibre. Lo-pro meal, produced by extracting the oil from soybeans without dehulling, has around 44% protein and 7% fibre.

The main interest of producing hi-pro meal is that it increases the amount of energy that is metabolised when it is given as feed to chickens and other nonruminants. This translates into about a 20% increase in weight gain per kilogram of meal fed to the animal. Poultry feed represents by far the largest market for soya meal. When the extracted meal is destined to feed birds, fish or monogastric animals in general, it is advantageous to separate the hull and process the kernel only.

The main destination of lo-pro soya meal is cattle feed. Ruminants are capable of digesting fibre. Cattle also represent a market for soya hulls, the byproduct of soybean dehulling.

The hull sticks to the kernel. To help their separation during dehulling, soybeans are first dried and then stored for some time. This storage time is called tempering, and for its duration the dry hull loosens itself from the kernel, making the later separation of hulls from kernels easy.

Before any drying operation, it is recommended that large impurities be separated on a screen in order to avoid any risk of blockage and to aspirate dust, loose hulls and other light impurities, so as to prevent risk of fire in the dryer. A full cleaning is sometimes preferred at this stage, separating large, small and light impurities.

Large impurities (pods) consist of leaves and sticks. They are recovered and sent to a hammer mill, where they are ground. The ground pods are added back to the hulls and sold as byproducts, for their fibre content.

The soybean is dried in a seed dryer, ideally to between 9.5 and 10.5%, without being heated beyond 65 °C. It is then held for tempering. A minimum of 12 hours' tempering is recommended, with 48 hours being optimum. The beans can also be stored for much longer after drying, then processed and dehulled without the need for further drying.

Separation of the soybean hull from the kernel is done after bean cracking and before conditioning and flaking. Drying and tempering loosens the link between the hull and the kernel, and cracking produces a mixture of broken kernels and pieces of hulls.

The dehulling process consists in aspirating the lighter hulls with a current of air, which separates them from the heavier kernel particles. The hulls are separated from the air and recovered in a cyclone or a bag filter. Some good product particles are aspirated with the hulls, and in order to minimise the amount of oil and protein lost with the hulls, screening and a second aspiration of the hulls fraction is usual (hulls purification).

The kernels go to conditioning and flaking. The hulls are generally ground in a hammer mill before storage and dispatch. When exported, they are sometimes pelletised.

3.4.6.2 Hot dehulling process

Hot dehulling is a newer process that is gaining ground. Preparation with hot dehulling combines soybean conditioning, separation of hulls and preparation in one process. It requires no prior drying and tempering.

The cleaned soybeans are elevated to the seed conditioner. This equipment is a vertical tower, built up of elements provided with special shaped tubes, heated by low-pressure steam. It is entirely filled with soybeans, which descend through these elements by gravity and are in direct contact with the tubes.

The soybeans are conditioned over 30 minutes and reach 65–70 °C at the outlet. Bean conditioning takes place in a saturated atmosphere – the beans are said to 'sweat'. The moisture from the internal kernel migrates to the interstitial space between the kernel and the hull. The magnitude of this effect depends to a great extent on properties of the seed, but there will always be a difference in moisture concentration between the kernel and the hull.

After conditioning, the beans are submitted to quick superficial heating in a fluid bed dryer. In this equipment, they are carried through an intensive stream of super-heated air. The heat dries the hull but also evaporates the moisture trapped between the hull and the kernel. This cracks the hull and detaches it from the kernel.

The beans are then cracked and dehulled, much in the same way as in the traditional process. Some adjustments are required, however, since the beans are soft and warm instead of cold and brittle. There is no additional conditioning before flaking.

3.5 Preparation and pressing of rapeseed (canola)

Rapeseed oil is obtained from *Brassica napus*, commonly called oilseed rape or colza, and from *Brassica rapa*, known as turnip rape. Rapeseed is the oilseed best adapted to cold and temperate climates.

Rape is a herbaceous plant whose central stem may reach 1 m in height and a diameter of 1–5 cm. Its leaves are of an elongated shape, with several lobes and a green-blue colour similar to cabbages leaves. The fruits are of the shape of an elongated husk and the peduncles contain 9–25 seeds. The seeds have a spherical shape, are very small (1.5–2.0 mm diameter), are coloured from black-brown to red-black and have a weight of 3–5 mg. They contain 35–45% oil, 19–23% protein, 10–15% cellulose fibre, 3–4% ash and carbohydrates. Moisture when arriving at the plant ranges from 7 to 10%. The seed bulk density is 0.60–0.65 tonnes/m^3.

Other than in Asia, most rapeseed crops grown today are 'double low' or 'double zero' ('00') varieties, meaning they contain low levels (<2%) of erucic acid in the oil portion and low levels (<30 μmol/g) of antinutritional compounds called glucosinolates in the meal portion. Canada pioneered the development of low-erucic acid, low-glucosinolate rapeseed varieties in the mid 1970s; the name 'Canola' was registered in 1978 by the Western Canadian Council Crushers Association to designate them. The trade name today belongs to the Canola Council of Canada.

Currently, the global production of rapeseed is in the order of 60 million tonnes per year, with approximately 20 million tonnes in the EU and 9 million tonnes in Canada. Other major producers are China, India and to a lesser extent Australia.

Oilseeds with high oil content, like rapeseed, are difficult to extract in one step without leaving a substantial percentage of oil in the final cake or meal. To achieve maximum yield, most industrial plants use a two-step extraction process for rapeseed and canola: mechanical pressing first (called prepressing), followed by solvent extraction. A 'prepress' is used to reduce the oil content to around 20%. The press cake is then treated with hexane to produce a meal with 0.7–1.5% oil. Justifying the capital and operating costs is generally not possible for small-capacity solvent plants. At small capacities, high-pressure pressing – either full or double pressing – is a potential alternative to solvent extraction. Depending on the seed type and the process used, residual oil contents of between 6 and 12% may be expected.

Cold pressing may also be used when a particular oil quality is desired. However, extraction rates are generally significantly poorer than with conventional processing routes.

3.5.1 Preparation

As with soybean, it is recommended to start the process by cleaning and weighing the incoming rapeseed.

The cleaned seeds were traditionally sent cold to flaking. Today, it is more usual to start by conditioning them. The purpose of conditioning is to heat the oilseeds to 50–60 °C and soften them before flaking. This is particularly recommended in regions where the incoming seeds can be very cold. The general observation is that conditioning improves the overall deoiling.

The conditioner is a horizontal rotary or vertical stack vessel, much the same as for soybean.

The warm, soft seeds are then sent to flaking. Here again, the flaking mills are of the same type as for soybean. Because rapeseeds are very small, special care must be taken to ensure that all seeds pass between the two rolls, and not through the side. Any seed that is not correctly flaked will cause the residual oil after solvent extraction to rise. Because of their small size, rapeseeds are usually not cracked before flaking.

3.5.2 Cooking

The flaked seeds are conveyed to the cooker–conditioner. The purpose of cooking–conditioning is to heat the flakes in order to soften the oil cells and facilitate the release of the oil. The cooking step has the following main objectives:

(1) Decrease the viscosity of the oil, making it easier to remove.
(2) Rupture the oil cells by flashing off intrinsic moisture as steam.
(3) Coagulate the proteins in the seed.
(4) Sterilise the seed by destroying enzyme activity and preventing the growth of moulds or bacteria.

Multistage vertical stack cookers or large horizontal conditioners can be used (Figure 3.4). In either case, the heating medium is steam at 6–10 bar gauge pressure. The seed is usually heated to 90–105 °C and dried down to 3–5% moisture content in order to ensure efficient operation of the screw press. The exact conditions will depend on the seed being processed and the pressing duty required.

A moisture adjustment is also made. Drying in the cooker helps transmit mechanical pressure to the product during mechanical extraction in the screw press.

3.5.3 Mechanical pressing

The extraction stage itself is carried out using a screw press, or as it is sometimes known, an expeller (Figures 3.5 and 3.6). The press will be fed by means of a variable speed conveyor within the feeder unit. The feeder regulates the flow of material into the press and thereby controls loading on the press main motor. Oil released along the length of the cage is allowed

Figure 3.4 Rotary cooker. Courtesy of Allocco. The full colour version can be found on the plates.

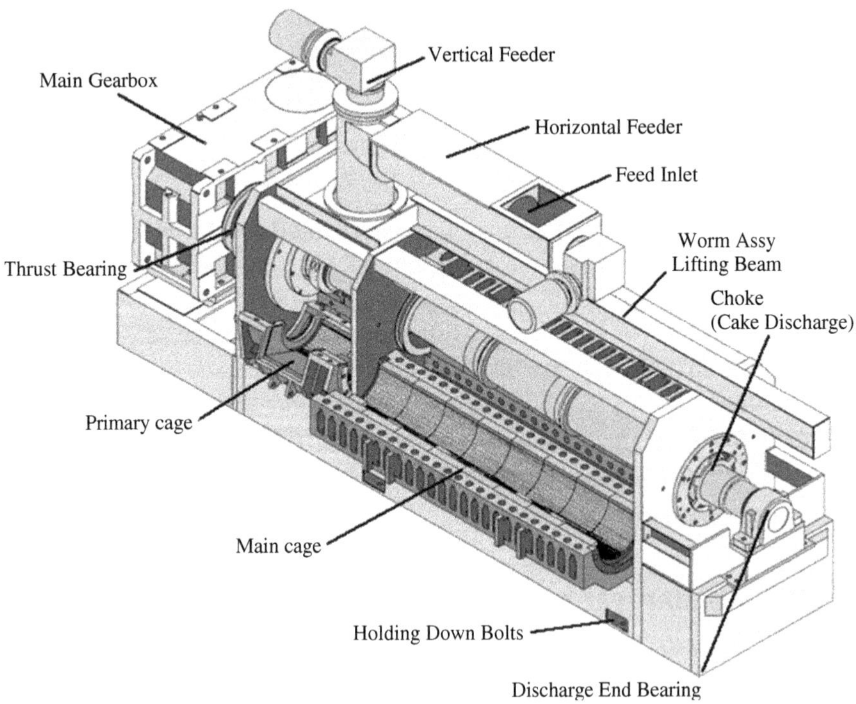

Figure 3.5 Elements of a typical press (opened cage). Courtesy of Desmet Rosedowns.

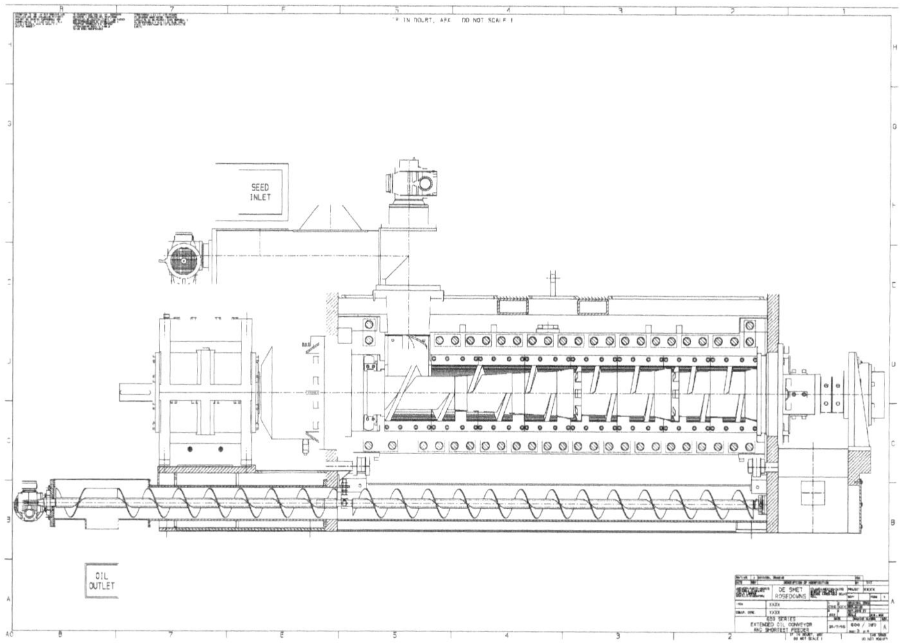

Figure 3.6 Technical drawing of an industrial press. Courtesy of Desmet Rosedowns.

to drain into the base of the press, where it is collected. The solid material remaining within the press is finally discharged into conveyors, to be removed for subsequent processing.

V.D. Anderson produced the first continuous screw press in 1902 and called it 'expeller'. This device was based upon a decreasing-volume Archimedean screw and enabled continuous operation rather than the batch processing imposed by hydraulic presses. Simon-Rosedowns of Hull, England and French Oil Mill of the USA started to produce screw presses at about the same time.

The mechanical screw press soon became the standard method for oil extraction for most materials, and since those first machines the basic design has been copied and improved upon many times over. The performance of a modern press has changed beyond recognition, with developments boosting the capacities and controlling the residual oil levels obtained from a press.

Today there are a great number of press manufacturers around the world, and while all presses share many features, the detail of the machines can vary significantly. These differences can have a great impact on the efficiency and durability of a press. The manufactures themselves are a mixture of local companies, operating in one country or one a region, and a handful of truly international companies, such as DeSmet Ballestra Rosedowns and Hardburg Freudenberger (ex Krupp), which operate on a global scale.

Figure 3.7 Typical worm assembly, showing worms, distance pieces, knife bars and a mixing device. Courtesy of Desmet Rosedowns. The full colour version can be found on the plates.

The fundamentals of a press are more or less independent of both the manufacturer and the press's duty. A mechanical screw press extracts oil by applying pressure to seed material by means of a decreasing-volume Archimedean screw contained within a drained barrel. This screw is generally made up of a number of 'flights' and 'distance pieces' built up on the main shaft of the press. The screw elements are collectively known as the 'worm assembly' (Figure 3.7). The drained barrel, known as the 'cage', is usually made up by lining the inside of a cylindrical frame with parallel bars, separated by shims.

The worm elements themselves consist of a cylinder, bored to fit on to the shaft and driven by means of a keyway. The flight is wrapped around the outside of the cylinder. The flight is an open-ended helix, and has the effect of pushing material along the barrel as the worm is rotated. Between each worm element are distance pieces. These are similar to the worms but without a flight. The inclusion of distance pieces allows 'knife bars' to be included within the cage. A knife bar is a stationary blade that is attached to the cage and projects into the barrel in order to stop the rotation of seed material along with the worm assembly. This helps ensure that the seed material is pushed forward. The worm assembly is arranged so that as material is transported along the press, the bosses of the worms becomes larger and the pitch of the flight smaller. This has the effect of progressively reducing the volume swept by the worm flights, applying pressure to the solid material and causing the oil to be released. Release of oil and a greater degree of milling of the material within the press can be achieved by the inclusion of other devices, such as pressure pieces and mixing rings, into the worm assembly design.

Figure 3.8 Oil flowing between cage bars. Courtesy of Desmet Rosedowns. The full colour version can be found on the plates.

The second major element of the press is the cage. The cage forms a drained barrel around the worm assembly, which allows the oil to be expelled while retaining the solid material within it. The cage consists of a series of semicircular 'barrel rings' linked together by longitudinal ribs, top and bottom. Into this half-cage are placed 'lining bars': parallel bars laid in the cage, separated by shims or 'spacers'. This has the effect of producing a series of parallel slots along the barrel, through which the oil may flow (Figure 3.8). The lining bars are clamped into the cage by 'shoe frame bars' and the knife bars.

3.5.4 Press oil clarification

Fine solids or 'foots' accompany the oil produced in a screw press. These solids must be removed from the oil. This clarification is generally a two-stage process.

Most of the solids can be simply removed by screening the oil over either a static screen or a vibratory screen.

The screened oil is then clarified using a filter – usually a hermetically sealed filter with stainless-steel leaves, in which the material itself is used to generate a precoat through which the oil is filtered.

On large prepress plants, centrifugal separators are used in place of filters.

The solids from the oil clarification process are generally quite high in oil content. They are normally returned to the press cooker feed stream for reprocessing.

Figure 3.9 Pressed cake at outlet. Courtesy of Desmet Rosedowns. The full colour version can be found on the plates.

The crude press oil is generally dried and cooled, and sometimes water degummed, before storage and dispatch or refining.

3.5.5 Press cake treatment

Prepress cake needs to be delivered to the extraction plant in a form suitable for the particular type of extractor in use. This can entail breaking the cake to ensure that the solvent percolates through it at the correct rate and/or cooling the cake to about 60 °C. Other than this, the cake should be produced in a form that can be sent directly to the extractor (see Figure 3.9).

Typical prepressed cake from rapeseed has an oil content between 18 and 22%, a moisture content of 5–7% and a bulk density of 0.4–0.5 tonnes/m^3. The average size of cake pieces going to the extractor should be around 6 mm, and the proportion of fines limited to 10%. Rapeseed press cake is very fragile and must be handled with great care before cooling.

3.6 Preparation and pressing of sunflower seed

As with rapeseed, the high oil content of sunflower seed requires a two-step process – mechanical pressing followed by solvent extraction – to fully recover the oil.

The seed is composed of a kernel protected by a shell, and it is customary in many markets to dehull the seed to a degree before extracting the oil.

Sunflower oilseed originates from Argentina, where it is still the second largest oilseed crop (behind soybean). Its main area of production is Eastern Europe, including Russia. It is also the major oilseed grown in the southern half of France, Spain and Turkey.

The sunflower is well known (as suggested by its name) by its heliotropism. The stem is tall, up to 4 m high, has a diameter of 2–8 cm and supports very wide leaves and a heavy flower (15–40 cm wide). The flower is surrounded by orange leaves and carries a quantity of seeds corresponding to 50–60% of its weight. Individual seeds have an elongated shape, with a length of 7–19 mm, are covered by a ligneous shell, which is very hard, abrasive and waxy (3% wax) and are generally of black colour.

The exact composition of sunflower seed varies according to its origin. The undecorticated seed may contain from 42 to 48% oil. The shells represent 25–30% of the total weight.

Sunflower seeds contain:

- 70–75% pure kernels, containing 55–65% oil, 5–10% moisture, 52–57% protein (on an oil- and water-free basis) or 16–20% as such, the balance being carbohydrates and ash.
- 25–30% pure hulls, containing 1.5–3.0% oil, 7–11% moisture, 4–6% protein, 50–60% fibre, 5–7% ash and carbohydrates.

When arriving at the plant, sunflower seeds are generally dried to 7–8% moisture – adequate for their storage and for their later processing. As with rapeseed, the extraction of the oil is carried out in a two-stage process involving mechanical pressing followed by solvent extraction.

If processed without separating the shell, the deoiled sunflower meal has a protein content of around 26–28% and fibre content of around 24–26%. This composition makes it suitable for feeding ruminants, but less so for feeding poultry and other monogastric animals. It is customary in regions where sunflower seed is a major source of seed oil to separate about 15% of hulls (on a seed basis) and produce feed meal with about 34% protein content, or to separate up to 22% of hulls and produce high-protein meal with 37–39% protein and 14–15% fibre. The dehulling consists in most cases in opening the seeds by impact, then separating hulls and meats by screening and aspiration. A two-step dehulling process is generally used to produce high-protein meal, in order to avoid entraining more than 1–2% oil (above botanical) with the hulls.

The hulls are in most cases conveyed to a specially designed boiler, where they are burnt to produce steam for the process. A kilogram of hulls produces between 3.5 and 4.5 kWh of heating energy. In large plants, the boiler can be coupled with a turbine to produce electricity. A sunflower seed crush and refining plant producing high-protein meal produces enough hulls to generate all the steam and electric power it needs to run.

A minimum of 8% of hulls are left with the meats. Without this, there would not be enough grip for the press to efficiently extract the oil, and the cake would lack the structure required to stay firm in the extractor.

After dehulling, the meats are flaked to 0.4 mm thickness, cooked and mechanically pressed to 18–22% oil, then sent to the solvent extractor.

When processed unhulled, seeds are cracked in a double-pair cracking mill, cooked and pressed before solvent extraction. Flaking the cracks before cooking improves deoiling, but this has to be balanced against the wear caused by the hulls to the flaking rolls. Processing undecorticated seeds will also substantially increase the power consumed by the press per tonne of product entering, and accelerate wear in the press.

Typical prepressed cake from sunflower seed has an oil content of between 18 and 20%, 5–7% moisture and a bulk density of 0.4–0.5 tonnes/m^3. Sunflower seed press cake is hard and must be broken into pieces. The average size of cake pieces going to the extractor should be around 6 mm, and the proportion of fines limited to 10%.

3.7 Full pressing

In large plants, full pressing is not a viable option as the cost per tonne of seed processed is far greater than that for solvent extraction plants. In addition, crude oil has a much higher value than cake or meal, and leaving too much oil with the meal is highly costly. However, there are reasons why the decision may be taken to build and run a small press plant.

In areas where poor infrastructure makes transport of seed over long distances difficult, processing seed locally means only the high-value products have to be transported, relieving the major problem. Additionally, the capital cost of a large extraction plant is beyond the means of all but the largest companies, and therefore many smaller and family-run businesses begin with small full pressing operations.

Cost and practicality are however not the only reasons for the existence of high-pressure pressing plants. Even in Northern Europe and North America, where solvent extraction remains the norm, there can still exist good economic reasons for operating a small high-pressure press plant. Usually these are centred on a small niche product that can command a premium price. This might be an oil or cake with a particular property or it might just be a small-scale speciality crop.

3.7.1 Cold pressing

A more recent phenomenon is the use of full pressing, as well as cold pressing, to process organically produced seed, creating an oil that has not been in contact with hydrocarbon solvents. These products are typically marketed as speciality products and command a premium price in the health food market.

The idea of cold pressing is not a new one. It probably started originally with the use of hydraulic presses in the 19th century, although there are examples of beam presses going back three or four hundred years; with these machines, there was little or no heat generation in the pressing operation. Today, this technology continues to be used for the production of the highest grades of very specific products, such as virgin olive oil, in which flavour is a desirable characteristic.

The use of the term 'cold pressing' has today become rather broad. It is generally used to describe a process in which no heat is applied to the raw material prior to its passing through a conventional screw press. In the press, heat is generated through friction. Modern worm assembly design is reducing the energy cost per tonne of material processed by a press, enabling temperatures to be better controlled.

Such technology has been used over many years with small presses in the range 40–200 kg/hour of seed. These presses are capable of processing whole seed that is fed directly into them. They have been used in several areas:

(1) For the development of village industry in developing countries. For a very small outlay, a village can extract its own oil for use locally in the crude form for cooking purposes. Machines are often funded through aid agencies.
(2) Research work in academic institutions.

Cold pressing was also attempted in the late 1970s as an alternative to conventional prepressing using technology similar to that employed in the mini presses. Scaling up the technology was difficult, and at that time it was difficult in practice to maintain a uniform prepress cake for feeding to the extraction plant, in which both percolation characteristics and extractability are critical parameters. Many such plants subsequently retrofitted conventional cookers or conditioners.

With various soft seeds, there is an advantage to producing cold-pressed oil for the specialised market. There is a growing demand for products that are perceived to be free of contact with any chemicals and can be viewed as natural. If you visit any supermarket, you will find not only the traditional cold-pressed virgin olive oil but also a whole range of other cold-pressed products being marketed at premium prices.

3.7.2 Double pressing

In the late 1970s, locally manufactured small machines with capacities of around 400 kg/hour were extensively used for the double pressing of palm kernel in Malaysia, where a single plant can have as many as 70 small machines. In this situation, it eliminated the need for steam generation. Maintenance in such an operation is very high, with many machines offline at any one time.

Current developments in this field fall into two distinct categories. The first is the processing of palm kernels or hard seeds. Using either single or double pressing operations, it is possible to reduce the oil content to as low as 7 or 8%, provided careful control of the moisture in the whole kernels is maintained. The second category is the processing of soft seeds, where it is not possible to achieve a low oil content in the cake through purely cold pressing. This is leading to the development of double-press plants, in which cold pressing is followed by hot pressing to enable maximum total oil recovery.

For the cold pressing stage, the seed is broken and/or rolled depending on its type. It is then fed to the press without any heat treatment. The result of the first stage is typically a high-quality crude oil, which is low in phospholipids and can be bottled and marketed with little downstream processing. This oil is processed separately from the second-press oil and sold at a premium. It is important to emphasise that this premium market is growing, but is still marginal.

The cake from the first press is still relatively high in oil. This material is generally given a conventional cooking, before being fed to the second press to produce normal hot-pressed oil as a secondary product. Several double pressing plants were installed, particularly in Germany, when the biodiesel boom began in the EU. These plants leave 8% or more oil in the cake, and became unable to compete once big players entered the biodiesel market with large prepress + solvent extraction operations.

3.7.3 Cake treatment

Full-press cake destined for use without further processing must be broken in order to aid mechanical handling. The cake produced from a high-pressure press can come in quite large pieces, and as it cools these become very hard. In smaller plants, the cake is typically broken in an integral cake breaker mounted in the discharge chute of the press. In larger, multipress plants, a separate set of breaking rolls may be added to handle all the cake.

The main hazard with full-press cake is spontaneous combustion during storage. In order to reduce the dangers of this, the cake, which may exit a press at well over 125 °C, must be cooled before storage. Cake is commonly cooled to within 10 °C of ambient temperature, but even after this the cake store should be checked regularly for temperature variations. It may also be necessary to move the cake about if it is stored for extended periods. One must be careful and admit that the high oil content of cake (>5%) prevents very long storage.

3.8 Oil from other seeds

Edible oil – and in some cases industrial – is also extracted from a number of other seeds. The list in this section gives some examples, without being exhaustive. In most cases, the oil production process involves unit operations that are similar to the ones described earlier; that is, cleaning, sometimes cracking, with or without dehulling depending on the case, conditioning/cooking and flaking, followed, depending on the oil content, the value of the oil, the plant size and location and the market, by full pressing, mechanical pressing plus solvent extraction or direct solvent extraction.

3.8.1 Cottonseed

Cotton oil is extracted from the cottonseed, a byproduct of the cotton ginning industry. Its world production ranges around 46 million tonnes annually.

The cotton fibre is used for textiles. It represents about 30–35% of the total seed weight and is separated from the seed in ginning factories. The seed from ginning remains covered with short fibres called 'lints', adhering to the shell. Before extraction, in most plants the cottonseed is delinted and decorticated – 100 kg of 'white' (undelinted) cottonseed yields approximately:

- 8 kg of lint;
- 28 kg of hulls with 1.6 kg of lint;
- 62 kg of decorticated seeds with 32% oil.

Delinting and decorticating of cottonseeds are jobs on their own, requiring extensive machinery. The seeds are then flaked, cooked and pressed before solvent extraction.

Alternatively, the seeds can be flaked, conditioned and expanded, and then direct solvent extracted. This process is the standard in the USA, whereas prepressing is the preferred method in countries where the extracted meal is given as feed to animals other than ruminants.

Cottonseeds may contain from 1 to 3% gossypol, a toxic pigment which is distributed in microscopic glands throughout the tissue of the meat. Mechanical operations will not rupture the cells, because of their small size. Heating, in the presence of moisture, will effectively rupture the pigment glands, but the degree of heating required will also denature the proteins. During cooking before pressing, the free gossypol binds with protein and is converted to a nontoxic form called 'bound' gossypol. The 'unbound' or toxic gossypol is called 'free' gossypol. Ruminants can digest the free gossypol but other animals feel its toxic effects (such as a growing shortage) when they eat meal containing more than 0.18%. The temperature reached during conditioning before direct extraction is not high enough to bind the gossypol.

The protein content of the solvent-extracted meal is 25–30% if the seed has not been decorticated before processing and in the range of 35–45% if it has.

Crude cotton oil has a red colour, coming from the gossypol pigment.

3.8.2 Corn germ

Maize is a complete food: it contains 70–80% starch, 7–8% protein and 1–2% fatty matter. It is used to prepare a corn flour which is the basic food of many different peoples around the world.

The starch, dextrose and breweries industries consume huge amounts of corn. After fermentation, corn may produce alcohol.

Corn oil is extracted from the corn germ, and its oil content varies widely with the degermination method. The germ represents approximately 12.5% of the grain weight. Degermination is carried out in one of two ways, depending on the industry:

- The dry process is used in feed milling. The corn, once cleaned, is humidified to 20% water content before entering the degerminator and having the germs separated by sieving. The germs are then dried to 14% moisture. Their oil content is usually only 18–20%, because part of the grain stays with them. The germs can be solvent extracted directly after a simple preparation consisting of cooking and flaking. The flakes are sometimes pelletised or expanded before extraction, but not always.
- The wet process is used in the starch and derivatives industry. The cleaned corn is left steeping in large baths for a period of 30–40 hours, during which acidulated water (0.2% sulfuric acid) causes the grain to soften. The corn then passes to the degerminator, where it is crushed and the germs are separated from the grains. The germs are washed in a large amount of water, in order to separate particles of seeds by floating. Finally they

are squeezed and dried to 5% moisture. This process produces a cleaner separation and the germs may contain up to 50% oil. With this process, the germs must be flaked, cooked and pressed before solvent extraction.

The crude oil has a colour from golden to dark yellow and a specific taste. The refined oil is used for salad or cooking oils and in the preparation of margarine. The meal is used as a component for animal feed. It contains about 20% proteins.

3.8.3 Coconut or copra oil

The coconut palm tree is found on all tropical coasts.

The coconut is made of a very hard shell containing white, sweet liquid called coco milk. When the fruit has matured, nearly all the milk has been transformed into albumen. It is this albumen that makes the copra, after appropriate drying.

The preparation of the endosperm in view of oil extraction varies from country to country, but is usually done in a primitive way. After a manual separation from the fibrous husk, the coconuts are cut in two to eliminate remaining milk; they are then dried, either by exposure to the sun or in a rotary drum dryer that uses the husks and shells as fuel.

The drying method has a large influence on the copra oil quality: it must not be overheated.

At 7% moisture or lower, the shell separates easily, and the dried endosperm is then cut with copra cutters to pieces of about 15–20 mm long. Copra contains about 65% oil and 7.5–8.5% protein. The copra pieces are cracked and flaked, cooked and dried to 4% moisture, then pressed to 23–25% residual oil.

The cakes are then solvent extracted. In many cases, the cakes are flaked again before extraction. The extracted meal is used as animal feed. The meal contains15–20% protein and over 40% carbohydrates.

3.8.4 Linseed (flaxseed)

Flax species used for oil production are different to and smaller (0.3 m high) than the ones used for the production of textile (1.5 m high).

The fruit contains about 10 flat, 5 mm-long seeds. The seeds must be cleaned and dried to 10% moisture for storage, or else be processed directly. Linseeds contain about 38–45% oil and about 25% proteins.

Cold pressing produces a light yellow oil, which may be used as food (Russia). Linseeds contain a small percentage of a cyanogenous carbohydrate called linamarin and an enzyme called linasis. At 40–50 °C and in the presence of moisture, the enzyme acts on the linamarin to release cyanhydric acid (HCN). During cold pressing, the linasis and the linamarin are not eliminated, and feeding the cake to animals may kill them.

The seeds are cracked on corrugated rolls, flaked, cooked and hot pressed. Hot pressing produces a cake with around 8% residual oil. Prepressing to around 18% then solvent extracting to 1% oil is also carried out.

The heat treatment in the cooker and in the press permits the elimination of toxic elements. Cakes or meals are thus no longer dangerous for livestock feed; they actually make excellent feed, as they contain 30–40% proteins and are easily digested. They can also be used as fertiliser.

The oil from hot pressing and that from solvent extraction is of dark brown colour and is used in the varnish, paint, linoleum and soap manufacturing industries.

Linseed oil's main characteristic is that it contains 45–58% linolenic acid, which makes it the best known siccative oil. It has the property of absorbing the oxygen from ambient air and forming a solid but elastic material known as linoxyne. It is this characteristic that makes it interesting to the paint and varnish industries.

3.8.5 Safflower

Safflower has a very long history of cultivation in the Middle East, India, Egypt, Ethiopia and China, primarily as a medicinal and dye source. It was only in the middle of the 20th century that it started to be cultivated as oilseed. The major producer today is the state of California. Safflower seed is also grown in some areas for bird feeding.

Safflower seed resembles sunflower seed, except for its ivory-like colour. Its oil content averages 35% but can be as low as 30% or can reach over 40% according to area of growth and its variety. The protein content is around 15%, but also varies widely. The hard shell represents 33–45% of the seed.

For the production of oil, seeds are cleaned, cracked, cooked and pressed, then solvent extracted. Flaking before cooking is sometimes added to the process. Dehulling was tried at first, but is not done industrially.

It is also possible to replace the press with an expander, preferably equipped with a drainage cage, then carry out solvent extraction.

The oil is highly polyunsaturated, containing 60–80% linoleic and around 15% oleic acid, but no linolenic. As a semidrying industrial oil, the high linoleic level, the absence of linolenic fatty acid and the low colour values give it the ability to produce nonyellowing white paint of excellent quality. Of all the commercially available edible oils, it has the highest content of polyunsaturated fatty acids.

3.8.6 Peanut (groundnut)

Groundnut or peanut is a herbaceous annual plant of the leguminous branch, originating from tropical America. The most important countries producing groundnut are India, the USA, Argentina, China and some tropical African countries. The fruit is made of an external shell (21–29%) and the nut (79–71%), consisting of:

- a thin hull surrounding the nut (2–3%);
- the nut (69–73%);
- the germ (2.0–3.5%).

The groundnut contains 40–55% oil, 30% protein and 12% hydrocarbon matter; its high vitamin B content makes the groundnut an essential part of a balanced diet in tropical countries.

A large number of peanuts produced are locally consumed; they are eaten crude or grilled, slightly roasted, and are used as appetisers and in the confectionery industry.

'Peanut butter' is made from a preparation of crushed groundnuts, roasted and mixed with 5–7% groundnut oil and salt.

Peanuts are available unshelled or decorticated. Once decorticated, however, it is difficult to store them, as the oil acidifies rapidly. The decorticating allows the transport volume to be reduced to a considerable extent. Generally, peanuts are not decorticated except when they are to be used in the production of industrial nonfood oil.

Decorticating is done either by corrugated rolls, by pounding or by centrifugation; the shells are then separated from the nuts by ventilation and the groundnuts are dehulled from their fine husks.

To obtain the oil from decorticated groundnuts, they are cleaned and cracked before pre-expelling, and the cakes are cracked, heated and flaked before solvent extraction.

Depending on whether they are dehulled or not, groundnut cakes are called 'white cake' or 'brown cake'. The cakes are called 'shelled' when they still contain a certain amount of shell. Shelled cakes are used as fertiliser. Decorticated, they are used for animal feed.

The bran made by the hulls is an excellent animal feed as it contains 14–19% oil and 22% nitrogen-containing matter.

The extracted meal is still very rich: 41–50% protein content. Shells and hulls are used as fertilisers or as combustible feed for the boiler.

Refined groundnut oil is an excellent food-grade oil.

Aspergillus flavus, a mould present in the soil, mostly in tropical countries, frequently contaminates peanuts. The mould generates toxic products, the most dangerous being aflatoxin B1. It grows particularly well on materials rich in carbohydrates. To develop, it needs a relative humidity of 80% and a temperature between 30 and 40 °C. It is especially active on seeds with 15–30% moisture and does not develop on seeds with moisture below 8%. For groundnuts, the most critical time is harvesting: any defect in drying before ensiling in tropical conditions and any increase in moisture afterwards may have extremely bad consequences. Therefore, drying to 8% or less is required.

Once present in a contaminated nut, aflatoxine is not eliminated in processing and will end up in the meal, where it represents a danger to animal health. This toxin is resistant to high temperature and is practically insoluble in hexane; basic, acidic and oxidising agents are the most appropriate means of breaking it down.

3.8.7 Rice bran

Rice is the staple cereal for at least half of the world population.

A very fine integument covers the grain; it is also called 'rice bran' or 'silvery skin'. The husks remain adhesive to this even after threshing. The grain plus husks and bracts is called 'paddy', the decorticated rice is named 'cargo rice' and the rice cleaned from its integument is called 'white rice' or 'polished rice'.

The crop is harvested with sickles and the sheaves are dried and put in stacks. Threshing is mechanical or manual.

The rice bran from the polishing of rice is a highly nutritive food: 35% starch, 10–20% fatty matter and 10–12% protein.

Rice bran oil can easily be obtained in direct solvent extraction of the rice bran. Due to the powdery nature of the rice bran, however, it requires preparation before extraction in order to ensure a good percolation of the solvent. It is thus heated and humidified before pelletising (to compress through a die).

Due to the acidity of rice bran oil, special care should be taken to select an appropriate construction material. Moreover, as sand is often present in rice bran, heavy wear can be expected.

The crude oil generally has a very high acidity (up to 50% free fatty acids (FFA)!). This is due to the fact that the oil contained in the rice bran rapidly deteriorates in humid atmosphere and under the action of a lypolitic enzyme (lipase); this enzyme favours the hydrolysis of triglycerides and makes fatty acids free. After 2–3 days of storage, the acidity of the oil rises to 10%; during the first hours, it may rise at 1% per hour! It is thus necessary:

- Either to process the rice bran immediately after the decorticating. This is rarely possible, due to the fact that the solvent plants are usually far from the fields where the threshing is done.
- Or to sterilise the rice bran at the rice plants, by heating it at 90–100 °C and drying it in order to stop the action of the lipase and thus allow storage before transport and extraction.

Rice bran oil, due to its high acidity, has often been used in making soap. As it is produced in countries with high populations, it has of course also been used for human food. Much research has already been done on the subject of refining rice bran oil.

Rice bran oil still contains 5–10% waxes and stearins, which can be removed easily in a miscella phase winterising.

3.8.8 Sesame seed

Sesame (*Sesamum indicum*) is the most representative of the pedialaceous branch and exists under many different species. Known since the highest antiquity, it has many different names: s.a. ajonjoli (Mexico), benne, till and simsia.

Sesame is an annual herbaceous plant, with a height of 1.0–1.5 m, which grows mostly in tropical and subtropical countries' hot climates. The main producers are China, India, Burma, Turkey, Egypt, Sudan and Mexico.

The fruit consists of a pod with an elliptic form, containing up to four longitudinal cells, each divided into two parallel cells, which may contain 15–20 seeds each. The seeds are white, yellow, dark red or black.

The sesame seed trade divides them into two categories:

- White sesame, which contains more than 95% white or yellow seeds.
- Black sesame, containing 15–25% of black seeds.

Sesame seeds contain 45–55% oil and 19–25% protein. The oil from white sesame seeds is of better quality.

The extraction of oil is either by cold pressing or by prepressing followed by solvent extraction. The seeds are cleaned, flaked and conditioned before hot pressing. The oil from cold pressing is yellow and may be consumed as such after filtration. The oils from prepress and solvent extraction are darker and require refining. Sesame oil is an excellent, semisiccative, edible oil. It is appreciated and known for its high resistance to rancidness and its agreeable taste.

Sesame meal contains 40–45% proteins; it may be used as feed for ruminants and poultry when mixed with soybean meal. Sesame seed, however, is mostly consumed as seed, in the bakery and pastry industry. Only a small quantity is crushed to produce oil.

3.9 Olive oil production

Olive oil and palm oil are not extracted from seeds, but from the pulp of fruits. Their production processes are entirely different from those previously examined.

The olive, the fruit of the olive tree, has been used for more than 6000 years for the production of olive oil. It is linked to Mediterranean culture and history. The Latin words *olea* (oil) and *olivum* (olive) are derived from the ancient Greek name *elaia*. From the Hebrew *zait* come the Arabian words *az-zait* (aceite in Spanish) and *zaitum* (aceituna in Spanish).

Olive trees are grown mostly in countries around the Mediterranean, but also to some extent in California, Argentina and Australia. The main producing countries are Spain, Italy and Greece, followed by Syria, Tunisia and Turkey. Well over 90% of the world's olive oil is produced in the Mediterranean region.

The yearly world production of olive oil is around 3 million tonnes. Its consumption is stable in the traditional producing countries, but consumption and imports have increased in spectacular ways in Northern Europe and the USA over the last 20 years. Today, these regions represent over 20% of world consumption. The main reason for this success is the dietetic quality of olive oil.

The olive tree is an evergreen, and can live several hundred years. It grows in rather arid areas and protects against desertification and soil erosion.

The olive fruit is oval shaped and has an average weight between 4 and 12 grams. It contains:

- A pulp representing 65–80% of the total weight and containing 50–60% water and 20–25% oil (40–60% on dry basis); the pulp is surrounded by a skin.
- A kernel or pit representing 15–30% of the total weight and containing about 30% water and 8–10% oil (12–17% on dry basis).

The harvest of olives is a delicate process. The quality of the olive oil depends on a number of factors, starting with the degree of ripening of the fruit. The fruits used for the production of first quality (extra virgin) oil are collected by hand or in another manner that does not damage them.

Harvesting in the northern hemisphere starts in November or December and lasts about 3 months. For optimum quality, the fruits should be processed immediately after harvesting.

Virgin olive oil is extracted from the pulp, and has the unique feature of being edible without further processing, if obtained from good-quality raw material.

Classification of olive oils according to the EU, International Olive Oil Council (IOOC) and the Codex Alimentarius is as follows:

- virgin olive oils:
  - obtained cold from the fruit by physical means; moisture and volatile $< 0.2\%$; impurities insoluble in petroleum ether $< 0.1\%$;

  - 'extra virgin olive oil' with organoleptic index above 6.5 and free acidity expressed as % of oleic acid up to 0.8%;
  - 'virgin olive oil' with organoleptic index above 5.5 and free acidity < 2%;
  - 'ordinary virgin olive oil' with organoleptic index above 3.5 and acidity < 3.3%.
- olive oils:
  - obtained from further treatments:
  - 'virgin lampante olive oil' was used long ago in oil lamps, is obtained from a further hot pressing, organoleptic rate < 3.5 and free acidity > 3.3%;
  - 'refined olive oil (*)' obtained from refining virgin oil in a way which doesn't lead to alteration in the natural triglyceridic composition, FFA up to 0.3%;
  - 'olive oil' obtained by blending refined olive oil with virgin oil, FFA up to 1%.
- olive kernel oils (olive pomace oils):
  - obtained by solvent extraction of the olive pomace or kernel, the solid residue left after virgin oil extraction;
  - 'crude pomace olive oil' crude solvent extracted;
  - 'refined olive residue oil' (refined oils cannot be traded in the EU) refined in a way that does not alter the natural triglyceridic composition, FFA up to 0.3%;
  - 'olive residue oil' obtained by blending refined residue oil with virgin oil, FFA up to 1%.

Parameters used for evaluating the quality of an oil include FFA content, peroxide value, absorbance in ultraviolet wavelength, organoleptic assessment, halogenated solvent content and heavy metals.

There are basically two methods for extracting the oil from the olive fruit: pressing and centrifugation.

3.9.1 Pressing

Pressing is the oldest method. The olives are first cleaned of leaves and washed, and then crushed, generally in a stone mill, in order to tear the flesh and release the oil from the cells that contain it. After crushing, the paste that is formed is slowly mixed at ambient temperature for 10–20 minutes. This step allows the oil droplets to merge and increase in size. Cylindrical mixers with a vertical or horizontal shaft equipped with blades are used for this purpose.

From the mixer, the paste is spread in layers in a hydraulic press. A single pressing step, lasting 1.0–1.5 hours, is generally applied. The press separates a water–oil liquid mixture from a solid phase, called pomace, containing stones and pulp residue. Oil and water are then separated by centrifugation.

Pressing is simple and efficient, but it is a labour-intensive, discontinuous process. Oil recovery yield is 86–90%. Its main drawback is labour cost and the risk of contamination in the press, if it is not kept spotlessly clean. Today, the traditional pressing method remains in use in very small plants only.

3.9.2 Centrifugation

Oil extraction by centrifugation is a more recent process. It allows the operating costs to be reduced and the production capacity to be increased. The separation of oil is performed in horizontal separators, called decanters.

The olives first go through a leaf-removal and water-washing process, followed by crushing. Crushing is generally done in a hammer mill, which produces a finely ground paste. Hammer mills are preferred to stone mills because they enable a higher oil yield to be attained in the centrifugation process. The paste is then mixed and heated, generally for an hour or more. This duration is necessary to break the oil–water mixture that is formed in the crusher.

For many years, separation was carried out in three-phase centrifuges. Water representing 40–60% of the weight of the fruits had to be added in order to produce a fluid paste, improve separation and increase the oil yield. The centrifuge produces three phases: an oil–water mixture, vegetable water mixed with the added water and the solid waste, or pomace. Oil and vegetable water are then separated in a vertical centrifuge. This process has the disadvantage of producing about twice the amount of wastewater as traditional pressing. In addition, some antioxidants are lost with the wastewater.

Two-phase centrifuges began to be used to take care of the wastewater problem. There is no wastewater phase, and the solid waste or pomace is very liquid. This liquid pomace is called *alpeorujo* in Spain. Its water content is about 60–65%, compared to 50% from the triphase centrifugation process and 35% from pressing.

Oil yields are similar with the two centrifugation processes, and slightly higher with pressing.

3.9.3 Olive pomace extraction

The olive pomace contains enough residual oil to make it worth extracting. The acidity of the oil is around 2% at the beginning of the harvesting campaign. It rises by 2% per month, due to hydrolysis and enzymatic action. The pomace should thus be processed as quickly as possible.

The olive pomace is first dried to 8–10% moisture. Rotary dryers employing hot air circulating co-current with the product are used.

The dried olive pomace contains:

- 45–50% kernel;
- 50–55% pulp, of which 5–7% is skin.

The pulp has 10% moisture and 10–15% oil content and the stone contains 1–2% oil.

The oil can be solvent extracted with hexane without separating the pulp and stone (batch process). After extraction, the pulp meal contains 4–5% oil, the stones 0.3–0.5%.

A residual oil content in pulp of less than 1% is achieved when using a continuous percolation extractor. In this process, the pomace pulp and stones are separated and the pulp is pelletised before extraction. The stones are not extracted and some oil is lost with the pulp sticking to the stone when they are separated. The stones are an excellent fuel, in the first instance for the pomace dryer. The spent pulp meal can be used as animal feed, fertiliser or fuel.

3.10 Palm oil production

Palm oil is obtained from the *Elaeis guineensis*, a tree native to Guinea, West Africa. Today, the oil palm is grown on plantations of the equatorial tropics (latitudes 15° north to 12° south) in South East Asia, Africa and Central and South America. Malaysia and Indonesia are the principal producers of palm oil, with Ivory Coast, Nigeria, Colombia, Thailand and Papua New Guinea the other main players.

Thc production of palm oil, which was slightly below that of olive oil in 1960 – 1.2 million tonnes – rockcted sky high in the early 1970s, to reach 2.8 million tonnes in 1975, 4.54 million in 1980, 10.75 million in 1990 and a whopping 51 million in 2011.

These very impressive figures are partly due to the exceptionally high productivity of the palm tree, which yields 6 tonnes of palm oil, corresponding to 30 tonnes of bunches per hectare. These figures definitely outstrip the productivity of all other oil-bearing materials. They also reflect the low cost of production and the oil palm's unique feature of yielding two qualities of oil, namely palm oil and palm kernel oil.

Another reason for the soaring production is demand from the world margarine market, as well as from nonproducing countries who find palm oil a suitably cheap product for refining and fractionating to suit their own requirements. The Middle East is perhaps the principal importer for this purpose, followed by a good number of East African countries.

The early oil palm grew in its wild form and was a source of oil for many inhabitants of the equatorial zones. The plant was very tall and could reach a height of 25 m. Harvesting was a difficult operation, because the fruit grows in large bunches at the top of the tree. Only the natives could pick the fruit, climbing up the trees to do so. Damage to the fruit resulted from improper handling and lack of mechanisation of the milling process. Under these conditions, the quality of palm oil was very poor, due to its very high FFA content (20–25%), caused by lipolytic enzymes developing in the bruised fruit. The oil was used in soap production and in traditional dishes.

Over the last 30 years, constant research, experimentation and husbandry have made it possible to plant short-stemmed varieties. At the same time, high-oil strains have also been developed. These new types of tree grow to a height of 4–5 m and have an extra 20% oil. Among these varieties, the tenera is predominant. It is a hybrid obtained by crossing the dura with the pisifera.

The fruit bunch of a dwarf oil palm bears from 800 to 2000 fruits (1200 on average) and weighs from 10 to 50 kg, according to the plant's age and climatic conditions.

The oil is contained both in the mesocarp (palm oil) and in the kernel (palm kernel oil), and hence the procedure of extraction must be carried out with due care so as to get the majority of the pulp with a minimum breakage of nuts and kernels.

The development of the new varieties, modern estate cultivation, harvesting and extraction techniques have changed the image of palm oil, which is today extensively used in the production of shortenings and margarine, for deep frying and as a liquid edible oil.

The palm fruit offers a range of products at the various processing stages:

- Pressing of the fruit produces crude palm oil.
- The press cake is separated into fibres and nuts.
- The fibres are used as boiler fuel.
- The nuts are processed to obtain palm kernels.
- The shells from the nuts are burnt in the boiler.
- After drying, the kernels are further processed to produce palm kernel oil and cake.
- The palm kernel cake is used as animal feed.
- The empty bunches can be used as boiler fuel after reduction of the water content.
- Empty bunch oil can be used for soap manufacturing.

3.10.1 Before reaching the mill

The actual processing of palm oil does not start at the mill, but in the plantation itself.

It is of the utmost importance to handle the bunches with great care, lest they should suffer rapid degradation. The efforts made to this end in many countries have provided great dividends, and very low-FFA crude oil has now become widespread.

At the time of cutting, the level of FFA in ripe and unbruised fruit is pretty low: some 0.3–0.8%. Improper handling between the tree and the mill causes the acidity to rise very rapidly, especially in the injured outer part of the fruit. This phenomenon is attributable to a very active lipase, which splits the moles of fats into fatty acids and glycerol, once the cell structure of the fruit has been adversely affected.

3.10.2 Sterilisation

Sterilisation is a heat treatment that stops the development of lipolytic enzymes in the fruit bunch. The process also serves two other purposes: it eases mechanical stripping and prepares the kernels for further processing.

Sterilisation is conducted in large steam autoclaves. Immediately after arrival at the oil mill, the bunches are unloaded from the trucks through special transfer hoppers into cages introduced into an autoclave, which will generally contain 3–10 cages of 1.5–7.0 tonnes of fresh fruit bunches (FFB). Once full, the doors of the autoclave are closed and saturated steam is injected until a pressure of 2.5–3.0 kg/cm^2 is reached. Each autoclave normally performs 12–14 operations per day. The bunches are processed for some 45–60 minutes, according to their size and ripeness.

3.10.3 Threshing

The sterilised bunches are then fed continuously into a rotary drum thresher or beater arm thresher, which strips and separates the fruits from the bunch stalks. Rotary strippers comprise a drum made up of equally spaced longitudinal U-bars, which revolve at low speed. Longitudinal flat bars are located radially inside the drum. The clusters are introduced at one end of the stripper and are raised by centrifugal force and by the flat bars, eventually falling on the U-bars. The impact frees the fruits, which pass through the U-bars and are finally removed by a screw conveyor. The empty bunches are discharged at the other end of the drum and sent to the empty bunch hoppers to be evacuated or further processed.

From the thresher, the fruits are conveyed into a digester. The purpose of digestion is to separate the fruit pulp from the nuts and to cause the oil-bearing cells to burst, which facilitates oil extraction by pressing. This operation is of the utmost importance as it affects the oil yield. Proper digestion ensures that most of the oil cells are ruptured, thus easing the task of the press, which will have to break a limited number of unruptured cells. The best digestion conditions are attained by mixing the fruits at a temperature of 90–100 °C for about 20 minutes. The factors influencing oil release are temperature and stirring/shearing, which cause the oil cells to break. The digester is a vertical cylindrical vessel provided with a double steam jacket, in which 3 atmospheres of steam is circulated, and a central paddle stirrer. For high mixing efficiency, it is extremely important that the vessel be always full.

3.10.4 Pressing

The mash, consisting of digested pulp and palm kernel nuts, is then transferred to the continuous screw press, which separates the solid portion, including the fibres and nuts, from the liquid phase (water and oil). Continuous screw presses are made up of a perforated cage or barrel of horizontal cylindrical design, in which the mass is subjected to increasing pressure by a variable pitch worm, which causes the oil to come out. The counter pressure is adjusted by means of a mobile cone, which regulates the press cake thickness. The press also maximally reduces nut breakage and maximises palm kernel recovery. The majority of manufacturers of these continuous screw presses are based in Malaysia.

Two products are obtained at the outlet of the press:

- A mixture of water, oil and solid impurities (sand and vegetable residue).
- A press cake, which contains fibres and nuts.

3.10.5 Crude oil clarification

In the clarification section, water and impurities are removed from the oil to yield a clean and dry product.

A clarification plant conventionally comprises a continuous decantation or settling tank, which separates oil, water and impurities by taking advantage of the difference in specific gravities of the crude oil components. Continuous decantation tanks are of rectangular horizontal or cylindrical vertical design.

Alternatively, three-phase decanters can be used on crude oil or sludge, reducing liquid effluent quantities and maintenance costs.

In some mills, the sludge is treated in centrifuges.

3.10.6 Oil drying

Finally, the oil is dried to below 0.1% moisture. Vacuum dryers are generally used, as the oil is maintained at a low temperature to avoid oxidation.

3.10.7 Fibre–fruit separation

The nuts are separated from the fibrous material in the depericarper, a vertical column through which a hot air current is circulated. The fibrous material is sucked into a duct, then separated from the air in a cyclone and finally conveyed to the boiler house to be used as fuel.

3.10.8 Nut conditioning

After separation from the fibre, the fresh nuts are dried to detach the kernels from the shells. The moisture content of the nuts is reduced from about 16 to some 10–12%.

3.10.9 Nut cracking installation

Large installations require two or more crackers, in front of which it is advisable to install a nut grading screen which can classify the nuts according to size.

3.10.10 Kernel separation

The cracked mixture consists of free kernels, shells, unbroken nuts, partly cracked nuts and dust (fibre, thin shells, bits of broken kernels). We recommend the use of a dry separation system followed by a clay bath or hydrocyclone.

The dry separation system consists ideally of two double-stage winnowing plants, in which separation is carried out by sucking air through a vertical column. This way, a maximum number of fibre, shell and dirt particles are removed without carrying the kernels away.

The clay bath separation system makes use of the difference in specific gravity between the kernels (1.06–1.16) and the shells (1.30–1.35).

In the hydrocyclones, separation is carried out in a water stream, rotating at high speed inside a cyclone.

After separation, the kernels contain about 20% moisture. Drying is therefore an absolute necessity. A water content of about 7% in the kernels appears to be ideal to maintain stability (low increase in FFA during storage).

3.10.11 Uses of secondary palm fruit products

3.10.11.1 Palm kernel meal

In addition to the oil, palm kernel leaves a meal with the following average content (percentage by weight):

- moisture: 10%;
- protein: 20%;
- fat: 0.9%;
- fibrous matters: 15%;
- ash: 4%;
- nonnitrogenous extracts: 50%.

This meal is normally used as animal feed, usually in the form of pellets.

3.10.11.2 Fibres and shell

After drying, these byproducts of the oil palm are utilised as fuel in the oil mill where the fruit is processed, with a view to generating steam and power. This is most advisable from an economic viewpoint because palm oil mills are usually located in the heart of a plantation, which will cover a very large area, and the mills are therefore far away from towns and power stations.

The ash of the empty bunches can be used as fertiliser thanks to its potash content.

4

Solvent Extraction

Timothy G. Kemper
Global Technical Director, Solvent Extraction, Desmet Ballestra, Marietta, GA, USA

4.1 Introduction

Solvent extraction is the preferred method for the final separation of oil from oleaginous materials containing less than 30% oil by weight. The oil content in the spent solids can be reduced to approximately 1% by weight, making maximisation of oil yield the primary economic driver for the solvent extraction process.

Solvent extraction can be employed for a variety of oleaginous materials. For oleaginous materials with less than 30% oil by weight, such as soybeans, cottonseed, dry process corn germ and rice bran, the material is mechanically and thermally prepared and then sent to the solvent extraction process for oil separation. This approach is commonly referred to as direct solvent extraction. For oleaginous materials with more than 30% oil by weight, such as rapeseed, sunflower, ground nuts, wet process corn germ and copra, the material is mechanically and thermally prepared, mechanically deoiled to approximately 20% oil by weight and then sent to the solvent extraction process for oil separation. This approach is commonly referred to as prepress solvent extraction.

The solvent extraction process dates back to 1855, when Deiss of Marseilles, France was the first to employ it, using carbon disulfide to dissolve olive oil retained in spent olive cakes (Kirschenbauer, 1944). This technology utilised batch solvent extraction, wherein the material was held in a common kettle for both the extraction process and the subsequent meal desolventising process. Deiss obtained a patent for batch solvent extraction of olive oil in 1856 (Kirschenbauer, 1944). In the early 1920s, with

Edible Oil Processing, Second Edition. Edited by Wolf Hamm, Richard J. Hamilton and Gijs Calliauw.

the availability of petroleum-based solvents, the German inventor Hildebrandt created the continuous countercurrent immersion extractor and the German inventor Bollman created the continuous two-stage percolation extractor. The first commercial-scale continuous solvent extraction plants were installed in Germany in the late 1920s, with the Hansa-Muhle facility in Hamburg, using Hildebrandt extractors, being the largest (Wan & Wakelyn, 1997). Numerous variations on continuous extractor designs were developed in the 1930s through the 1950s as the industry rapidly expanded. Today's leading solvent extraction process equipment suppliers, Crown Iron Works and Desmet Ballestra, both got their start with unique continuous extractor designs developed in the 1940s.

Various solvents have been utilised in the solvent extraction process over the years, including carbon disulfide, petroleum napthas, benzene, trichloroethylene, alcohols, pentane, supercritical carbon dioxide and hexane. The solvent used in the vast majority of solvent extraction processes around the world today is commercial hexane, a mixture of hydrocarbons generally boiling in the temperature range of 65–69 °C. Most commercial hexane contains approximately 65% normal hexane, with the remaining 35% of the composition consisting of cyclopentane and hexane isomers. Commercial hexane is the preferred solvent today, due to its wide availability, relatively low cost, excellent diffusivity through oilseed cell walls, high solubility with edible oils, low solubility with water, low latent heat of vaporisation, low specific heat and moderate boiling range. Commercial hexane also has its downsides. Hexane vapour is three times heavier than air and slight amounts of hexane vapour mixed in air can create an explosive mixture. Special care must be taken in constructing and operating commercial hexane solvent extraction processes. Atmospheres Explosive (ATEX) is the recognised guide for safe construction and operation of solvent extraction processes in Europe, and the National Fire Protection Association bulletin *NFPA-36 Solvent Extraction Plants* is the recognised guide for safe construction and operation of such facilities in the USA. Due to special safety considerations, the solvent extraction process is constructed in a structure isolated from the seed receiving area, seed preparation process, meal finishing process, oil refining process and any other important facility structures.

Steam, electricity, labour and initial capital expense are the major cost centres for the solvent extraction process. Large-scale facilities have significantly lower initial capital expense and labour cost per tonne of seed processed. As a result, larger plants have an economic advantage over smaller facilities up to the point where the freight rate basis on seed, meal and oil counteracts the economy of scale. These economic issues in a commodity industry have driven up the throughput rate of solvent extraction processes over the years. Most rapeseed solvent extraction plants being built today are in the 1500–3000 tonne per day range. Most soybean solvent extraction plants built today are in the 3000–5000 tonne per day range, with some in Argentina reaching 20 000 tonnes per day of seed processed in a single facility.

Solvent extraction is an integrated process involving five key unit processes: the solvent extractor, meal desolventiser, meal dryer/cooler, miscella distillation and solvent recovery.

4.2 Solvent extractor

The extractor is the apparatus within the solvent extraction process in which the vegetable oil fraction of the oleaginous material is separated from the meal fraction of the oleaginous material by dissolving the oil fraction in a solvent.

The prepared oleaginous material is conveyed from the seed preparation process to the solvent extraction process and enters the solvent extractor. The solvent extractor conveys the prepared material from its inlet to its exit, providing the prepared material approximately 30–120 minutes of residence time. While the material is being conveyed forward, miscella (solvent and oil solution) is washed down through the bed of material to extract out the edible oil. Each miscella wash is of a decreasing concentration of vegetable oil. After 4–10 miscella washes of descending concentration, the material is washed once more by fresh solvent, ending the extraction process. Before the material exits the extractor, it is allowed to gravity drain in order to reduce its solvent retention. The extracted, spent material then falls into the extractor discharge and exits the apparatus. The miscella with the highest concentration of oil also exits the apparatus to a full miscella tank.

In order to understand the extraction process on a macroscale, it is helpful to understand it on a microscale. Figure 4.1 indicates the microstructure of a soybean cotyledon parenchyma (mcats fraction) cell. This transmission electron micrograph is at 9000 : 1 magnification and represents a 0.020 mm tall by 0.023 mm wide cross-sectional view of a soybean flake: the approximate size of a single cell. As clearly seen, the oil within the cell exists as thousands of spherical oil bodies clinging to the inside surface of the cell walls and to the exterior surface of the protein storage vacuoles.

In the solvent extraction process, the miscella at the surface of the oleaginous material diffuses through the cell walls to the oil bodies located within the cells. The miscella quickly goes into solution with the oil bodies. As miscella continues to enter and go into solution, internal pressure builds within the cell and concentrated miscella diffuses back out of it. This concentrated miscella diffuses through the adjacent cell walls and eventually reaches the surface of the particle. Once the more concentrated miscella reaches the path of miscella outside the oleaginous material, it quickly goes into solution with

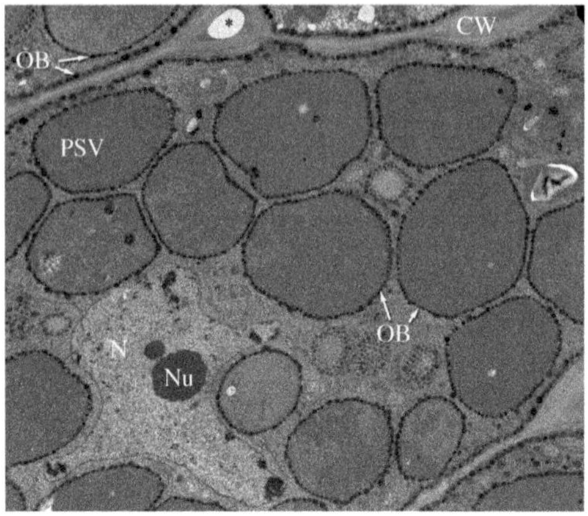

Figure 4.1 Soybean cellular structure. CW, cell wall; PSV, protein storage vacuoles; OB, oil bodies; N, nucleus; Nu, nucleolus; *, intercellular spaces. Special thanks to Dr Robert Yaklich and Dr Charles Murphy at the Soybean Genomics and Improvement Laboratory in Beltsville, MD for creating this electron transmission micrograph. Courtesy of USDA-ARS.

the miscella, incrementally increasing its concentration. This process continues until the concentration of the miscella inside the cells of the oleaginous material comes into equilibrium with the concentration of miscella outside the oleaginous material.

All oleaginous materials have a somewhat different cell structure and therefore a different time required for the miscella in the cells of the oleaginous material to come into equilibrium with the miscella outside the oleaginous material. Soybean flakes have a cellular structure (17) which allows equilibrium to occur within approximately 5 minutes at each extractor wash, while sunflower cake requires approximately 9 minutes and rapeseed cake approximately 12 minutes.

There are six parameters which affect the performance of the solvent extractor apparatus: contact time, particle thickness, extractor temperature, miscella flux rate, number of miscella stages and solvent retention.

4.2.1 Contact time

The total time that the oleaginous material spends in the extractor is its residence time. Residence time can be subdivided into wash time and drain time. Wash time is the time the oleaginous material spends under the washing nozzles of the extractor, and drain time is the time between the last washing nozzle and discharge.

Wash time can be further subdivided into contact time and dormant time. Contact time is the time a particle of oleaginous material spends in the washing zone of the extractor during which it is in contact with miscella. Oil extraction only takes place during contact time. Dormant time is the time the particle spends in the washing zone of the extractor during which it is not in contact with miscella. The ratio of contact time to dormant time varies with extractor design.

Extractors with high material bed depth and small bed surface area are generally operated by immersing the oleaginous materials in miscella. This is accomplished by providing a sufficient miscella flow rate per unit of material bed surface area in order to ensure that miscella fills all the voids around the oleaginous material particles as it passes down through the material bed. This type of extractor operation provides a very high ratio of contact time to dormant time in the washing zone.

Extractors with low bed depth and large bed surface area are generally operated by percolating the miscella down through the oleaginous materials. With most oleaginous materials, the oleaginous material particles occupy 40–50% of the material bed, with the remaining 50–60% composed of void spaces between the particles. In percolation, the oleaginous material particles are surrounded by both solvent vapour and miscella as the miscella rains downward. This type of extractor operation provides a lower ratio of contact time to dormant time in the washing zone.

In comparing two different extractor designs, the contact time can be similar despite very different configurations:

(1) Deep-bed extractor example: 50 minutes' residence time = 30 minutes' wash time + 20 minutes' drain time; 30 minutes' wash time – 5 minutes' dormant time = 25 minutes' contact time.
(2) Shallow-bed extractor example: 50 minutes' residence time = 45 minutes' wash time + 5 minutes' drain time; 45 minutes' wash time – 20 minutes' dormant time = 25 minutes' contact time.

Regardless of extractor design, adequate contact time is critical for maximising extraction efficiency and minimising the amount of residual oil remaining in the oleaginous material. Increased contact time requires a larger extractor. Even though this represents a higher initial investment, the long-term economic benefits of reduced residual oil content as a result of surplus contact time are significant and justify oversizing the extractor.

4.2.2 Particle thickness

Various oleaginous materials are prepared for extraction using different process steps, but one process step that virtually all oleaginous materials have in common is flaking. The principal purpose of flaking is to reduce the thickness of the oleaginous material in order to reduce the distance and the number of cell walls that miscella needs to diffuse through in order to reach the oil bodies. Note that for soybeans, a typical 0.38 mm-thick flake is approximately 20 cells thick. By reducing the particle thickness, the time required for the miscella within the cellular structure of the oleaginous material to reach equilibrium with the miscella surrounding the oleaginous material is reduced; also, desired results can be achieved with less contact time. If all other extraction parameters remain constant, reduced particle thickness will allow a smaller extractor to be used.

Reducing particle thickness represents additional cost. For example, with soybeans, reducing particle thickness from 0.38 mm to 0.30 mm will increase flaking mill electricity requirements by 1–2 kwh/tonne of soybeans processed. Since this is a significant ongoing operating expense, it is not economically feasible to undersize an extractor and reduce particle thickness. Conversely, by increasing particle thickness, the desired results will demand more contact time. An extractor can be oversized in order to obtain the desired results with increased particle thickness, thereby reducing the ongoing operating cost.

For all oleaginous materials, the economic balance between the initial cost of the extractor and the ongoing electricity costs required for flaking can be analysed and the optimum particle thickness determined.

4.2.3 Extractor temperature

As the temperature of the miscella increases, its rate of diffusivity through the cell walls of the oleaginous material increases. Since the prepared oleaginous material enters the extractor at approximately 60 °C and both the oil and meal fractions are heated in excess of 100 °C in subsequent process steps, there is no extra energy required to operate the extractor at a warm temperature. As a result, optimising extraction results requires operating the extractor at as high a temperature as is possible.

There is a practical upper limit for the extractor operating temperature. The solvent must remain safely in a liquid state. Since the boiling range of commercial hexane is typically 64–69 °C at sea level, the maximum possible temperature without boiling is 63 °C. Operating on the edge of the boiling range may cause rapid evaporation during an upset condition. Rapid evaporation can cause pressurisation of the extractor, leading to excessive solvent loss: a safety hazard. Therefore, most processors operate the extractor at 60 °C in order to provide several degrees Celsius of safety margin below the lower end of the solvent boiling range.

If the prepared material temperature is too low, or heat loss in the extractor is too high, then it may not be possible to achieve an extractor temperature of 60 °C. Additional contact time will be required in order to achieve desired results. Insulating the conveying system and extractor to prevent heat loss and enable operation at 60 °C is typically less expensive than oversizing the extractor to compensate for a low operating temperature.

4.2.4 Miscella flux rate

The miscella flux rate is the maximum volumetric flow rate of miscella that can flow down through the bed of material per unit of material bed surface area. In SI units, it is commonly expressed as m^3/hour per m^2, while in Imperial units it is commonly expressed as gpm per ft^2. By simplification of units, the miscella flux rate can also be expressed as the downward velocity of the head of miscella before entering the material bed (in m/minute or ft/minute). Miscella flux rates for various prepared oleaginous materials vary widely (see Table 4.1).

The miscella flux rate is determined by the screen below the bed of material. As stated earlier, the material bed is approximately 40–50% solids and 50–60% void space. Therefore, as the miscella is moving downward, it has 50–60% open area to pass through. The screen under the material bed has less open area, and therefore the material interface with the screen creates the greatest restriction to flow. Most deep-bed extractor screens have approximately 30% open area, whereas most shallow-bed extractor screens have less than 10% open area. In both cases, the screens have less open area than the material bed itself, and the material interface with the screen provides the greatest flow restriction.

As downward miscella flow reaches the miscella flux rate, the material–screen interface reaches its maximum flow rate and begins restricting the flow of miscella. All void spaces between the oleaginous material particles fill with miscella as the solvent vapours are pushed out the top of the material bed. Eventually, the entire material bed becomes immersed in miscella, with no void spaces. At this point, miscella breaks through the top of the miscella bed and forms a pool. This phenomenon is often referred to as 'flooding' of the material bed. Once the material bed is

Table 4.1 Miscella flux rates.

	m^3/hour/m^2	gpm/ft^2	m/minute	ft/minute
Cottonseed flakes (0.38 mm)	10	4	0.17	0.53
Cottonseed extrudate	30	12	0.50	1.60
Rapeseed cake	20	8	0.33	1.07
Soybean flakes (0.30 mm)	15	6	0.25	0.80
Soybean flakes (0.38 mm)	20	8	0.33	1.07
Soybean extrudate	45	18	0.75	2.40
Sunflower cake (ground)	25	10	0.42	1.33

flooded, the rate of flow passing down through the material bed cannot be increased.

At each washing stage of the extractor, miscella needs to have an opportunity to wash the material bed, pour through the screen and enter the proper miscella collection receptacle beneath the material bed. For a given extractor and prepared oleaginous material, each miscella collection receptacle is carefully calculated to be located a specific distance from its washing nozzle, in order to maintain separation between washing stages. For example, if there is a deep-bed extractor operating on 0.38 mm-thick soybean flakes with a 3.0 m bed depth and a forward material velocity of 0.3 m/minute, the distance that the miscella collection receptacle needs to follow the washing nozzle by can be calculated as follows:

$$\text{Downward head miscella velocity for 0.38 mm} - \text{thick}$$
$$\text{soybean flakes} = 0.33\,\text{m/minute}$$
$$\text{Void space between 0.38 mm} - \text{thick soybean flakes} = 57\%$$
$$\text{Downward flake bed miscella velocity} = 0.33\,\text{m/minute} \div 0.57$$
$$= 0.58\,\text{m/minute}$$
$$\text{Time for miscella to pass through material bed} = 3.0\,\text{m}$$
$$\div 0.58\,\text{m/minute} = 5.2\text{ minutes}$$
$$\text{Wash nozzle to miscella receptacle distance} = 5.2\text{ minutes}$$
$$\times 0.3\,\text{m/minute} = 1.6\,\text{m}$$

As another example, if there is a shallow-bed extractor operating on 0.38 mm-thick soybean flakes with a 0.75 m bed depth and a forward velocity of 1.2 m/minute, the distance that the miscella collection receptacle needs to follow the washing nozzle by can be calculated as follows:

$$\text{Downward head miscella velocity for 0.38 mm} - \text{thick}$$
$$\text{soybean flakes} = 0.33\,\text{m/minute}$$
$$\text{Void space between 0.38 mm} - \text{thick soybean flakes} = 57\%$$
$$\text{Downward flake bed miscella velocity} = 0.33\,\text{m/minute} \div 0.57$$
$$= 0.58\,\text{m/minute}$$
$$\text{Time for miscella to pass through material bed} = 0.75\,\text{m}$$
$$\div 0.58\,\text{m/minute} = 1.3\text{ minutes}$$
$$\text{Wash nozzle to miscella receptacle distance} = 1.3\text{ minutes}$$
$$\times 1.2\,\text{m/minute} = 1.6\,\text{m}$$

If the miscella flux rate is significantly reduced for a given prepared oleaginous material, the miscella can partially discharge into a later, undesired miscella collection receptacle. This will cause concentration contamination due to mixing of the extractor washes and reduce the efficiency of the extractor.

Miscella flux rates can reduce as a result of thinner-than-normal flakes, surface moisture or an abundance of fine particles. Flake thickness is a normal operator-controlled parameter, but surface moisture and an abundance of fine particles are more difficult to control.

Since hexane solvent is not soluble in water, the liquids repel each other. If surface moisture exists on the material in the extractor, the solvent will have difficulty penetrating the particle surface. Also, the moisture can collect on the screen at the bottom of the material bed, building a protein layer and narrowing the screen slots, further reducing the miscella flux rate. To prevent this from occurring, it is very important to adequately aspirate flakes from flaking mills, cake from screw presses or extruded pellets from extruders in order to remove all water vapour created when the material evaporatively cools from preparation temperature to extraction temperature. As a secondary precaution, it is ideal to have an extractor designed such that the oleaginous material moves with respect to the screen surface so as to constantly keep the screen surface brushed clean, so that the impact of surface moisture on miscella flux rate is minimised.

The material bed is approximately 40–50% material particles and 50–60% void space. If there is an abundance of fine material particles, these particles can sift down through the material bed and settle in the void spaces just above the screen. This causes an additional flow restriction and the miscella flux rate will be reduced. An abundance of fine material at the extractor is generally caused by overdrying of the material at some point in the seed preparation stage or by rough handling of the friable material during a seed preparation step. Both should be avoided, so as to ensure a uniform material shape at the extractor and uniform miscella flux rates.

4.2.5 Number of miscella stages

In most extraction applications, the prepared material has approximately 20% oil by weight and the goal is to reduce the oil content to approximately 0.5% by weight. If an extractor has one miscella stage, the miscella concentration exiting the extractor (1.15% oil) will be in equilibrium with the miscella concentration remaining in the material cells (1.15% oil). A mass balance for a single-stage extractor can be calculated (see Table 4.2). An extractor which has only one miscella stage will require 17.2 parts of solvent per 1.0 part of material to be extracted. The energy required to evaporate the solvent in the

Table 4.2 Single-stage extractor mass balance.

Prepared material entering		Solvent entering	
Solids:	800 units	Solvent:	17 200 units
Oil (20%):	200 units		
Total:	**1000** units	Total:	**17 200** units
Spent material exiting		**Miscella exiting**	
Solids:	800 units	Solvent:	16 856 units
Oil (0.5% residual):	4 units	Oil:	196 units
Solvent (30% retention):	344 units		
Total:	**1148** units	Total:	**17 052** units
Miscella concentration		**Miscella concentration**	
4 / (4 + 344) = **1.15% oil**		196 / (196 + 16 856) = **1.15% oil**	

miscella will be tremendous. As a result, countercurrent multistage extractors are required.

Through iteration of the mass balance, the minimum number of miscella stages can be calculated for a given solvent–material ratio. For an energy-competitive distillation system, the solvent–material ratio should be maintained below 1 : 1 in order to maintain a full miscella concentration of 23% oil by weight or greater. The minimum number of stages required to achieve 0.5% residual oil with 23% miscella concentration can be calculated as four.

A four-stage extractor with sufficient contact time to allow the miscella concentration in the material cells to come into equilibrium with the miscella concentration in the surrounding miscella bath at each miscella stage can be designed to extract to less than 0.5% residual oil using a solvent–material ratio of 1 : 1. The resultant outgoing miscella concentration will be approximately 23% oil. A four-stage extractor is sufficient theoretically, but leaves no contingency for failure to achieve equilibrium at each miscella stage.

The more miscella stages, the greater the theoretical extraction efficiency. In practice, however, if too many miscella stages are designed into an extractor, causing the individual stages to have insufficient contact time to reach equilibrium, residual oil will not be further reduced by the addition of new stages. In this case, more miscella stages simply leads to more pumps, and therefore more pumping energy and more potential points for solvent loss. The number of miscella stages is generally determined by the total washing zone time and the number of stages that can theoretically come to equilibrium within the washing zone time. Commercially, most extractors have somewhere in the range of five to nine miscella stages.

Increasing the solvent flow to the extractor in order to increase the solvent–material ratio and reduce the full miscella concentration can reduce the number of miscella stages required, although this is not cost-effective due to the distillation energy required to evaporate the additional solvent. Very high solvent–material ratios (full miscella concentrations well under 23%)

are only justified in specialty oil extractions with extremely high incremental oil values.

4.2.6 Solvent retention

After leaving the washing zone of the extractor, the extracted material is left to drain by gravity. This gravity drain time is generally in the range of 5–20 minutes. Extractors designed with low material bed depths will generally have a drain time closer to 5 minutes, while extractors with high material bed depths will generally have a drain time closer to 20 minutes. After gravity drainage, the solvent retained with the extracted material will be in the range of 25% (fast-draining extrudate) to 32% (slow-draining flakes).

The 'solvent retention' of the drained material is something of a misnomer, and could be more accurately defined as the 'weak miscella retention'. The weak miscella retained in the drained material contains approximately 1% oil. In the meal desolventiser, the solvent is evaporated, leaving behind traces of oil, often referred to as the residual oil. In order to minimise the residual oil left in meal, it is important to minimise the amount of weak miscella carried forward to the meal desolventiser.

An adequate extractor drain time is the most economical means by which to minimise weak miscella retention. Maintaining the desired miscella flux rate is also important. Thus, once again, there is a need to maintain proper flake thickness, minimise surface moisture and minimise fine material particles in the material bed in order to minimise weak miscella retention.

Today there are two major suppliers of solvent extractors: Crown Iron Works, headquartered in the USA, and Desmet Ballestra, headquartered in Belgium. The Crown Model III Extractor is a shallow-material bed extractor which utilises a chain conveyor design to convey the material over fixed screens in a loop pattern (see Figure 4.2). Desmet Ballestra supplies two major types of extractor: a Reflex® Extractor and an LM™ Extractor. The Desmet Ballestra Reflex® Extractor is a deep-material bed extractor which uses a cylindrical rotating set of baskets to convey the material over a fixed circular screen (see Figure 4.3). The Desmet Ballestra LM™ Extractor is provided in both shallow and deep material bed versions, both utilising a belt conveyor made of screens to convey the material along a linear path (see Figure 4.4). Extractors of these three types process somewhere in the range of 25 tonnes per day of specialty oilseeds to 10 000 tonnes per day of soybeans through a single unit.

4.3 Meal desolventiser toaster

After the prepared material has had its oil extracted in the solvent extractor, it is conveyed to the meal desolventiser toaster, commonly referred to as the

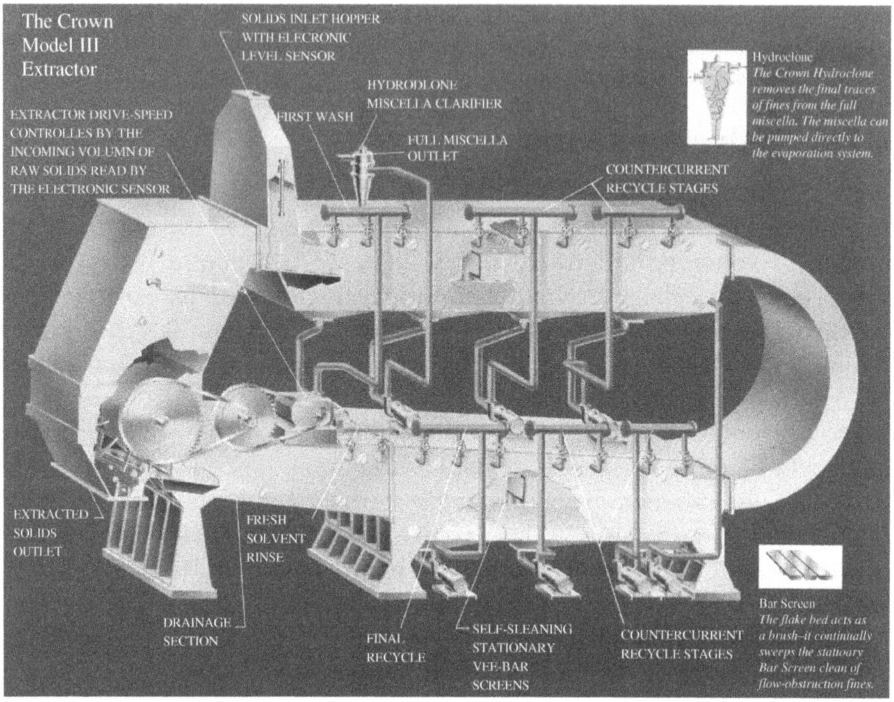

Figure 4.2 Crown Extractor. Courtesy of Crown Iron Works.

DT. The material entering the DT is typically at the extractor temperature of 60 °C, and contains 25–32% by weight solvent. The primary purpose of the DT is to remove the solvent from the meal fraction so that the solvent can be recovered.

DTs are vertical, cylindrical vessels with a multitude of horizontal trays. The extracted material enters at the top and is supported by the first tray. The material is mixed above each tray and then conveyed downward from tray to tray, by agitating sweeps anchored to a central rotating shaft. The heat used to increase the meal temperature and evaporate the solvent is supplied by steam, introduced directly and indirectly into the meal via the trays. Figure 4.5 illustrates a typical DT.

The trays of the DT are designed with an upper plate, lower plate and structural members between used to hold pressurised steam. The DT has three different types of tray: predesolventising trays, countercurrent trays and a sparge tray.

4.3.1 Predesolventising trays

The predesolventising trays have as their sole purpose the provision of conductive heat transfer through their upper surface to the solvent-laden

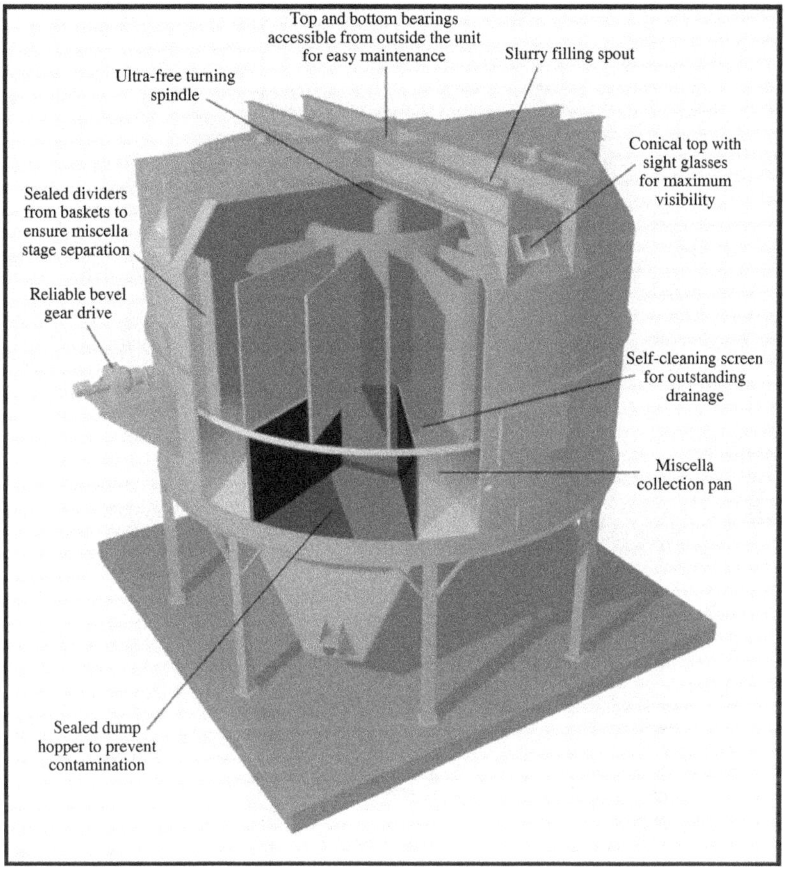

Figure 4.3 Reflex® Extractor. Courtesy of Desmet Ballestra.

material supported above. The steam is typically held at 10.5 kg/cm^2 pressure within the predesolventising trays, providing a surface temperature of 185 °C. The steam condenses within the trays, providing its latent heat to maintain the 185 °C tray surface temperature and allowing heat to be conducted into the solvent-laden meal layer above.

A DT may have as many as seven predesolventising trays, or as few as one. The predesolventising trays are located in the upper portion of the DT and must allow ascending vapours from below to pass around them to the vapour exit at the top of the DT. Some manufacturers design disc-shaped trays, providing space for the ascending vapours to pass between the outside perimeter of the tray and the shell wall, while others design donut-shaped trays, providing space for ascending vapours to pass between the inside perimeter of the tray and the central shaft. As an alternative to a large number of predesolventising trays, the upper section of the DT is often expanded in diameter so as to enable fewer, larger-diameter trays.

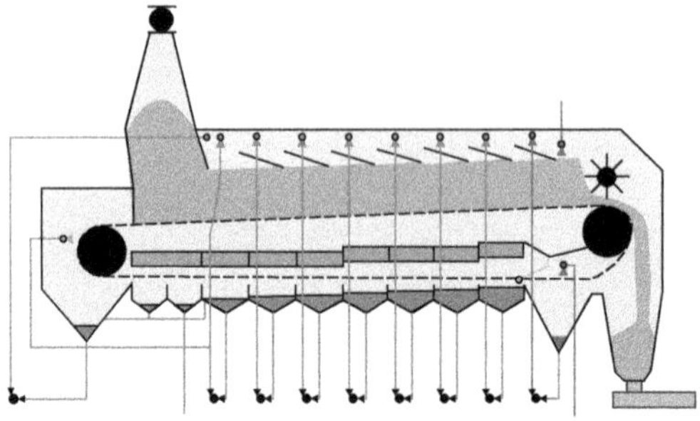

Figure 4.4 LM™ Extractor. Courtesy of Desmet Ballestra.

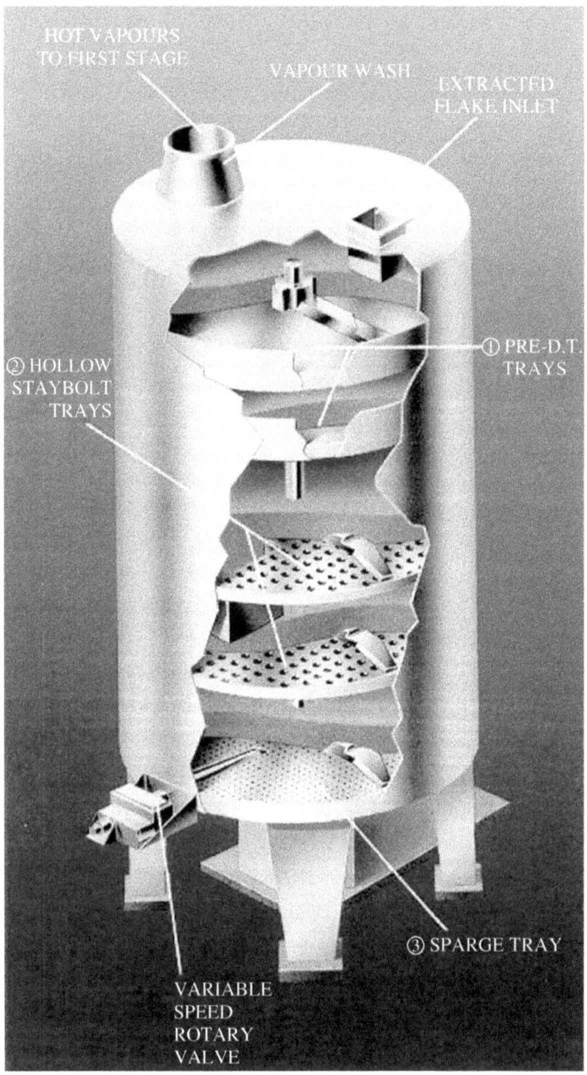

Figure 4.5 Desolventiser toaster. Courtesy of Crown Iron Works.

4.3.2 Countercurrent trays

The countercurrent trays have three purposes: to provide conductive heat transfer through their upper surface in order to warm wet material supported above; to provide convective heat transfer through their lower surface in order to superheat direct steam swirling below; and to provide apertures to allow direct steam to evenly ascend up through the tray and into the meal supported above. The steam is typically held at 10.5 kg/cm^2 pressure within the countercurrent trays, providing a surface temperature of 185 °C. Steam condenses within the trays, providing its latent heat to maintain the 185 °C tray surface temperature and allowing heat to be conducted into the meal layer above and transferred into the direct steam swirling below.

A DT will have from one to four countercurrent trays. The countercurrent trays are located directly under the predesolventising trays in the centre of the DT. The apertures in the trays must allow the direct steam swirling below to pass through and into the meal supported above. There are three designs of countercurrent tray, with different apertures: initial modern DTs by Schumacher, circa 1982, utilised hollow stay-bolts as apertures and generally had a 1–2% open area to allow the vapours to ascend (Schumacher, 1985); a later modern DT design by Mason, circa 1985, utilised hollow stay-pipes capped by a perforated plate, with generally 2–4% open area to allow vapours to ascend; while the latest modern DT design by Kemper, circa 1997, utilises hollow stay-pipes capped with stainless-steel slotted screens, with approximately 10% open area to allow the vapours to ascend (Kemper & Farmer, 1999). A greater open area in the countercurrent trays allows for more uniform steam distribution and has been the trend in modern DT design.

4.3.3 Sparge tray

The sparge tray has the dual purpose of providing a uniform means of introducing direct steam into the meal layer and providing conductive heat transfer through its upper surface to the wet material supported above. The direct steam introduced through the sparge tray provides approximately 75% of the total heat required to desolventise and heat the meal in the DT. The sparge tray is typically designed with a plurality of apertures across its entire upper surface, in order to evenly introduce direct steam into the meal. The size and quantity of apertures is calculated based upon the anticipated direct steam flow rate so as to provide a pressure drop of 0.35–0.70 kg/cm^2. The direct steam supply is 10.5 kg/cm^2 pressure saturated steam (185 °C). After passing through a flow-control valve, its quality changes to 0.35–0.70 kg/cm^2 pressure superheated steam (150–160 °C). Therefore, the upper surface of the sparge tray is maintained at approximately 155 °C.

Solvent-laden meal enters the DT at a temperature of 60 °C and contains 25–35% by weight solvent. The solvent-laden meal is stirred across the surface of the predesolventising trays by the rotating sweeps. Since the heat is transferred into the meal layer by conduction, a shallow layer of 150–300 mm meal depth is held above each tray. The solvent-laden meal temperature is increased to approximately 68 °C, and approximately 10–25% of the solvent is evaporated on the predesolventising trays.

The material exits the predesolventising trays of the DT and falls on to the top countercurrent tray. This is perhaps the most critical tray of the DT. Since most of the heat is transferred into the meal layer by condensation of direct steam, a deep layer of 1000–1200 mm meal depth is held above the tray. The solvent-laden meal is stirred above the top countercurrent tray by the rotating sweeps. The direct steam passes from below up through apertures in the countercurrent tray. As the direct steam penetrates the upper meal layer, it reaches the solvent-laden meal and condenses, providing direct latent heat to evaporate the solvent, which exits the meal layer as vapour. The condensation of steam causes the meal exiting the tray to be wet, typically in the range of 17–21% moisture. After the majority of the solvent evaporates, the meal temperature increases via direct and indirect steam heat, surpassing 100 °C before the material exits the tray. The protein solubility of soybean meal is reduced from approximately 90 to 45 PDI as a result of the elevated moisture and temperature conditions.

Once the wet meal exits the top countercurrent tray, it has had over 99% of its solvent removed. On the remaining countercurrent trays and the sparge tray, the meal is typically held in a 1000 mm-deep layer, in order to provide residence time to allow solvent stripping and toasting. The wet meal is stirred above each tray by rotating sweeps. The final desolventising takes place as the ascending steam passing through the meal slowly strips out the final traces of residual solvent, down to 100–500 ppm. The meal temperature increases from 100 to 105–110 °C and the meal moisture decreases approximately 1% before the meal discharges from the sparge tray. The meal colour darkens slightly, providing the meal with a toasted colour. For soybeans, antinutritional factors such as trypsin inhibitors and urease are reduced on these trays by maintaining the meal moisture and temperature elevated for a period of time. The protein solubility drops approximately 1% PDI for every minute the meal spends in the remaining countercurrent trays and the sparge tray. Ideal feed for monogastric animals (poultry and swine) is high in protein solubility, and ideal feed for ruminant animals (cattle) is low in protein solubility (high in rumen bypass protein). Meal residence time on the remaining countercurrent trays and the sparge tray is dictated by both the degree of solvent recovery required and meal quality parameters.

An important parameter in the energy efficiency of the DT is the exit vapour temperature. The condensing sparge steam provides a plentiful supply of

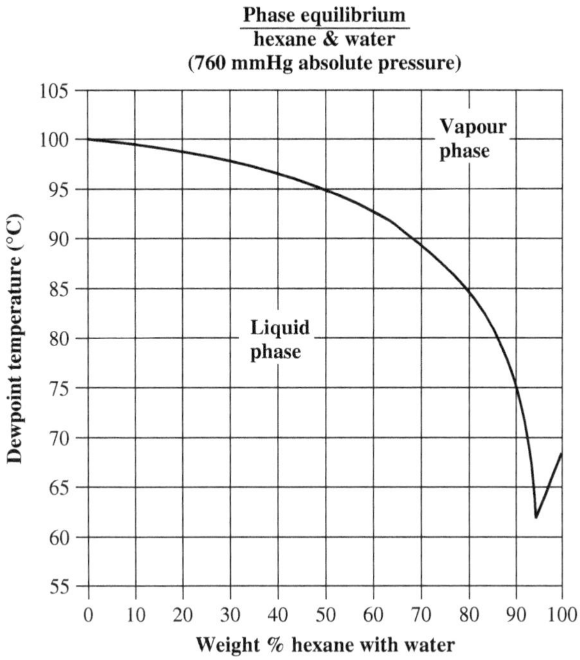

Figure 4.6 Solvent–water equilibrium boiling curve.

surface moisture, allowing the solvent and water to evaporate as an azeotropic mixture. Figure 4.6 indicates the solvent–water equilibrium boiling curve. As the chart shows, a mixture of 94% solvent with 6% water can boil at as low as 62 °C. Therefore, the lowest possible DT exit vapour temperature is 62 °C. In practice, in order to maintain low solvent loss and a safety margin, modern DTs are operated with exit vapour temperatures ranging from 66 to 78 °C, with the most typical temperature being 71 °C. As the DT vapour temperature increases, the ratio of water vapour to solvent vapour increases. Therefore, to minimise total DT energy, it is very important to maintain a vapour temperature as low as is safely possible.

Determining the optimum DT configuration for a given process application is a rather complex process. It requires the determination of all input parameters and calculation of the mass and heat balance of both the DT and the follow-on meal dryer cooler (DC). The mass and heat balance of the DC will determine the maximum allowable DT exit moisture, which will minimise meal drying energy. This moisture is generally in the range of 18–20%. With the DT exit moisture determined, the amount of direct steam introduced into the meal can be calculated. The DT diameter is generally determined by the direct steam flow rate per unit area. It is important to have a sufficiently high direct steam flow rate per unit area to allow adequate solvent stripping. The number of countercurrent trays is determined by the residence time required to balance meal quality with residual solvent objectives. By calculating the total DT heat demand and subtracting the heat supplied by live steam, the total heat supplied by indirect steam can be determined. Subtracting the heat supplied by countercurrent tray indirect steam from the total heat supplied by indirect steam will provide the amount of indirect steam heat that must be supplied by the predesolventising trays. With this information in hand, the diameter and quantity of predesolventising trays can be selected. Major manufacturers of DTs utilise process simulation tools to assist processors in optimising the DT configuration for a given application.

4.4 Meal dryer cooler

After the solvent-laden material is desolventised, it is conveyed to the DC. The material entering the DC is typically at the DT exit temperature of 108 °C, and contains 18–20% moisture (for soybeans). The primary purposes of the DC are to reduce the moisture in the meal to within trading rule limits and to lower the meal temperature prior to storage.

DCs are vertical, cylindrical vessels with a multitude of horizontal trays. The desolventised material enters at the top and is supported by the tray. The material is mixed above each tray and conveyed downward from tray to tray by agitating sweeps anchored to a central rotating shaft. The DC has three different types of tray: steam-drying trays, air-drying trays and air-cooling trays.

4.4.1 Steam-drying trays

The steam-drying trays of the DC are designed with an upper plate, lower plate and structural members between them designed to hold pressurised steam. The steam-drying trays have the purpose of providing conductive heat transfer through their upper surface to wet meal supported above. The steam is typically held at 10.5 kg/cm^2 pressure within the steam-drying trays, providing a surface temperature of 185 °C. Steam condenses within the trays, providing its latent heat to maintain the 185 °C tray surface temperature and allowing heat to be conducted into the wet meal layer above.

A DC may have as many as five steam-drying trays, or as few as none. The water vapour evaporated from the meal can be compressed in an ejector and have its heat recovered within the solvent extraction plant.

4.4.2 Air-drying trays

The air-drying trays of the DC are designed with an upper plate, lower plate and structural members between them designed to hold low-pressure air. The air-drying trays are designed with a plurality of apertures across their entire upper surface in order to evenly introduce air into the meal. The size and quantity of apertures is calculated based upon the design air flow rate so as to provide a pressure drop of 0.02–0.03 kg/cm^2. These apertures are generally small round holes, but Desmet Ballestra has recently introduced patent-pending technology that utilises narrow slots.

The air supplied to each air-drying tray is first filtered to remove dust and then pressurised using a centrifugal blower. The air for the dryer trays is passed through a steam-heated coil between the blower and the entrance to the trays. After the air enters the trays, it flows upward through the meal at a nominal velocity of 14–21 m/minute, partially fluidising the meal. The meal evaporatively cools and the released moisture is transferred to the ascending air. The warm, damp air exits the top of the meal layer and then moves through the sidewall of the DC to a cyclone collector in order to remove dust prior to discharge to atmosphere. Figure 4.7 illustrates a meal DC with two air trays and related air-handling equipment.

The major source of heat for evaporation of the moisture in the meal is the high temperature of the meal exiting the DT or the DC steam-drying trays. When the meal drops in temperature from 108 to 38 °C, the heat provided is adequate to reduce the meal moisture by 6.5%. For soybean meal, the trading rule moisture limit is 12.5%; therefore, if the incoming moisture from the DT, or the DC steam-drying trays, does not exceed 19.0%, the DC will typically require no additional evaporative heat source in order to dry the meal. If additional heat is required to evaporate moisture from the meal, the air entering the meal dryer trays can be heated to up to 150 °C prior to entering the air-drying trays. The heat source may be recovered flash steam, hot glycol–water solution (oil cooler) or fresh steam.

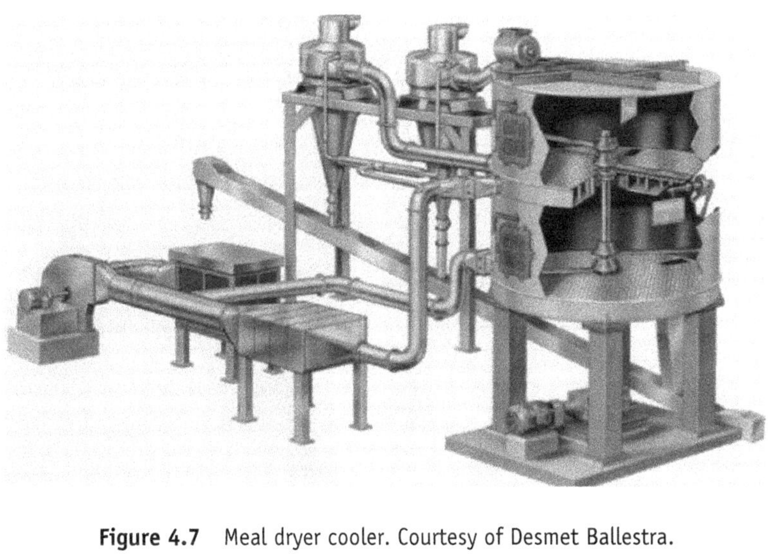

Figure 4.7 Meal dryer cooler. Courtesy of Desmet Ballestra.

The air must have adequate capacity to carry the moisture released from the meal without becoming saturated. Cold air can hold less moisture than warm air, so winter conditions may limit the moisture-carrying capacity of the air. If additional heat is required to increase the dew point of the incoming air, the air entering the air dryer trays can be heated to temperatures up to 150 °C.

The energy required to heat the incoming drying air is largely dictated by the meal moisture coming into the DC. Adequate predesolventising heat transfer area in the DT or steam-drying tray heat transfer area in the DC is the key to minimising meal moisture to the DC air-drying trays and therefore to minimising DC heater coil steam consumption.

4.4.3 Air-cooling trays

The air-cooling trays of the DC are designed with an upper plate, lower plate and structural members between them designed to hold low-pressure air. The air-cooling trays are designed with a plurality of apertures across their entire upper surface so as to evenly introduce air into the meal. The size and quantity of apertures is calculated based upon the design air flow rate to provide a pressure drop of 0.02–0.03 kg/cm^2.

The air supplied to each air-cooling tray is first filtered to remove dust and then pressurised using a sparkproof centrifugal blower. After the cool air enters the trays, it flows upward through the meal at a nominal velocity of 14–21 m/minute, partially fluidising the meal. The meal continues to evaporatively cool, and also convectively cools. The cool, damp air exits the top of the meal layer and then moves through the sidewall of the DC to a cyclone collector in order to remove dust prior to discharge to atmosphere.

Ambient air is heated by approximately 5 °C by the energy of the blower, before it blows into the air-cooling trays of the DC. The meal cools to within approximately 5 °C of the air temperature passing through the meal. Therefore, the temperature of the meal exiting the DC is typically cooled to within 10 °C of ambient air. A suction blower fitted downstream of the DC can thus reduce exit meal temperatures by up to 5 °C by avoiding the heating energy of the upstream blower.

The dry cool meal is conveyed from the DC outside the solvent extraction plant to allow size reduction and then on to meal storage. It is important to properly dry and cool the meal in order to prevent continued evaporative cooling in storage or transport, which will cause reduced flowability, solidification and bridging of the meal inside storage and transport vessels.

4.5 Miscella distillation system

The miscella exiting the extractor contains up to 1% by weight particles of meal. Therefore, the first step in miscella distillation is meal particle separation. Meal particles can be separated from the miscella by filtration or centrifugal separation. The large meal particles (+80 mesh) need to be removed in order to prevent them from settling out in the distillation equipment. The fine meal particles (−80 mesh) are removed from the oil after solvent extraction, typically in the oil degumming process or oil refining process. For those plants producing food-grade lecithin, the fine meal particles must be completely filtered out prior to degumming.

Some extractor designs include an internal miscella filter (100 mesh), enabling the miscella exiting the extractor to be sufficiently free of meal particles to go directly to distillation. Other extractor designs require external meal particle separation, accomplished by pumping the miscella through a liquid cyclone. The liquid cyclone spins the miscella at high velocity and utilises centrifugal force to separate the larger meal particles (over 80 mesh) from the miscella. The larger meal particles, along with 5–10% of the miscella flow, exit the liquid cyclone underflow orifice and return to the extractor. Separation of meal particles by external filtration of the miscella is not recommended due to the safety hazards associated with opening the filter to remove the solvent-laden meal fines.

Once the large meal particles are separated, the clean miscella is stored in a surge tank, generally referred to as a full miscella tank. The full miscella tank has several purposes: it separates the continuous extraction process from the continuous distillation process, it provides miscella storage capacity during a power outage and it provides surge capacity so that fluctuations in miscella flow rate from the extractor can be absorbed prior to distillation. The miscella in the full miscella tank is generally 25–30% oil and 70–75% solvent by weight, and exists at the typical extractor temperature of 60 °C.

The miscella is pumped from the full miscella tank to the first of two rising film evaporators. The first-stage evaporator, often referred to as the economiser, utilises the waste heat from the DT as its heating source. Miscella enters through tubes at the base of the evaporator, at a temperature of approximately 60 °C. Since the tube side of the evaporator is held at approximately 300–400 mmHg absolute pressure, the miscella temperature will drop to approximately 43–48 °C temperature upon entry into the tubes. Solvent will begin evaporating and solvent vapour bubbles will rise up through the centre of the tubes. Additional DT vapour heat is transferred through the tubes into the miscella and additional evaporation takes place. When sufficient solvent vapour is formed, the vapour velocity through the centre of the tubes will become sufficiently high to drag a thin film of miscella up the inner walls of the tubes, creating high heat transfer rates. At the tops of the tubes, the high-velocity solvent vapour and remaining miscella contact an impingement plate to break foam and are then centrifugally separated in the evaporator dome. Solvent vapours exit the top of the dome and concentrated miscella exits the base of the dome. The concentrated miscella exiting the first-stage evaporator is generally 75–85% oil and 15–25% solvent, and approximately 48 °C.

Since the temperature of the miscella exiting the first-stage evaporator is low, it is a good heat sink for heat recovery. In various facilities, heat from hot finished oil, heat from steam ejector exhausts or recovered flash steam is used to preheat the miscella to approximately 75 °C. The preheated, concentrated miscella is then typically heated to 110 °C in a steam-heated exchanger prior to entering the second rising film evaporator.

The preheated, concentrated miscella is pumped into the second-stage evaporator, which utilises low-pressure steam as its heating source. Since the tube side of the evaporator is held at approximately 300–400 mmHg absolute pressure, the solvent temperature will drop to approximately 43–48 °C upon entry into the tubes, providing sufficient heat to immediately begin vigorous evaporation. The vapour velocity through the centre of the tubes will be sufficiently high to drag a thin film of miscella up the inner walls of the tubes at relatively high velocity. This is very important for preventing phospatides and fine solid particles in the miscella from baking to the lower, inner surface of the tubes. Additional low-pressure steam heat is transferred through the tubes into the miscella and additional evaporation and heating takes place. At the tops of the tubes, the solvent vapour and remaining miscella contact an impingement plate to break foam and are then centrifugally separated in the evaporator dome. Since velocities exiting the tubes are insufficient to break all foam, the dome must be sufficiently large in diameter to allow the remaining foam to collapse and not discharge with exiting vapours. Solvent vapours exit the top of the dome and concentrated miscella exits the base of the dome. The concentrated miscella exiting the second-stage evaporator is generally 95–98% oil and 2–5% solvent, and approximately 95–110 °C.

Miscella from the second-stage evaporator is pumped or gravity-fed into the oil stripper. The oil stripper is a tall, thin, cylindrical vessel and is commonly operated at 150–300 mmHg absolute pressure. The hot oil passes downward through the vessel across trays. Simple, robust disc-donut trays or sieve-type trays are typically utilised, since fouling due to baking of gums and fine particles on to stripper trays is common. Live steam is introduced into the oil at the top of the vessel to initiate evaporation, and again at the base of the vessel to provide countercurrent stripping. The steam and solvent vapours exit the top of the oil stripper through an enlarged-diameter dome to prevent entraining of oil mist. The oil typically exits the base of the oil stripper with 0.1–0.3% moisture and 5–200 ppm solvent, at a temperature of approximately 95–110 °C.

The oil leaving the oil stripper, particularly in soybean plants, is often water-degummed. In these facilities, the oil temperature is reduced to 70–80 °C, and 1–2% soft water is injected and mixed into the oil inline. The oil is then held for approximately 30–60 minutes in an agitated tank in order to allow gums to hydrate. The gums (water, phosphatides, fine meal particles and some neutral oil) are centrifugally separated from the oil using a high-speed centrifugal separator. The gums are often pumped back into the DT and mixed into the meal fraction. Alternatively, the gums may be dried for food-grade lecithin, or feed-grade lecithin for animal feed applications. The degummed oil, at a moisture level of approximately 0.5% and a temperature of approximately 70 °C, is then heated in a heat exchanger to approximately 110 °C.

Whether or not the oil is degummed, it is typically pumped to an oil dryer. The oil dryer is a vertical cylindrical vessel that is commonly operated at 50–80 mmHg absolute pressure. The hot oil is sprayed downward into the vessel, with or without trays. The solvent and water vapours exit the top of the oil dryer and the oil exits the bottom. The oil typically exits the base of the oil dryer with 0.05–0.10% moisture and 5–100 ppm solvent, at a temperature of approximately 105 °C.

The dried oil must have its temperature reduced from 105 to 50 °C in order to prevent degradation in storage and transport. The hot oil is commonly cooled in two stages. First, the oil is cooled from 105 to approximately 70 °C in a heat exchanger using concentrated miscella or a water–glycol solution (for preheating DC air) as the cooling medium. Second, the oil is cooled from 70 to 50 °C in a heat exchanger using cooling water as the cooling medium. The cool oil is then pumped to storage.

4.6 Solvent recovery system

Modern solvent extraction plants recover over 99.9% of the solvent pumped to the extractor. The solvent recovery system includes solvent and water vapour condensation, solvent–water separation, stripping of solvent from water and air effluent streams and heating of the solvent prior to reuse in the extractor.

The water and solvent vapours from the first- and second-stage evaporators as well as the mineral oil stripper are typically condensed in a common medium–vacuum condenser. The medium–vacuum condenser is a shell-and-tube vessel, with the vapours typically on the shell side and the cooling water on the tube side. The noncondensable vapours are removed from the condenser by a steam ejector to maintain the 300–400 mmHg absolute pressure on the shell side and are typically discharged into the first-stage evaporator for heat recovery.

The water and solvent vapours from the edible oil stripper are typically condensed in a high-vacuum condenser. The high-vacuum condenser is a shell-and-tube vessel, with the vapours typically on the shell side and the cooling water on the tube side. The noncondensable vapours are removed from the condenser by a steam ejector to maintain the 150–300 mmHg absolute pressure on the shell side and are typically discharged into the wastewater evaporator for heat recovery.

The water and solvent vapours are evacuated from the oil dryer by a steam ejector in order to maintain the 50–80 mmHg absolute pressure on the oil dryer and are typically discharged into the base of the edible oil stripper. These vapours, along with the ejector motive steam, serve as the edible oil stripper's source of countercurrent stripping steam for heat recovery.

The solvent and water vapours from the DT typically pass through a vapour scrubber to remove meal particles. Two types of vapour scrubber are commonly used: water-spray scrubbers and centrifugal scrubbers. Water-spray scrubbers utilise a heavy spray of hot water droplets sprayed through the vapours in the duct exiting the DT to entrap meal particles. The meal particles and hot water are collected in a tank and recirculated, with a small stream of dirty water being returned to the DT. Centrifugal scrubbers utilise centrifugal force to separate meal particles from the vapours, allowing the particles to fall out the bottom of the scrubber directly into the DT. Some centrifugal separators use a solvent wash to keep the walls of the centrifugal scrubber clean.

The clean vapours exiting the DT vapour scrubber, along with the much smaller vapour streams from the steam ejectors and the wastewater stripper, are partially condensed in the shell side of the first-stage evaporator. The vapours enter the top of the evaporator shell side at approximately 70 °C and quickly condense water vapour until the temperature drops to 62 °C, the minimum hexane–water equilibrium temperature. The vapours continue condensing at the mixture of 94% solvent and 6% steam and temperature of 62 °C as they progress downward through the shell of the first-stage evaporator. The remaining solvent vapour and water vapour exiting the first-stage evaporator shell (approximately 20–30% of what entered) are further condensed in either a vapour contactor or a solvent preheater. In a vapour contactor, cool condensate from condensers is pumped and sprayed into the top of a vertical tank. The remaining solvent and water vapours from the first-stage evaporator pass through the cool liquid stream and condense on the droplets, causing the liquid spray temperature to rise. In a solvent preheater, cool fresh solvent en route to the extractor is pumped through the tubes of a shell-and-tube heat exchanger and the solvent and water vapours from the first-stage evaporator pass through the shell side, condensing on the tubes while warming the fresh solvent passing through the tubes. Whether using a vapour contactor or a solvent preheater after the first-stage evaporator, the remaining DT vapours are reduced to less than 10% of what exited the DT, recovering 90% of the waste heat from the DT. The remaining vapours pass on to the vent condenser.

Up to 1 m^3 of air enters the extractor with each 1 m^3 of material. To maintain the extractor under a slight vacuum, the air must be continuously vented from it. At the extractor temperature of 60 °C, an equilibrium condition will occur, where the vent gas exiting the extractor will contain approximately 10 parts of solvent vapour for every 1 part of air (see Figure 4.8). The solvent vapour and air exit the extractor and are typically condensed in an extractor condenser. The extractor condenser is a shell-and-tube vessel, with the vapours typically on the shell side and the cooling water on the tube side. The noncondensable vapours exiting the extractor condenser pass on to the vent condenser.

The vent vapour streams from the extractor condenser, vapour contactor or solvent preheater, and atmospheric tanks pass on to the vent condenser. The normal heat load on the vent condenser is quite low. However, if the miscella is not passing through the first-stage evaporator, or solvent is not

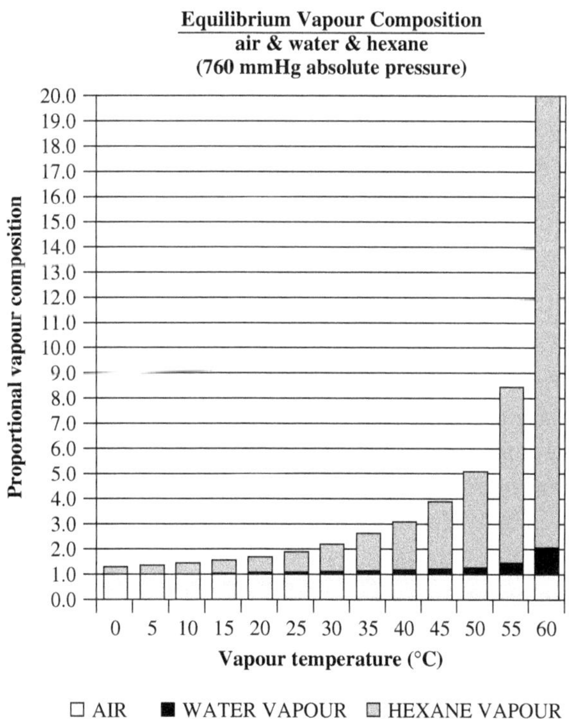

Figure 4.8 Equilibrium vapour composition.

passing through the vapour contactor or solvent preheater, the load on the vent condenser can be very high. Therefore, the vent condenser is generally sized for these contingency conditions and has an excess of heat transfer area for normal operating conditions. The vent condenser is a shell-and-tube vessel, with the vapours typically on the shell side and the cooling water on the tube side. The noncondensable vapours exiting the vent condenser pass on to the mineral oil absorption system.

The composition of solvent vapour with the air exiting the vent condenser is a result of an equilibrium condition determined by the vent gas temperature. Figure 4.8 illustrates the equilibrium vapour compositions at various temperatures. For example, at a vent gas temperature of 30 °C, the vent gas composition will be slightly greater than 1 part solvent for every 1 part air; if the vent gas temperature increases to 40 °C, the vent gas composition will be slightly less than 2 parts solvent for every 1 part air. Therefore, to minimise the load on the mineral oil system, it is important to minimise the vent gas temperature exiting the vent condenser. Some facilities, particularly in hot and humid climates, install a refrigerated vent gas cooler following the vent condenser in order to minimise the solvent vapour load on the mineral oil system.

The vent gas from the process enters the bottom of the mineral oil absorber. The mineral oil absorber is a tall, small-diameter packed column. Cold mineral oil cascades down through the column, absorbing solvent vapour from the vent gas as the vent gas rises up through the packing. When the vent gas exits the mineral absorber, it generally contains less than 10 g solvent per 1 m^3 of air. The vent gas is pulled from the mineral oil absorber via a spark resistant suction fan and is discharged to atmosphere through a flame arrestor.

The mineral oil enters the mineral oil absorber at approximately 30 °C and contains 0.1–0.4% moisture plus solvent. When the mineral oil exits the mineral oil absorber, its temperature rises slightly due to the heat of absorption and it contains 3–5% moisture plus solvent. The cool solvent-rich mineral oil is then heated to 65 °C by hot/cool mineral oil heat recovery and then further heated to over 100 °C using a steam-heated heat exchanger. The hot, solvent-rich mineral oil enters the mineral oil stripper: a packed column vessel similar in construction to the mineral oil absorber. Since the mineral oil stripper is typically maintained under 300–400 mmHg absolute pressure, much of the solvent evaporates as soon as the mineral oil enters the mineral oil stripper's top. Most of the remaining solvent is removed as the mineral oil cascades down across the packing countercurrent to the rising stripping steam. The water and solvent vapour exit the dome of the mineral oil stripper. The hot, solvent-lean mineral oil exits the mineral oil stripper and has its temperature reduced to 65 °C by hot/cool mineral oil heat recovery, and is then further cooled to 30 °C using a heat exchanger with water as the cooling medium. The cool, solvent-lean mineral oil is then recirculated to the top of the mineral oil absorber.

All water and solvent that drains from the various condensers in the solvent extraction plant enters a decanting tank. Since solvent is immiscible with water, the lighter solvent (0.65 specific gravity) floats above the water. The key to the decanting tank performance is to minimise turbulence within the tank, so as to allow sufficient time for gravity decanting to take place. The elevation of the interface between solvent and water is established by the highest elevation of the water drain pipe. Water, typically containing 0.01% solvent, exits the decanting tank to the wastewater stripper. Solvent, typically containing 0.05% water, exits the decanting tank to the solvent work tank.

The wastewater stripper is a small tank used to increase the temperature of the wastewater to approximately 95 °C, in order to evaporate residual solvent prior to discharge to the plant sump. The water exiting the wastewater stripper is typically less than 10 ppm solvent. The hot water exiting the wastewater stripper is often interchanged with the cool water entering the wastewater stripper for heat recovery.

Solvent enters the work tank from the decanting tank. The work tank is a surge tank used to hold solvent prior to the extractor, in order to ensure that there is always ample solvent available to be pumped to the extractor. If a large surge of solvent flows into the work tank, the work tank will automatically overflow to solvent storage. If the level in the work tank becomes low, additional solvent is pumped from solvent storage to the work tank. The temperature of the solvent in the work tank is typically in the range of 52–57 °C in plants equipped with a vapour contactor, and 43–49 °C in plants without a vapour contactor.

In plants with a vapour contactor, solvent is pumped from the work tank at 55 °C to a steam-heated solvent heater in order to increase its temperature to 60 °C prior to entering the extractor. In plants with a solvent preheater rather than a vapour contactor, solvent is pumped from the work tank at 45 °C to the solvent preheater, where its temperature is increased to 55 °C. The solvent is then further heated in a steam-heated solvent heater from 55 to 60 °C prior to its return to the extractor.

4.7 Heat recovery

With solvent extraction being a very established technology, many incremental improvements have occurred over the years in the optimisation of heat recovery between unit processes. Tools such as 'pinch analysis' are utilised today to chart and evaluate the available hot streams and cool streams in the process. Each stream can be plotted on a chart, with temperature on the y-axis and heat content on the x-axis.

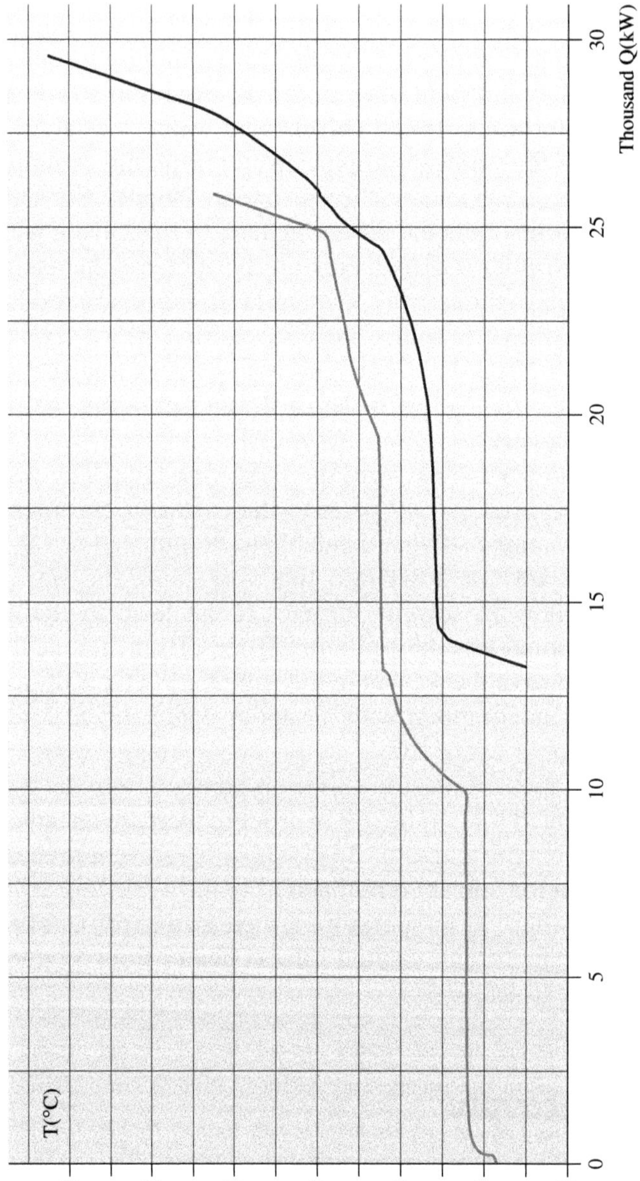

Figure 4.9 Composite pinch curves.

The vapours from the DT represent a hot stream with high heat content and moderate temperature, while the vapours from the evaporators have high heat content and very low temperature. Several hot streams exist with low heat content and high temperature, such as edible oil stripper vapours, mineral oil stripper vapours, wastewater evaporator vapours, waste water and edible oil.

The three large cool streams are miscella from the extractor, solvent to the extractor and air to the DC. A small cool stream is water from the solvent–water separator.

A composite hot stream and cool stream curve can be plotted (see Figure 4.9). The temperature range at which the two curves near each other (the 'pinch' range) is then determined. Ideally, heat recovery devices should be employed such that cooling water will not be used to cool any individual hot stream above the upper end of this pinch temperature range and steam will not be used to heat any individual cool stream below the lower end of this pinch temperature range. Using this tool, modern solvent extraction plants are achieving up to 96% of maximum potential heat recovery in the solvent extraction process.

References

Kemper, T.G. & Farmer, R.D. (1999) US patent 5 992 050 (to French Oil Mill Machinery Company).

Kirschenbauer, H.G. (1944) *Fats and Oils*. New York, NY: Reinhold Publishing Corp., pp. 122–123.

Schumacher, H.O. (1985) US patent 4 503 627 (to H.O. Schumacher).

Wan, P.J. & Wakelyn, P.J. (1997) *Technology and Solvents for Extracting Oilseeds and Nonpetroleum Oils*. Champaign, IL, USA: AOCS Press, pp. 14–18.

5

Edible Oil Refining: Current and Future Technologies

Wim De Greyt

R&D Manager, Desmet Ballestra Oils and Fats, Zaventem, Belgium

5.1 Introduction

Several factors, including a growing world population, higher fat intake per capita and increasing use for technical applications, have resulted in a steep increase in the demand for oils and fats. The annual growth rate of the world demand has virtually doubled since the late nineties and is currently estimated at 6.3 million tonnes per year (Mielke, 2011). To meet this increasing demand, the total production volume of 12 vegetable oils has more than tripled, from 40.8 million tonnes in 1980 to 146.2 million tonnes in 2010 (FAOSTAT, 2012). Production of palm oil (from 5 to ~40 million tonnes) and soybean oil (from 13 to ~38 million tonnes) have particularly increased.

Since most vegetable oils need to be (at least partially) refined for edible or technical applications, increased production volumes have resulted in a serious expansion of the edible oil refining industry. The required refining capacity is currently estimated at 400 000 tonnes per day.

In edible oil refining, the continuous effort to reduce overall production costs (including capital investment and operating costs) is mainly achieved by increasing plant capacities (up to 4000 tonnes per day today, to maximise economies of scale), installation of mono feedstock plants (for palm or soybean oil) and increasing the degree of automation (to reduce required manpower). As a result, edible oil refining has turned from a locally orientated

Edible Oil Processing, Second Edition. Edited by Wolf Hamm, Richard J. Hamilton and Gijs Calliauw.

operation with many small-scale plants into a large-scale industry dominated by a number of global companies.

Improving overall cost-efficiency has also been an important driver for new developments. Over the years, more energy-efficient processes and technologies, resulting in a higher refined oil yield, have gradually been introduced. The growing importance of the (nutritional) oil quality and the sustainability aspect of the refining process itself (minimal use of processing aids and chemicals) are new challenges for oil processors. To reach these new objectives, 'next-generation' oil refining process technologies will have to developed and implemented.

In this chapter, an overview is given of some process improvements and new developments that have recently been introduced in the edible oil refining industry. Some potential new innovations which are currently not yet applied are also briefly discussed.

5.2 Next-generation chemical refining with nanoneutralisation

Edible oils can be refined by either a chemical or a physical refining process. Chemical refining is still the most widely applied process for soft oils with low free fatty acid (FFA) content (soybean oil, rapeseed oil, sunflower oil etc.). The main byproduct of chemical refining is the so-called soapstock, which is a mixture of fatty acid soaps, salts, phospholipids, impurities and entrained neutral oil. Soapstock is usually split with sulfuric acid, resulting in a low value, difficult-to-valorise 'acid oil' and a difficult-to-treat wastewater stream. The high neutral oil losses in the soapstock (especially when crude oils with higher FFA and phospholipid contents are chemically refined), the low value of the resulting acid oil and the stricter environmental legislation (making wastewater treatment more expensive) are the main reasons for oil processors to consider physical refining. On the other hand, chemical refining is quite forgiving towards crude oil quality and it usually gives a good refined oil quality. For these reasons, it is still the preferred refining process for many processors, and it is not expected that chemical refining will disappear. Hence, there will remain a serious interest in new developments that make chemical refining more attractive.

At the end of the 1990s, several new neutralisation processes, such as soluble silicate refining (Hernandez & Rathbone, 2002), dry refining with CaO (Meyers, 2000) and chemical refining with KOH, were developed. All these developments aimed at the (partial) elimination of the washing step and a better valorisation of the soapstock. Unfortunately, none of them were finally implemented in industrial practice as the valorisation potential of Ca/K soaps was lower than expected, and soapstock-related problems thus remained unsolved.

In the last decade, process improvements in chemical neutralisation focused on increasing process automation and the use of better, more powerful mixing systems. This resulted in an overall better process control and the need for less (excess) chemicals. However, these developments did not have a significant positive impact on neutralised oil yield, and the need for acid pretreatment and excess caustic still remains.

In the search for a new neutralisation process that could further reduce the use of (excess) chemicals and oil losses in soapstock, the potential of so-called Nano Reactor® technology was investigated. Nano Reactors® are hydrodynamic cavitation reactors. Their working principle and possible applications in the chemical industry (for process intensification), biotechnology (cell disruption) and drinking water treatment (microbial disinfection and degradation of contaminants) are well described in recent literature (Cogate, 2010).

The use of ultrasound cavitation (created by a cavitational effect) for edible oil degumming was studied by Moulton & Mounts (1990). Although the results were promising, this process was never industrially applied due to some inherent drawbacks: (1) no uniform cavitational effect; (2) very high energy requirement; and (3) applicability only as a batch process.

Hydrodynamic Nano Reactors® are inherently more suitable for use in large scale oil processing as these can be used in continuous operation and require less energy. As a first industrial application, nanoneutralisation was recently developed and successfully introduced in edible oil processing (Svenson & Willits, 2012). A typical process flow diagram is given in Figure 5.1. Crude or water degummed oil is blended with the caustic solution and then transferred by a high-pressure feed pump through the Nano Reactors® at a typical pressure of 40–80 bar. The combination of this high pressure and the unique internal design of the Nano Reactors® creates a high turbulence and strong shear forces, resulting in a very good mixing of the crude oil and the caustic solution in the Nano Reactor®. Discharge pressure is 3–4 bar, which allows direct feeding of the nanotreated oil to the centrifugal separator. Afterwards, the nanoneutralised oil can flow on to the water washing or silica treatment process.

The proven industrial advantages of the nanoneutralisation process are a significant reduction (up to 90%) in phosphoric/citric acid consumption and a corresponding significant reduction (over 30%) of caustic soda use. The latter is due to the lower acid consumption and the very good mixing effect in the Nano Reactors®, which render nonhydratable phospholipids more

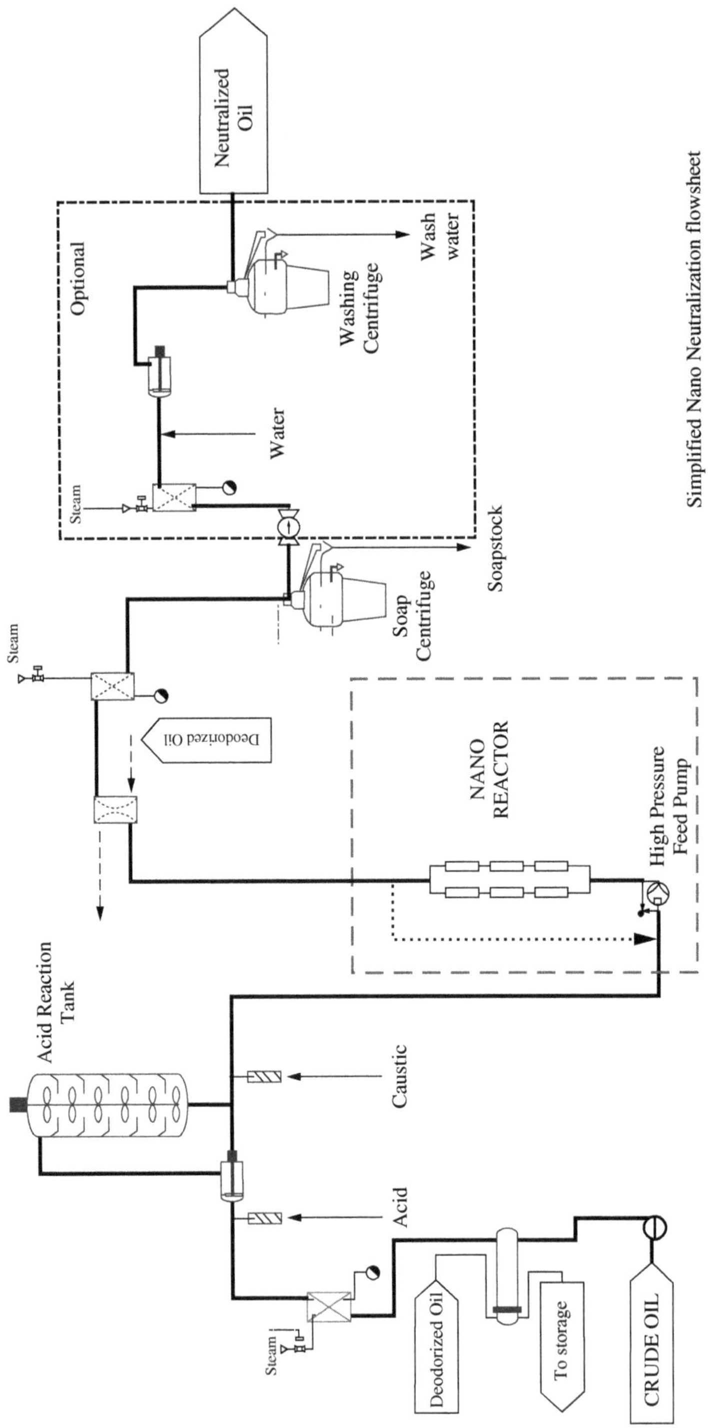

Figure 5.1 Typical nanoneutralisation process flow diagram. Courtesy of Desmet Ballestra.

easily removed and mean that almost no excess NaOH is required for complete FFA neutralisation.

The lower acid and caustic consumption results in a decreased formation of unwanted salts, with a resulting better centrifugal separation of the soapstock from the oil. As a consequence, an overall neutralised oil yield increase of at least 0.2% is observed in industrial operation. Overall oil increase can be even higher due to the lower soap content in the nanoneutralised oil. This reduces the need for silica treatment or water washing, which directly translates into lower oil losses.

The positive effects of Nano Reactors® in the neutralisation process can mainly be explained by the superior mixing effect created during the passage of the oil/caustic solution through them. However, Nano Reactors® are more than just superior mixers and a better understanding of their working principle will result in further applications in edible oil processing.

5.3 Enzymatic degumming: a missing link in the physical refining of soft oils?

Physical refining was originally developed for high(er) FFA oils (such as palm oil) for which chemical refining is not economically attractive. Physical refining results in more easily valorised side products (e.g. deodoriser distillate), but generally requires better quality crude oil. It is therefore more suitable for integrated crushing–refining plants with better control over the incoming crude oil quality.

The broader industrial application of physical refining first requires an efficient degumming process that can ensure a very good degummed oil quality ($P < 10\,ppm$) even when applied to lower quality crude (soft) oils.

The traditional classification of phospholipids into so-called hydratable and nonhydratable components is well known in the literature. Hydratable phospholipids can easily be removed during water degumming, which is generally applied as first refining step in the oilseed extraction plant. The resulting gums can either be added back to the deoiled meal or valorised separately as lecithin.

Nonhydratable phospholipids are removed during so-called acid degumming. This is usually the first stage of physical refining and can be considered the equivalent process to alkali neutralisation in chemical refining. Important developments in acid degumming date from the 1980s, driven by the first real interest in physical refining. New features such as improved dosing systems, more powerful mixing systems (to get finer dispersion of the degumming acid), addition of caustic and oil cooling for gum hydration were successfully implemented and resulted in a significant improvement in degumming efficiency. Processes such as TOP degumming (Vandemoortele) and Super- and Uni-degumming (Unilever), which are still used today in edible oil refining, were developed during that period.

First-generation enzymatic degumming (Enzymax process), soft degumming (Tirtiaux) and membrane degumming (Cargill, Desmet) were developed in the 1990s. The need for a milder but still efficient degumming process requiring less chemicals was the main driver. Unfortunately, these degumming processes were never broadly implemented on an industrial scale. Miscella membrane degumming (Lin & Koseoglu, 2004) was applied industrially for a short time but was soon abandoned due to excessive problems with irreversible membrane fouling. Industrial application of soft degumming (Deffense, 2002) was hindered by the fact that ethylenediaminetetraacetic acid (EDTA) was used as a chelating agent, which raised some acceptability issues. The main drawbacks of the Enzymax process (Clausen, 2001) were the high enzyme cost, the relatively poor stability and selectivity of the enzyme and the fact that a porcine pancreas lipase was used.

A renewed interest in enzymatic degumming has been observed in recent years. This is mostly due to the commercial availability of several new, cost-efficient and stable phospholipases with sufficiently high enzyme activity, developed and guaranteed by various suppliers (Table 5.1). In addition, there is the new market approach of the enzyme producers, who no longer present enzyme degumming as an efficient degumming process but rather as a process that results in a significantly higher refined oil yield. With the current high edible oil prices, oil refiners are very sensitive to this feature, making it the most important driver for the wider application of 'new-generation' enzymatic degumming.

Current commercial phospholipases are all of microbial origin. Their mode of action is illustrated in Figure 5.2. Phospholipase A1 (PL-A1, e.g. Lecitase Ultra from Novozymes) and phospholipase A2 (PL-A2, e.g. Rohalase MPL from AB Enzymes, GumZyme from DSM) both release a fatty acid from the phospholipid molecule, resulting in a lysophospholipid and an FFA.

Table 5.1 Commercially available phospholipases for enzymatic degumming.

Enzyme trade name	Producer	Activity
Lecitase Ultra	Novozymes	Phospholipase A1
Rohalase MPL	AB Enzymes	Phospholipase A2
GumZyme	DSM	Phospholipase A2
Lysomax	Danisco	Lipid Acyltransferase (type A2)
Purifine®	DSM	Phospholipase C

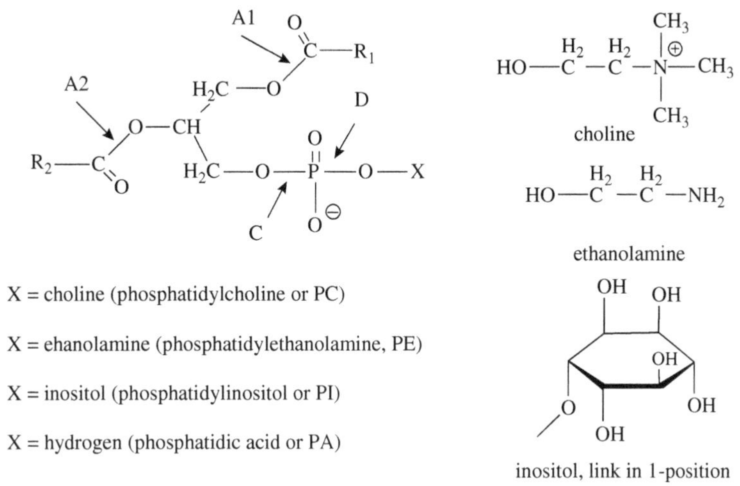

Figure 5.2 Specific activities of the various commercial phospholipases. A1, phospholipase A1; A2, phospholipase A2; C, phospholipase C; D, phospholipase D.

Theoretically, conversion of 0.1% PL (40 ppm P) leads to formation of 0.036% FFA. With sufficient reaction time (depending on enzyme dosing), phospholipases A1 and A2 are relatively unselective and will degrade nearly all phospholipids. LysoMax (Danisco) is a lipid acyltransferase (PL-A2 type) which transfers FFA released from phospholipids to free sterols, resulting in the formation of sterol esters. Unlike FFA, sterol esters are not removed during the refining process and thus represent a limited but real increase in the refined oil yield. Phospholipase C (PL-C, e.g. Purifine® from DSM) releases the P-containing part of the phospholipid molecule, with formation of diacylglycerols and phosphate esters as degradation products. Conversion of each 0.1% phospholipids results in the formation of 0.084% diacylglycerols. Phospholipase C will only react with phosphatidylcholine (PC) and phosphatidylethanolamine (PE) and has virtually no effect on phosphatidic acid (PA) or phosphatidylinositol (PI) (Hitchman, 2009).

A general flow sheet of an enzymatic degumming process (basically independent of the type of enzyme being used) is given in Figure 5.3. The first step is the acid conditioning/pH adjustment of the crude or water degummed oil. This step is required to make the nonhydratable phospholipids more accessible for enzyme degradation at the oil–water interface and to bring the pH closer to the optimal pH of the enzyme. Afterwards, the enzyme is added – either pure or diluted in water. High shear mixing is required to ensure optimal distribution in the oil. Enzyme dosing depends on the type of enzyme and on the phospholipids content of the oil, but usually varies between 50 and 200 ppm. The optimal reaction temperature is 50–60 °C, while the required

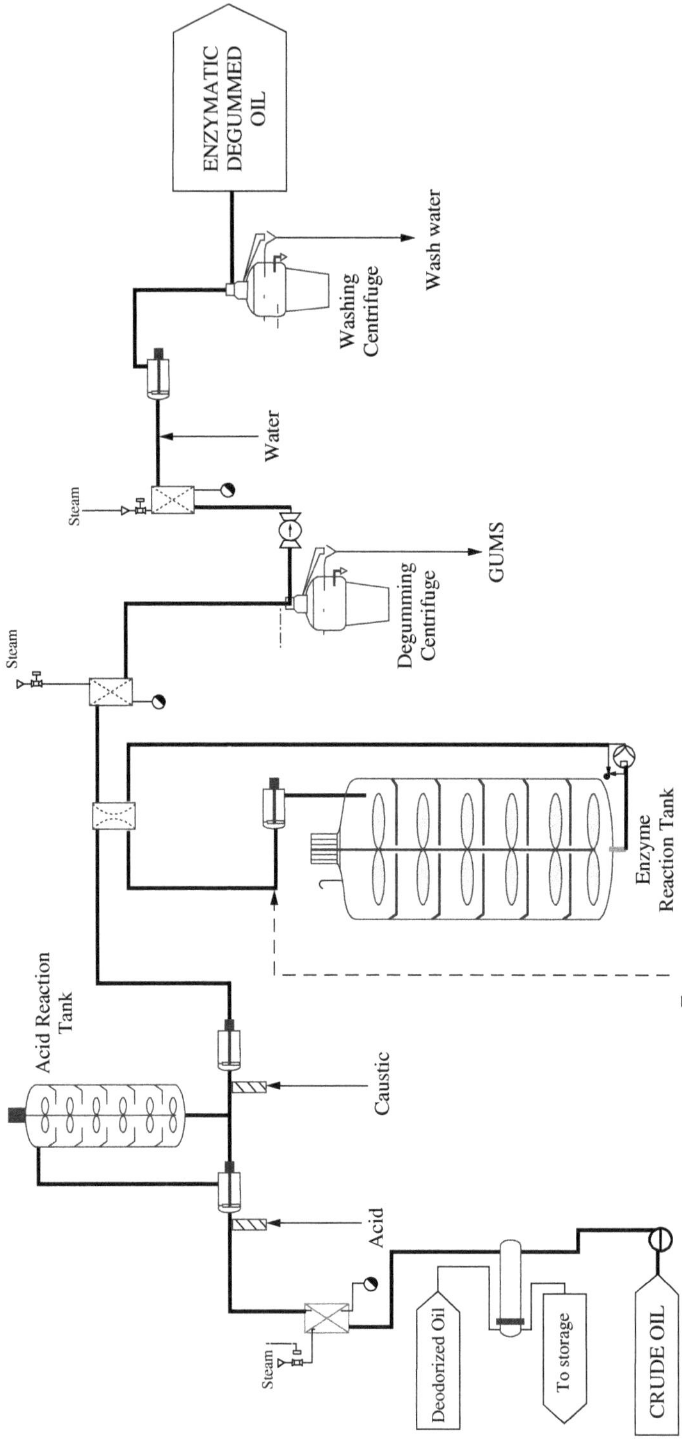

Figure 5.3 Typical process flow diagram of a deep enzymatic degumming process. Courtesy of Desmet Ballestra.

reaction time mainly depends on the enzyme dosing. While in the past it was common practice to apply a longer reaction time with a low enzyme dosage (e.g. 30 ppm enzyme for 6 hours' reaction), preference is now given to a shorter reaction time with higher enzyme dosage (e.g. 100 ppm enzyme for 2 hours' reaction). This practice is preferred because it increases the flexibility of the process while keeping the operating (enzyme) cost at an acceptable level. Finally, the heavy phase (consisting of water and lyso gums or phosphate esters) is separated by centrifugation from the degummed oil.

Two different types of enzymatic degumming can be distinguished: so-called enzymatic water degumming and deep enzymatic degumming. Enzymatic water degumming is typically applied in (soybean) crushing plants. Several large-capacity plants in South America (Argentina, Brazil etc.) are already running in this mode. Increased oil yield is the main driver for its implementation. The expected yield increase depends on the type of oil (P content) and the type of enzyme used. The highest increase (up to 1.8%) can be expected when crude soybean oil is enzymatically degummed with PL-C (Hitchmann, 2009; Kellens, 2009); in this case, the oil yield increase is the sum of the diacylglycerols formed and the lower neutral oil entrainment in a smaller heavy phase (gums fraction). A lower yield increase (1.0–1.5%) will be obtained from PL-C degumming of crude rapeseed oil or when phospholipase A1 or A2 is used on crude soybean oil (Kellens, 2009). In the latter case, the net oil yield increase is due to the lower neutral oil entrainment in the gums fraction alone.

An increase in refined oil yield is obviously a very attractive feature of enzymatic (PL-C) water degumming, but by itself it is not enough to lead to implementation in all crushing plants. In the overall cost/benefit analysis of the process, the enzyme cost and side-stream valorisation are also taken into account. Depending on the value of (lyso-) lecithin, it may be more profitable for a crusher to apply simple water degumming or enzymatic water degumming with PL-A1/PL-A2. The latter gives a lower net oil yield improvement compared to PL-C enzymatic degumming but yields a lysolecithin side stream that may have value for specific applications.

PL-C enzymatic degummed soybean oil typically still has 100–150 ppm residual P (mainly present in PA and PC). A significantly better degumming efficiency (P $<$ 10 ppm) can be obtained when crude or water degummed vegetable oils are enzymatically degummed with commercial PL-A1 or PL-A2. This so-called 'deep enzymatic degumming' is already applied in several industrial plants. In addition to the increased oil yield, the very efficient phospholipid removal – making the degummed oil suitable for physical refining – is of great interest to refiners. As an alternative option, a combination of PL-C and PL-A1/PL-A2 can be used for deep enzymatic degumming (Dayton, 2011; Galhardo & Dayton, 2012). The two enzymes can be added either separately or as a cocktail, depending on the plant design. Although the potential advantages of the latter process are well described in the (patent) literature (Dayton & Galhardo, 2008; Gramatikova *et al.*, 2011), it is still rarely applied on industrial scale.

A potential alternative to enzymatic degumming is the direct enzymatic deoiling of the lecithin fraction resulting from the water degumming of crude oils. In this patented process (De Greyt & Kellens, 2010), a phospholipase (e.g. Lecitase Ultra) is added to the wet lecithin and the phospholipids are degraded into much less hydrophobic lysophospholipids. As a result, 80–90% of the entrapped neutral oil can be recovered by simple static decantation or centrifugation (Kellens, 2009; Kellens *et al.*, 2010). The recovered neutral oil (FFA content: 25–30%) can be recycled to the crude or degummed oil or can be used as such as biodiesel feedstock, while the lysolecithin can be added back to the deoiled meal. The main advantages of the enzymatic lecithin deoiling process over enzymatic degumming are the lower enzyme consumption (~50% less) and the fact that it is applied on a small side stream, with no impact on the oil degumming/refining process. The process has been tested successfully on a pilot scale but is currently not yet applied on an industrial scale.

5.4 Bleaching: from single-stage colour removal to multistage adsorptive purification

Bleaching was introduced in edible oil refining at the end of the 19th century to improve the colour of cottonseed oil. Originally, it was a batch process at atmospheric pressure, in which natural bleaching clay was added to hot oil with the sole objective of removing colouring pigments. Today this is no longer the case, and bleaching has become a critical process in edible oil refining. It has gradually turned from a single-stage 'bleaching' into a multistage adsorptive purification process in which a wide range of unwanted components (soaps, phospholipids, oxidation products, trace metals, contaminants etc.) are removed prior to deodorisation.

In order to reach this point, a whole series of process improvements was gradually introduced, with the aim of reducing the overall processing cost and improving the bleached oil quality. Vacuum bleaching was implemented first, in order to avoid oxidation and related colour fixation and improve the oxidative stability. Later, as the capacity of refining plants increased, bleaching evolved from a batch to a (semi-) continuous process. This evolution further improved the bleached oil quality and made the process more energy efficient. Another process improvement was the implementation of (horizontal/vertical) pressure leaf filters. Initially, plate and frame filters were used, but these lost favour over the years due to the too high residual oil content in the spent bleaching earth (typically 35–40%) (Veldkamp, 2012).

With pressure leaf filters, the residual oil content in the spent bleaching earth varies between 25 and 30%, depending on the cake blowing efficiency.

Reducing oil losses in spent bleaching earth is very important as it will both directly increase the bleached oil yield and reduce the quantity of spent bleaching earth that must be disposed of. In the past, spent bleaching earth still had a certain value as it was sold mostly to the animal feed market. However, this practice became prohibited in a growing number of countries due to stricter feed safety regulations. Only in integrated crushing–refining plants it is still possible to incorporate spent bleaching earth in the deoiled meal. Refiners therefore have to look to other outlets for their spent bleaching earth. Some can dispose of it through biomethanisation, but for most disposal as landfill is the only (costly) option.

The increasing disposal costs, together with the oil losses in the spent bleaching earth, have a big impact on the overall operating cost of bleaching. The most efficient way of lowering the operating cost is to reduce the bleaching earth consumption. Much effort has been made to design more efficient bleaching processes and develop more efficient bleaching clays and other specific adsorbents.

The bleaching earth efficiency can be significantly improved by acid activation. Acid activation is carried out with H_2SO_4 or HCl. Its main effect is a significant increase in the specific surface area, by a factor of 3–6 to 250–350 m^2/g. Today, a wide range of acid-activated bleaching clays are commercially available from many (local) suppliers. This has resulted in a very price-competitive market, with products that are differentiated according to degree of activation, type of activation acid (H_2SO_4 or HCl) and particle size distribution. The latter is an important and sometimes underestimated characteristic as it greatly affects the filterability of the bleaching earth.

Although activated bleaching clays still have the highest market share, oil refiners are showing a growing interest in nonactivated bleaching earths, especially for the bleaching of palm oil. The main reason is the possible catalytic effect of highly (HCl) activated bleaching earths on the formation of potentially toxic 3-monochloropropane-diol (3-MCPD) esters during palm oil refining (De Greyt, 2012). This catalytic effect is not observed when natural (or less activated) bleaching clays are used. Although the effect of acid-activated bleaching earth on the formation of 3-MCPD esters is not yet fully understood and contradictory research data have been presented, it is clear that the growing demand for refined palm oil with low levels of 3-MCPD esters may have a serious impact on the palm oil refining process (De Greyt, 2012; Ramli *et al.*, 2011; Schurz, 2010). More specifically for the bleaching/dry pretreatment process, it may result in the use of less strong acids and other grades (more natural, less activated) of bleaching earth.

The active surface of bleaching earth can also be significantly improved by a drastic reduction of its particle size (<10 μm). Such bleaching earths can be produced, but the problem is that they cannot be separated from the oil by conventional filtration techniques. In order to overcome this filtration problem, the electrofiltration process was developed in the mid 1990s (Transfeld, 1998). In this process, fine electrically charged particles are agglomerated on an electrode at the end of the bleaching process and removed via conventional filtration. The oil, being a nonconductive liquid, will not interfere in this process. Unfortunately, electrofiltration was only tested on a pilot scale.

Bleaching earth consumption can also be reduced by increasing the efficiency of the bleaching process itself. Nowadays, bleaching is still mostly operated as a single-stage process, with only one bleaching reactor and one bleaching filter in operation. In such a process, the adsorption capacity of bleaching earth is largely underutilised, and therefore multistage bleaching processes (multistage co-current, countercurrent, prefiltration over spent bleaching earth etc.) have been investigated. The least complicated multistage bleaching is the co-current process, with two (or more) bleaching reactors and the same number of filters in operation. However, this only gives a limited bleaching earth saving, which for most oils does not compensate for the extra capital investment required. More bleaching earth saving can be expected from a countercurrent bleaching process. The first real countercurrent process was developed about 15 years ago by Ohmi Engineering (Transfeld & Schneider, 1996). This process features two bleaching reactors and two sets of filters. The spent bleaching clay of the second filtration is reused by mixing it back into the degummed/neutralised oil. Although bleaching earth savings up to 40% are guaranteed, this countercurrent process is not widely applied, mainly because of its technical complexity. Another alternative is the 'countercurrent' process in which the incoming oil is prefiltered over a filter loaded with spent bleaching earth before fresh bleaching earth is added. Prefiltration removes all solid impurities and most of the phospholipids and soaps, which increases the bleaching earth efficiency in the bleacher. Typically, 10–15% bleaching earth can be saved in this way. Although savings are lower than in the real countercurrent bleaching process, prefiltration is nevertheless preferred by most refiners as it only requires one additional filter.

Removal of interfering components from the oil (e.g. phospholipids) prior to effective bleaching will also have a positive effect on bleaching earth consumption. This is partially achieved by prefiltration, but can be done even more efficiently when silica hydrogels are used as an additional adsorbent.

Silica hydrogels (also named 'silica') are free-flowing white powders consisting of silicon dioxide (all or not acid activated) and high amounts of water (50–65%). These adsorbents have little or no affinity for (oil-soluble) colour pigments but are very efficient for the removal of polar impurities like phospholipids and soaps. Silica hydrogels were introduced in edible oil refining in the mid 1980s (Taylor, 2004). Their first application was in chemical refining, where they were used for the removal of soaps, phospholipids and trace elements as an alternative to the second washing (or sometimes even first soap) centrifuge. Today, this is still their main application in edible oil refining, but they can also be used in (multistage) bleaching. Silica and bleaching earth can be added to an oil together, but this has little or no beneficial effect as silica is only effective when applied at 70–80 °C and at atmospheric pressure (to avoid evaporation of the matrix water). A sequential addition of silica and bleaching earth is therefore more efficient. Silica (0.05–0.1%) is added first to neutralised/degummed oil at 70–80 °C. After 15–30 minutes' contact time at atmospheric pressure, the oil–silica mixture can be sent either straight to the bleacher or first to a separate drier. In the first, most straightforward option, silica and bleaching earth are removed from the oil on one filter. A better synergistic effect of silica and bleaching earth can be obtained when the oil–silica mixture is first dried and then transferred over a filter coated with spent bleaching earth (prefiltration with silica). This process (also known as the Trisyl® Silica Tri-Clear process) claims a bleaching earth reduction of more than 40% when applied in the physical refining of rapeseed oil (Figure 5.4) (Jalalpoor, 2008). This lower consumption is partially due to the enhanced ability of the fresh bleaching earth to adsorb colouring pigments from the silica pretreated oil. In addition, use of silica seems to improve bleaching earth filterability, resulting in longer filter cycles and a higher press bleach effect.

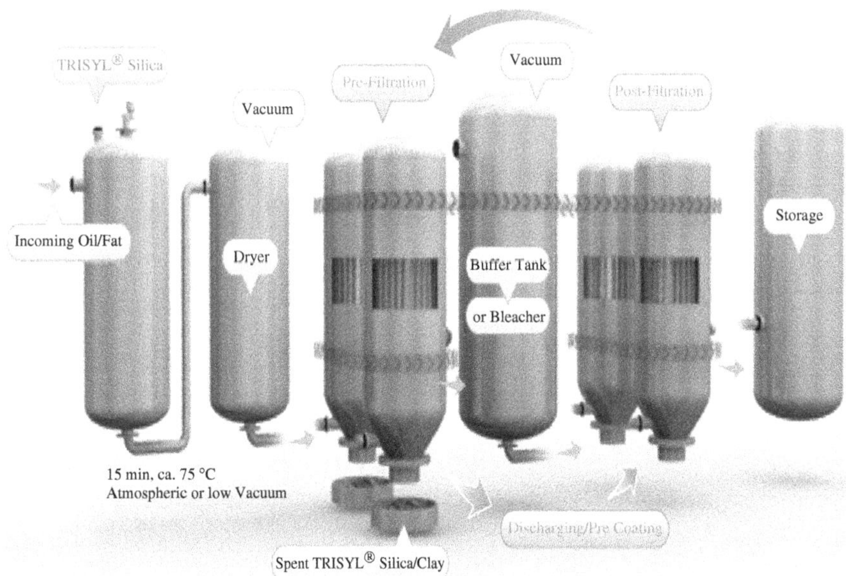

Figure 5.4 Schematic process diagram of the Trisyl® Silica Tri-Clear process. Published with the courtesy of W. R. Grace & Co.

Another adsorbent that has long been used in the bleaching of edible oils is activated carbon. This is produced from a carbon-rich material (nutshells, peat, wood, coal etc.), steam activated at a very high temperature (up to 1000 °C) to give it its typical structure with a high surface area (up to 1500 m^2/g). The traditional use of activated carbon in edible oil refining is in effective bleaching (colour removal), as a complementary (lipophilic) adsorbent to bleaching clay. However, as more efficient (activated) bleaching earths have become available, the use of activated carbon for bleaching has been substantially reduced (Taylor, 2004). Today, activated carbon is mainly used for the removal of heavy polycyclic aromatic hydrocarbons (PAHs) from vegetable oils (coconut oil, palm kernel oil, olive pomace oil etc.) that have been contaminated by smoke drying or direct heating of the raw materials (De Greyt, 2010; Kemeny *et al.*, 2011). At the end of the 1990s, it was also introduced in fish oil refining for the adsorption of dioxins and polychlorinated biphenyls (Maes *et al.*, 2005). More recently, activated carbon treatment has begun to be applied more systematically as 'best proven practice' for the decontamination of edible oils (like sunflower and rapeseed oil). This practice arose with the growing attention paid to the removal of contaminants from edible oils imposed by stricter legislation and driven by stricter trading specifications.

Activated carbon is still mostly added together with bleaching earth. If used for the removal of colour pigments, the bleaching earth–activated carbon ratio is typically 80 : 20 or 90 : 10. Typical dosing for the removal of dioxin/PCB from fish oil is 1–3 kg/tonne, while for the removal of PAH from coconut oil it can be 5 kg/tonne or higher. The recent, more strict EU regulation on the maximum allowable PAH content in vegetable oils will probably require higher dosing rates or even new, more efficient grades of activated carbon (Kemeny *et al.*, 2011).

Apart from its higher cost compared to bleaching earth, the main disadvantages of activated carbon are its higher oil retention and poor filtration characteristics. To overcome the latter, activated carbon is mostly filtered together with bleaching earth. In some cases (e.g. decontamination of fish oil), this is however not possible as either no bleaching earth is used or refiners want to keep the contaminated spent active carbon separate from the noncontaminated spent bleaching earth. For these applications, special activated carbon powders with improved filtration characteristics have been developed. These so-called high filterability (HF) grades can be filtered with classical pressure leaf filters or pulse tube filters, and membrane press filters are also used. The latter allows squeezing of the spent activated carbon cake at the end of the filtration cycle, which results in lower oil losses and a reduction in spent cake (De Kock, 2006).

The possibility of removing specific colour bodies (e.g. chlorophyll) from oils through enzymatic degradation has been under investigation for many years. Recently, new developments in enzymatic bleaching were presented (Carlson *et al.*, 2011). Chlorophyllases that can operate at pH 4.5–6.0 and at 55–65 °C have been identified (Mikkelsen, 2011). When added during water degumming, these enzymes degrade chlorophyll components in canola and soybean oil to very low residual levels (< 50 ppb), eliminating the need for bleaching earth addition (Carlson *et al.*, 2011). An indicative enzyme cost of US$3/tonne oil was mentioned, without further information on enzyme stability and required dosing rate. This process was successfully tested at pilot scale but is not yet ready for implementation on an industrial scale.

By combining silica pretreatment, bleaching with prefiltration over spent bleaching earth and decontamination with activated carbon, bleaching effectively becomes a multistage modular adsorptive purification process. However, the complete process is still rarely applied in industry. Not all refiners (need to) apply an activated carbon treatment, and if they do then it is still mostly integrated in the effective bleaching step. Bleaching and active carbon treatment are only separated when it brings significant advantages for the disposal costs of both solid waste streams. When applied separately, activated carbon treatment is best applied after bleaching. At the same time, the integration of silica pretreatment and/or prefiltration is not common practice either. The (apparent) advantages of these additional steps (lower oil losses, lower disposal cost, less wastewater etc.) have to be weighed carefully against the higher investment costs (additional equipment) and the additional process steps and material handling. In efficient (chemical) refining plants that consistently obtain low phospholipids and low soaps after neutralisation or degumming, silica pretreatment or prefiltration may only give minimal cost savings. Although oil quality and sustainable processing are growing in importance, in the end a potential cost saving remains the prime driver for the implementation of a new process.

5.5 Deodorisation: much more than just a process for the removal of off-flavours

Deodorisation is usually the last stage of the refining process of edible oils. It was introduced at the end of the 19th century to improve the taste and smell of refined oils. Today, the process is still commonly named 'deodorisation', but the objectives have become much broader than just the removal of off-flavours. In fact, the current deodorisation process has three main objectives: (1) stripping of volatile components such as FFA (in the case of physical refining), valuable minor components (tocpherols, sterols etc.) and contaminants (pesticides, light polycyclic aromatic hydrocarbons etc.);

(2) actual deodorisation by removal of different off-flavours; and (3) thermal destruction of pigments (so-called heat bleaching).

Deodorisation is obviously a crucial refining stage with a big impact on the refined oil quality. Apart from the desired effects, some unwanted side-reactions (like formation of *trans* fatty acids and polymeric triacylglycerols) may also occur during deodorisation. The effects of process conditions (temperature, time, pressure and stripping steam) on the standard quality parameters and the nutritional quality of the refined oil are well understood (see Table 5.2) (De Greyt, 2010; De Kock & De Greyt, 2009), so deodoriser design and process conditions have been optimised to ensure minimal formation of *trans* fatty acids, maximal removal of volatile contaminants and a controlled stripping of valuable minor components (tocopherols, sterols etc.).

New developments in deodorisation technology are driven by the continuous need for more efficient processes (lower operating cost, higher refined oil yield and better valorisation of side streams) and the increased attention paid to the (nutritional) quality of food oils and fats. Recent trends and developments are summarised in Table 5.3 (De Kock & De Greyt, 2009). As profit margins in edible oil deodorisation are small, further reduction of the operating cost has always been an important driver for new innovations. Fixed costs are primarily reduced by installation of higher capacity deodorisers. Today, single deodorisers with a capacity of more than 1500 tonnes per day have become more or less standard, especially for the deodorisation of commodity oils e.g. soybean oil, palm oil etc. Variable processing costs are mainly determined by energy consumption in the heating of the oil, generation of the vacuum and production of the stripping steam. The introduction of optimised oil–oil heat exchangers in continuous deodorisers resulted in a higher heat recovery (up to 90%), giving a significantly reduced net fuel consumption. Improved heat-recovery systems were also introduced in semi-continuous deodorisers. Installation of a double thermosyphon system gives a typical heat recovery of 65%, while the combination of one thermosyphon and the generation of low-pressure steam increases the heat recovery further,

Table 5.2 Effect of process variables on deodorised oil quality.

Quality parameter	Temperature	Time	Pressure	Steam
Taste	+	++	+	++
Colour (heat bleach)	++	+	–	–
FFA stripping	++	–	++	+
Trans fatty acid formation	++	++	–	–
Tocopherol/sterol stripping	++	–	++	+
Contaminant removal[a]	++	–	++	+

[a]Pesticides, PAH, dioxins.

–, little or no effect; +, significant effect; ++, large effect.

Table 5.3 Trends and developments in edible oil deodorisation.

Trend	Development
Higher capacities	>1500 tonnes per day, no exception
Higher energy efficiency	Improved heat recovery
Higher stripping efficiency	Improved tray design and integration of packed columns
Lower neutral oil losses	Improved scrubber design
Lower heat load	Application of dual-temperature deodorisation and integration of packed column
Lower pressure	Ice-condensing vacuum systems of closed loop with chilled water
Higher distillate value	Application of dual condensation

to 75%. More efficient stripping steam distribution systems (special steam lift pumps) have been designed, and, for specific applications, packed columns are integrated to improve stripping efficiency and reduce stripping steam consumption.

Lowering the heat load (residence time at high temperature) during edible oil deodorisation is another clear trend driven by the need to minimise unwanted thermal degradation reactions and the desire for maximum retention of the natural characteristics of an oil. Negative thermal effects during deodorisation can be minimised by the integration of packed columns (only for dedicated purposes) or by application of dual-temperature deodorisers (Figure 5.5). These deodorisers operate at two different temperatures in order to reach the best compromise between required residence time for actual deodorisation (longer time at lower temperature) and heat bleaching and stripping of volatile components (shorter time at higher temperature). The dual-temperature concept has been successfully introduced on an industrial scale. Both the low/high temperature and the high/low temperature concept can be applied.

In view of a further reduction of the heat load, the implementation of more powerful vacuum systems (chilled barometric vacuum system or dry-ice condensing) is important as it will allow a reduction of the deodorisation temperature without affecting the stripping efficiency in a negative way. Most conventional vacuum systems consist of a combination of steam ejectors (boosters), vapour condensers and mechanical (liquid-ring) vacuum pumps. These quite robust systems typically reach pressures in the deodoriser between 2.5 and 5.0 mbar, but the motive steam consumption required to generate the vacuum is high (up to 85% of the total steam consumption). Motive steam consumption can be significantly reduced (by a factor of 2–3) by

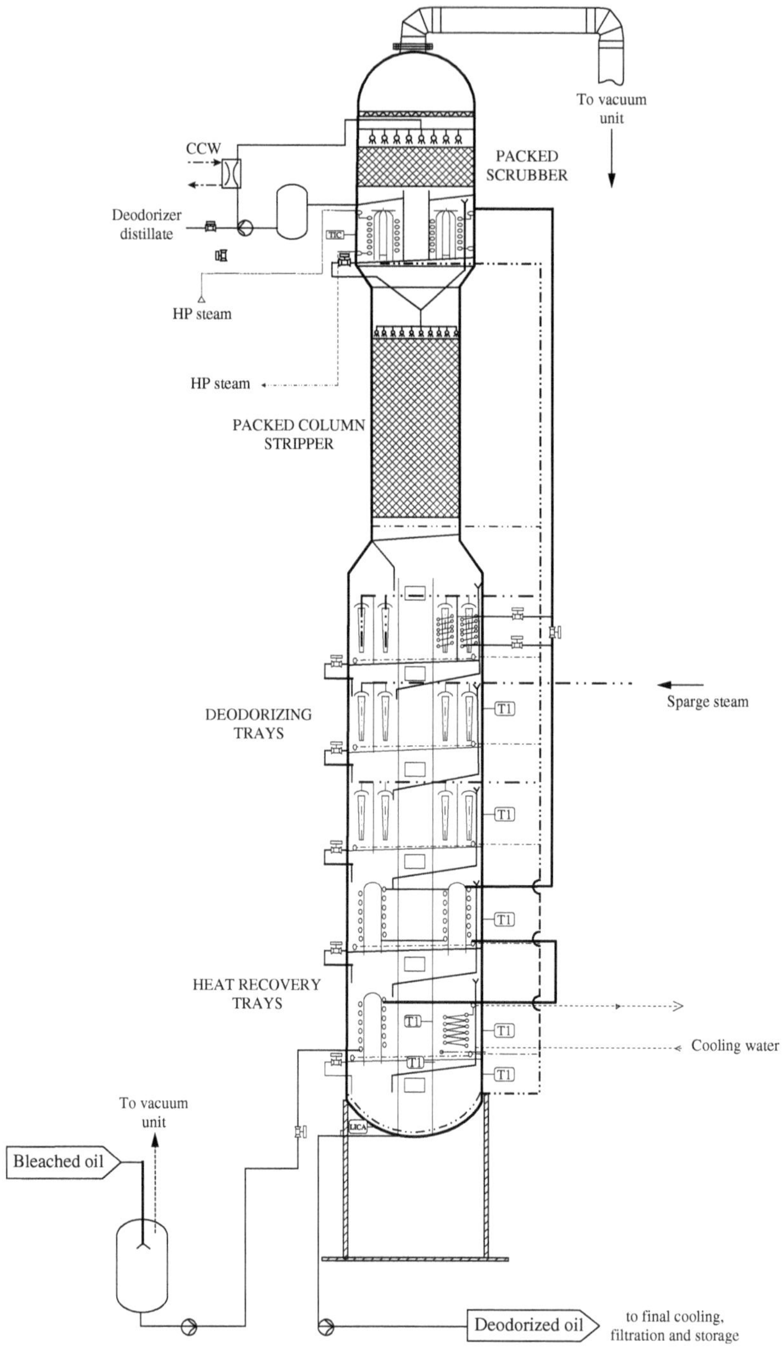

Figure 5.5 Dual-temperature deodoriser with integrated packed column stripper. Courtesy of Desmet Ballestra.

cooling the barometric condenser water. The benefit of a lower motive steam consumption must be weighed against the extra chilling capacity (higher electricity consumption) required to cool the barometric condenser water. Another benefit of using a chilled barometric vacuum system is a better condensation of the volatile matter, which gives a lower pressure in the deodoriser (e.g. 1.5 mbar). These classical vacuum systems are increasingly being replaced by dry (ice) condensing systems (Figure 5.6). With such systems, the stripping steam is condensed on surface condensers operating alternately at very low temperature (−30 °C). The efficient sublimation of steam and other volatile matter will give a very low pressure in the deodoriser (<1.5 mbar) and will strongly reduce odour emission. Dry–ice condensing systems strongly reduce the motive steam consumption but require extra electrical energy. Commercially available systems consist of two or more freeze condensers with horizontally or vertically orientated straight tubes, a refrigeration plant for the generation of the cold refrigerant evaporated in

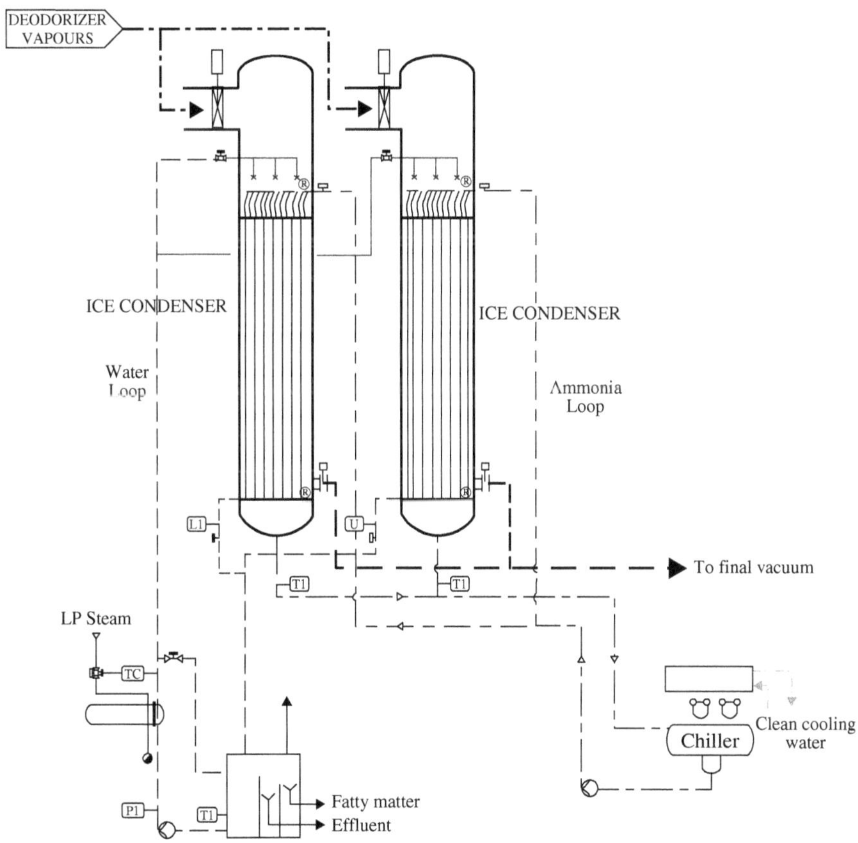

Figure 5.6 Typical process flow diagram of an ice condensing vacuum system. Courtesy of Desmet Ballestra.

the tubes and a vessel for the defrosting and cleaning of the tubes after a certain period of freezing.

Design improvements are also introduced in the vapour scrubber system. The main objectives are a further reduction of the neutral oil losses (due to mechanical entrainment) and an increase of the value of the deodoriser distillate. Most deodorisers have only one vapour scrubber, from which one single deodoriser distillate is collected. The amount and composition of this side stream is determined by a number of factors, including the processed oil composition, the applied refining mode (chemical or physical) and the deodorisation conditions (Verleyen *et al.*, 2001). Deodoriser distillate always contains a certain amount of 'neutral oil' (triacylglycerols, partial acylglycerols etc.) which has been mechanically entrained from the oil. This refining loss can be reduced by integration of a so-called 'neutral oil recovery system' (NORES) in the first part of the scrubber section. The purpose of this system is to recover the mechanical entrained neutral oil from the vapour phase before the volatile matter (FFA, tocopherols, sterols, odour components, contaminants etc.) is effectively condensed. Deodoriser distillate from physical refining is characterised by a very high FFA content (>85%) and is mostly used for technical applications (oleochemistry). Recently, it has also begun to be used as feedstock for biodiesel production (Echim *et al.*, 2009). Deodoriser distillates from the chemical refining of vegetable oils (e.g. soybean oil) have a significantly higher added value, due to their high concentration of valuable minor components such as tocopherols and sterols. Depending on the FFA content of the incoming oil and the amount of tocopherols stripped, a single scrubber can yield a deodoriser distillate containing 10–15% tocopherols (Table 5.4). The growing interest in and demand for tocopherol-rich distillates (as a source of natural vitamin E) with even higher tocopherol concentrations (up to 20%) has created a momentum for the broader implementation of the so-called 'double scrubber'. This concept was already developed by the end of the 1990s, but has been continuously improved since (Figure 5.7). The vapour phase leaving the deodoriser is first partially condensed at a higher temperature, giving a so-called 'hot distillate' in which the least volatile components (e.g. tocopherols and sterols) are concentrated. Complete condensation of the remaining, more volatile substances (mainly FFA) is then achieved in the second so-called 'cold scrubber', giving an FFA-rich 'cold distillate'. Provided that the condensation temperatures of the hot and cold scrubbers are properly set, this concept gives a very good separation between the FFA and tocopherols. Combining NORES with a double scrubber in the physical refining of soybean oil (for which processors seem to have a growing preference) gives a 'hot distillate' with 21.2% tocopherols (Figure 5.8). Today, the commercial value of such a deodoriser distillate is high (>US$8000 per tonne), which corresponds to a potential extra revenue of US$15–25 per tonne of deodorised oil.

Table 5.4 Tocopherol concentration in soybean deodoriser distillate from a single scrubber.

	NB oil[a]		DB oil[b]
FFAs (% C18 : 1)	0.05	0.1	0.6
Tocopherols (ppm)	1240	1200	1200
		Fully refined oil	
FFAs (% C18 : 1)	0.015	<0.03	0.6
Tocopherols (ppm)	900	511	515
		Deodoriser distillate	
FFAs (% C18 : 1)	39.2	31.3	71.2
Tocopherols (%)	**20.4**	**16.2**	**6.9**
Yield (kg/ton)	1.23	3.6	8.52

[a]Neutralised, bleached soybean oil.
[b]Degummed, bleached soybean oil.

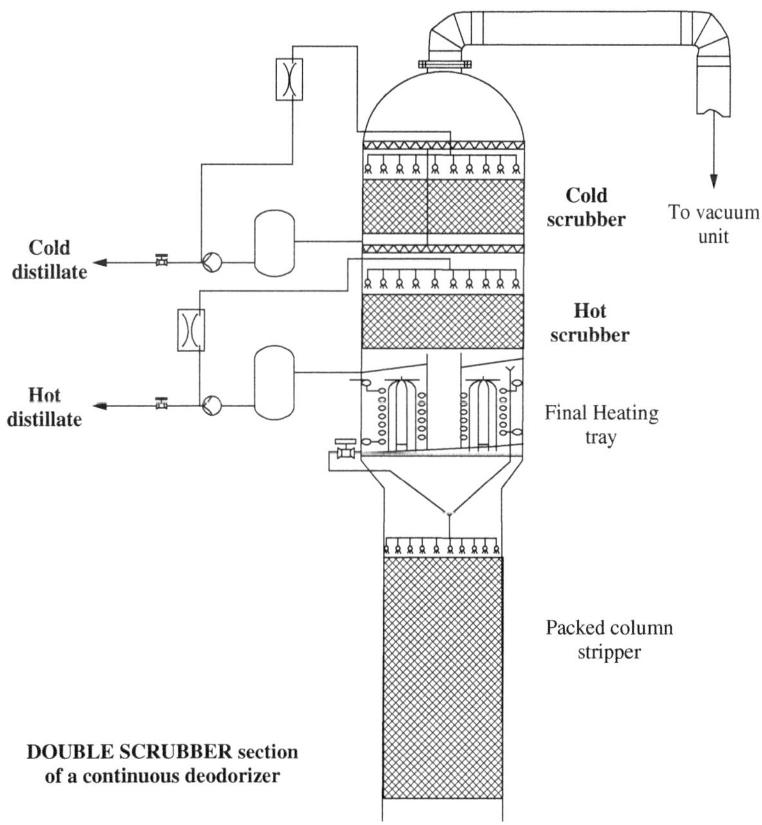

Figure 5.7 Double scrubber for the selective condensation of FFA and tocopherols. Courtesy of Desmet Ballestra.

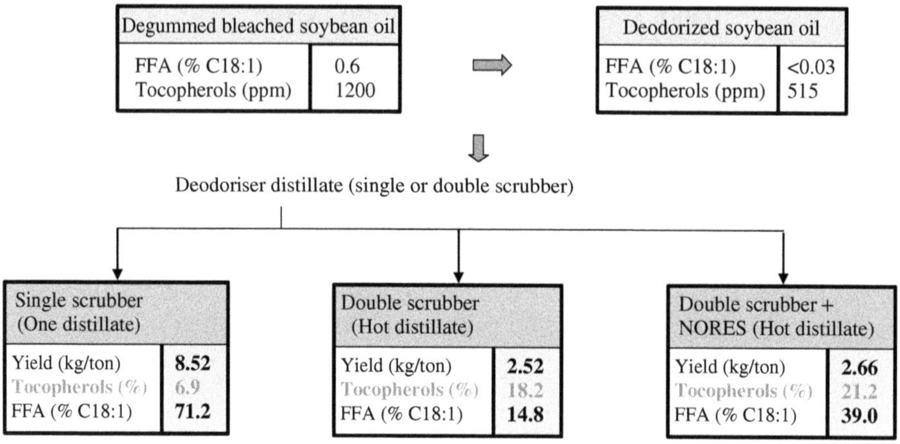

Figure 5.8 Expected tocopherol concentration in deodoriser distillate from physical refining of soybean oil (single- versus double-scrubber system). Courtesy of Desmet Ballestra.

5.6 Short-path distillation and supercritical processing: refining technologies for the future?

Short-path distillation is a well-known process that is characterised by the short residence time of the product in the evaporator (<10 seconds) and the very low operating pressure (<0.01 mbar). The incoming product flows in a thin liquid film on a vertical cylinder (falling film evaporator) or on a rotating surface (centrifugal film evaporator). The short distance (typically 10–50 mm) between the internal condenser and the evaporator causes minimal pressure drop, which results in a very low operating pressure (Figure 5.9). Short-path distillation is especially suitable for the treatment/purification of heat-sensitive products. It is established in lipid processing for the production of high-purity monoacylglycerols and the concentration of omega-3 fatty acids from fish oil (ethyl esters). It is also part of the downstream processing of deodoriser distillates for the concentration of tocopherols and sterols (Fabricius, 2009). Applications of short-path distillation in edible oil refining are still scarce. At the end of the 1990s, it was introduced on an industrial scale for the production of red palm oil (rich in carotenoids). However, as investment and operating costs are high, it never became broadly implemented. More recently, a short-path distillation process was developed for the removal of contaminants from fish oil (Oterhals *et al.*, 2010). This has been successfully applied by a limited number of fish oil processors, who prefer it over the more broadly implemented combined active carbon treatment/low-pressure stripping (Maes *et al.*, 2010). Finally, short-path distillation can also be used

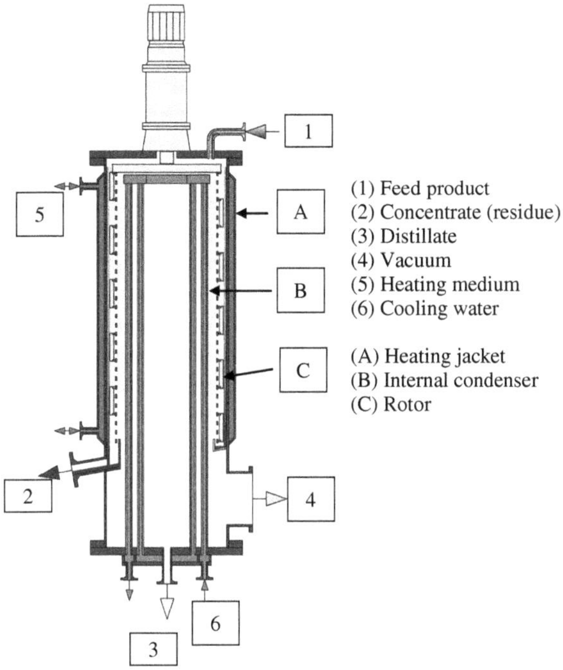

Figure 5.9 Schematic view of a short path evaporator. Courtesy of Buss-SMS-Canzler GmbH.

for efficient FFA stripping at low temperatures (<200 °C), which makes it an interesting process for the physical refining of heat-sensitive oils (fish oil, algae oil etc.). These oils are still mostly chemically refined, since complete FFA removal without significant degradation of the omega-3 fatty acids is very difficult to achieve using classical stripping technology. However, it has to be emphasised that short-path distillation is not an effective 'deodorising' process. As with packed column stripping, the residence time is much too short to produce a stable, odourless and bland refined oil: for fish oil in particular, this requires a long 'deodorisation' time of several (2–4) hours at moderate temperature (<190 °C), with the injection of sufficient sparge steam, which can still best be achieved in a 'classical' batch or (semi-) continuous deodoriser.

Supercritical CO_2 processing is another technology that has been investigated for use in edible oil refining. It was tested successfully at laboratory scale some years ago for the degumming of crude soybean oil (List *et al.*, 1993), the deacidification of rice bran oil (Dunford & King, 2001) and the mild refining of palm oil (Ooi, 1996). Nevertheless, it has never been industrially applied, due to the high investment and operating costs involved, which make it commercially viable only for high-value products.

References

Carlson K., Mikelsen R. & Soe J.B. (2011) New approaches for chlorophyll removal in oil processing. Paper presented at the 102nd AOCS Annual Meeting and Expo, Cincinnati, OH, USA.

Clausen, K. (2001) Enzymatic oil degumming by a novel microbial phospholipase. *Eur. J. Lipid Sci. Technol.*, **103**, 333–340.

Cogate, P. (2010) Cavitational reactors for process intensification of chemical processing applications: a criticial review. *Chemical Engineering and Processing*, **47**, 515–527.

Dayton, C. (2011) Enzymatic degumming for physical refining. Paper presented at the SCI workshop on enzymatic processing and modifications, Ghent, Belgium.

Dayton, C. & Galhardo, F. (2008) US patent application 2008/0182322.

De Greyt, W. (2010) Developments in edible oil refining for the production of high quality, contaminant-free food oils. *Proceedings of the POMREQ Conference*, Kota Kinabalu, pp. 147–156.

De Greyt, W. (2012) How to minimize 3-MCPD and glycidyl esters during oil processing. Paper presented at the 103rd AOCS Annual Meeting and Expo, Long Beach, CA, USA.

De Greyt, W. & Kellens, M. (2010) US patent 7 696 369.

De Kock, J. (2006) Technologies for cleaning oils and meals. Paper presented at the EWOS workshop, Byrkjedalstunet, Norway.

De Kock, J. & De Greyt, W. (2009) Recent developments in edible oil deodorization. Paper presented at the 7th EuroFedLipid conference, Graz, Austria.

Deffense, E. (2002) US patent 6 407 271.

Dunford, N. & King, J. (2001) Thermal gradient deacidification of crude rice bran oil utilizing supercritical carbon dioxide. *J. Am. Oil Chem. Soc.*, **78**, 121–125.

Echim C., Verhé, R., De Greyt, W. & Stevens, C. (2009) Production of biodiesel from side-stream refining products. *Energy Environ. Sci.*, **2**, 1131–1141.

Fabricius, R. (2009) Applications of short path evaporators in the oils and fats and oleochemical industry. Paper presented at the 28th ISF Congress, Sydney, Australia.

FAOSTAT (2012) Edible oil production volumes. Available from http://faostat.fao.org, last accessed 8 January 2013.

Galhardo, F. & Dayton, C. (2012) Enzymatic degumming. *AOCS Lipid Library*, available from http://lipidlibrary.aocs.org/processing/degum-enz/index.htm, last accessed 8 January 2013.

Gramatikova, S., Hazlewood, G., Lam, D. & Barton, N.R. (2011) US patent 7 977 080.

Hernandez, E. & Rathbone, S.J. (2002) US patent 6 448 423.

Hitchman, T. (2009) Purifine® PLC: industrial application in oil degumming and refining. *Oil Mill Gazetteer*, **115**, 2–4.

Jalalpoor, M. (2008) PCT patent application, 2008/02552 A2.

Kellens, M. (2009) Oil processing challenges in the 21st century: enzymes key to quality and profitability. Paper presented at the 28th ISF Congress, Sydney, Australia.

Kellens, M., De Greyt, W., Kemper, T. & Willits, J. (2010) Enzymatic oil processing: current status and future developments. Paper presented at the 101st AOCS Annual Meeting and Expo, Phoenix, AZ, USA.

Kemeny, Z., Hellner, G., Radnoti, A. & Ergomoa, T. (2011) Polycyclic aromatic hydrocarbon removal from coconut oil. Paper presented at the 9th EuroFedLipid conference, Rotterdam, Netherlands.

Lin, L. & Koseoglu, S.K. (2004) Membrane processing of fats and oils. In: F. Shahidi (ed.) *Bailey's Industrial Oil and Fat Poducts*, Volume V. New York, NY, USA: Wiley Interscience, pp. 437–457.

List, G., King, J., Johnson, J., Warner, K. & Mounts, T. (1993) Supercritical CO_2 degumming and physical refining of soybean oil. *J. Am. Oil Chem. Soc.*, **70**, 473ff.

Maes, J., De Meulenaer, B., Van Heerswynghels, P., De Greyt, W., Eppe, G., De Pauw, G. & Huyghebaert, A. (2005) Removal of dioxins and PCB from fish oils by activated carbon treatment. *J. Am. Oil Chem. Soc.*, **82**, 593–597.

Maes, J., Vila Ayala, J., De Meulenaer, B., Verhé, R., Gibon, V. & De Greyt, W. (2010) Process optimization for the removal of environmental contaminants from fish oils. *Oléagineux Corps gras Lipides*, **17**, 81–85.

Meyers, H. (2000) US patent 6 111 120.

Mielke, T. (2011) Rapidly increasing global dependence on rapeseed/canola and oil. Paper presented at the 15th International Rapeseed Congress, Prague, Czech Republic.

Mikkelsen, R. (2011) Biotechnological approaches to remove chlorophyll components in plant oils. Paper presented at the 102nd AOCS Annual Meeting and Expo, Cincinnati, OH, USA.

Moulton, K.J. & Mounts, T.L. (1990) Continuous ultrasonic degumming of crude soybean oil. *J. Am. Oil Chem. Soc.*, **69**, 443–446.

Ooi, C. (1996) Continuous supercritical carbon dioxide processing of palm oil. *J. Am. Oil Chem. Soc.*, **73**, 233–237.

Oterhals, A., Kvamme, B. & Berntssen, M. (2010) Modelling of a short-path distillation process to remove persistent organic pollutants in fish oil based on process parameters and quantitative structure properties relationships. *Chemosphere*, **80**, 83–92.

Ramli, M.R., Siew, W.L., Ibrahim, N.A., Hussein, R., Kuntom, A., Razak, R. & Nesaretnam, K. (2011) Effects of degumming and bleaching on 3-MCPD esters formation during physical refining. *J. Am. Oil Chem. Soc.*, **88**, 1839–1844.

Schurz, K. (2010) Development of the bleaching process under the perspective of classic and new applications. Paper presented at the 3rd DGF Symposium, Leipzig, Germany.

Svenson, E. & Willits, J. (2012) Nano-neutralization™. In: W.E. Farr & A. Proctor (eds) *Green Vegetable Oil Processing*. Champaign, IL, USA: AOCS Press, pp. 157–168.

Taylor, D.R. (2004) Bleaching. In: F. Shahidi (ed.) *Bailey's Industrial Oil and Fat Poducts*, Volume 5. New York, NY, USA: Wiley Interscience, pp. 285–339.

Transfeld, P. (1998) Electrofiltration cuts bleaching earth use. *INFORM*, **9**, 502–506.

Transfeld, P. & Schneider, M. (1996) Countercurrent system cuts bleaching costs. *INFORM*, **7**, 756–767.

Veldkamp, F. (2012) From fundamentals to new filter technologies. Paper presented at the Edible Oil Refining Update short course at the 103rd AOCS Annual Meeting and Expo, Long Beach, CA, USA.

Verleyen, T., Verhé, R., Garcia, L., Huyghebaert, A., Dewettinck, K. & De Greyt, W. (2001) Gas chromatographic characterization of vegetable oil deodorization distillate. *Journal Chromatography A*, **921**, 277–285.

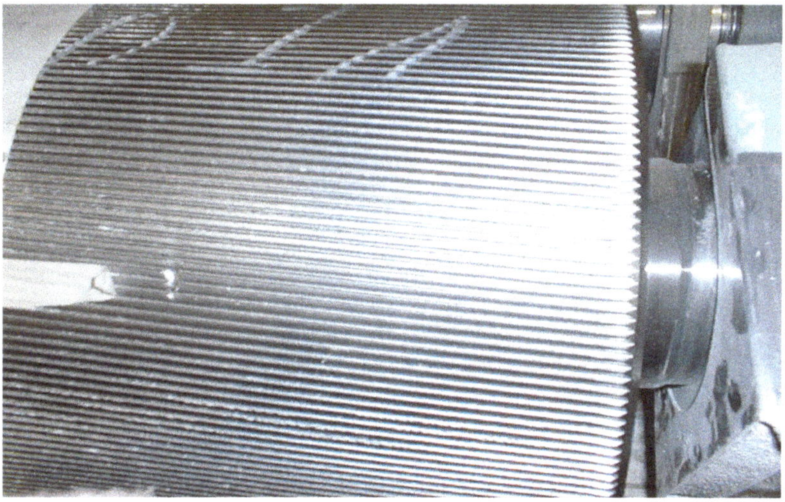

Plate 3.1 Detail of a corrugated roll in a cracking mill. Courtesy of Allocco.

Plate 3.2 Industrial flaking mill. Courtesy of Allocco.

Edible Oil Processing, Second Edition. Edited by Wolf Hamm, Richard J. Hamilton and Gijs Calliauw.

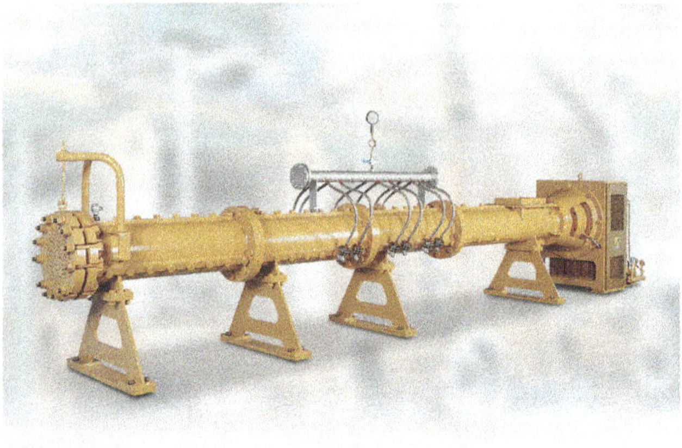

Plate 3.3 Expander with feeder and steam injection nozzles. Courtesy of Allocco.

Plate 3.4 Rotary cooker. Courtesy of Allocco.

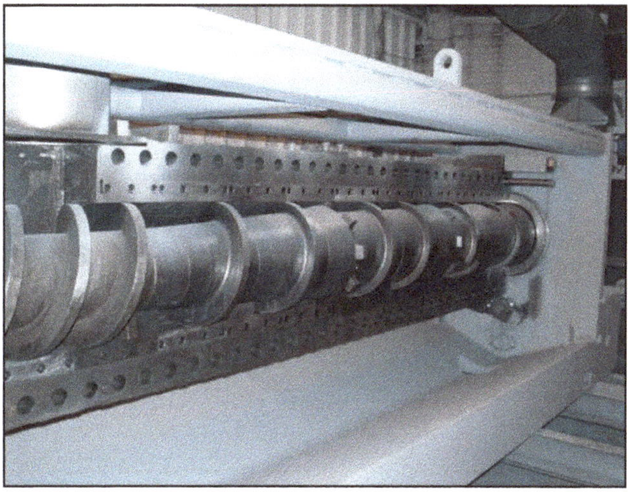

Plate 3.7 Typical worm assembly, showing worms, distance pieces, knife bars and a mixing device. Courtesy of Desmet Rosedowns.

Plate 3.8 Oil flowing between cage bars. Courtesy of Desmet Rosedowns.

Plate 3.9 Pressed cake at outlet. Courtesy of Desmet Rosedowns.

Plate 9.3 Crude oil risk matrix. This shows the risk classification for contaminant presence in a crude oil. It also shows the recommended frequency of analysis if an oil is of unknown origin.

	Pesticides	PAH	Mineral oil in edible oil	Dioxins and PCBs	Aflatoxins	Zearalenone
Limit	MRL or LOD	BaP < 2ppb				
Soybean oil	[yellow]	[yellow]	[blue]	[blue]		
Sunflower oil	[red]	[red]	[yellow]	[blue]		
Rapeseed oil	[yellow]	[yellow]	[blue]	[blue]		
Corn oil	[yellow]	[yellow]	[blue]	[blue]		[red]
Palm oil	[blue]	[blue]	[yellow]	[blue]		
Palm kernel oil	[blue]	[yellow]	[yellow]	[blue]		
Coconut oil	[blue]	[red]	[yellow]	[blue]	[yellow]	
Groundnut oil	[blue]	[blue]	[blue]	[blue]	[red]	
Fish oil	[blue]	[yellow]	[blue]	[red]		
Linseed oil	[yellow]	[yellow]	[blue]	[blue]		
Cottonseed	[yellow]	[yellow]	[blue]	[blue]		
Grape seed	[blue]	[red]	[blue]	[blue]		
Olive	[yellow]	[yellow]	[blue]	[blue]		

PAH, polycyclic aromatic hydrocarbon; PCB, polychlorinated biphenyl; MRL, maximum residue limit; LOD, level of determination (see Figure 9.1); BaP, benzo pyrene. [red] high risk (regular occurrence (> once a year), monitor every batch); [yellow] medium risk (occasional occurrence (every 1–5 years), monitor at least once a quarter); [blue] low risk (infrequent occurrence (< once every 5 years), monitor once a quarter at most).

Plate 6.10 Detail of MoBulizer™ cooling tubes in vegetable oil.

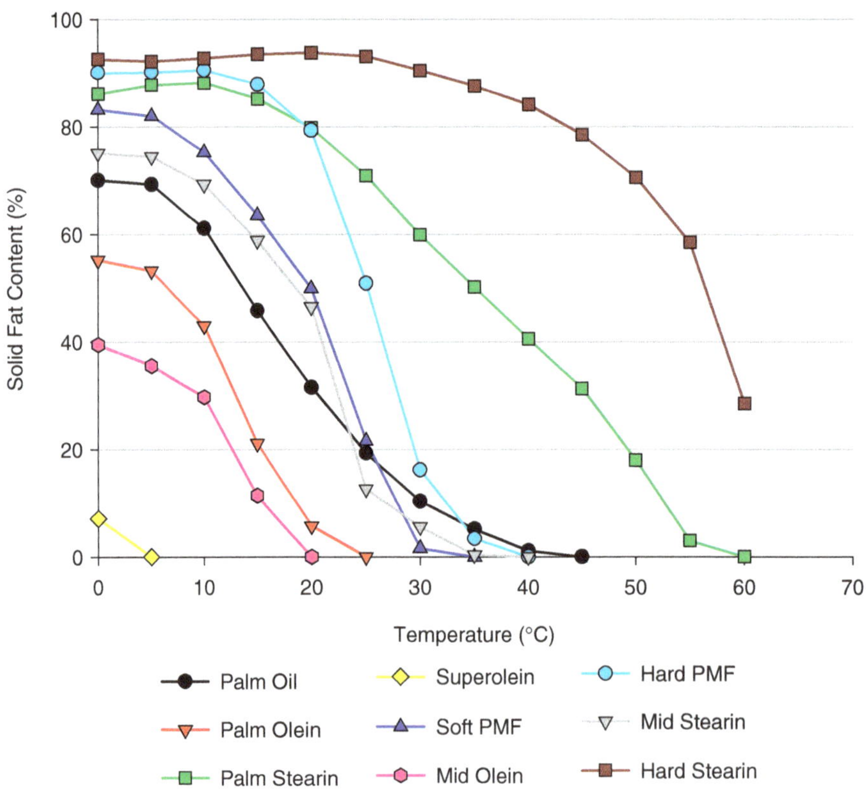

Plate 6.11 Overview of palm oil fraction melting curves in multistage fractionation.

6
Oil Modification Processes

Marc Kellens[1] and Gijs Calliauw[2]
[1]*Group Technical Director, Desmet Ballestra Oils and Fats, Zaventem, Belgium*
[2]*Development Manager Modification, Desmet Ballestra Oils and Fats, Zaventem, Belgium*

6.1 Introduction

All oil modification processes involve a substantial change of the physical behaviours and structural properties of an oil. This differentiates the modification processes from the refinery processes discussed in Chapter 5, in which the processing effect is more orientated towards improved organoleptic properties and nutritional value. There are three main modification technologies available in the edible oils industry at present: hydrogenation (reducing the degree of unsaturation on the acyl chains), interesterification (intermolecular redistribution of the acyl groups on the glycerol backbone) and fractionation (a fractional crystallisation of the oil, followed by a phase separation).

The drive for healthier edible oil products has substantially changed the face of the modification technologies as they were originally conceived in the 20th century. The use of catalysts, solvents and chemicals, which is in any case preferably kept to a minimum for simple cost reasons, is increasingly being discouraged due to these compounds' possible harmful and polluting effects. Moreover, the formation of undesirable side products – undesirable in terms of their health effects – such as the *trans* fatty acids, or more recently the monochloropropanediol (MCPD)-esters during deodorisation, has led to a complete rethinking or even total abandonment of well-established, proven industrial practices, as they are being traded for 'greener', more sustainable and healthier technologies.

Edible Oil Processing, Second Edition. Edited by Wolf Hamm, Richard J. Hamilton and Gijs Calliauw.

6.2 Hydrogenation

6.2.1 Historical perspective

The Nobel Prize winner Paul Sabatier was the first to work out the hydrogenation chemistry of volatile triglycerides, in the 1890s. Shortly after that, Wilhelm Normann demonstrated that the same principle could be applied to liquid oils. He filed several patents on the matter in 1902 and 1903 (Normann, 1903), and promptly built a hardening factory with his acquired knowledge. So, although technically the patenting of catalytic hydrogenation on oil was preceded by a fractionation patent a good 30 years earlier, hydrogenation was the first process to be applied globally on an industrial scale. It can therefore be considered the oldest of the three main oil modification technologies, for in both Europe and the USA, hydrogenation was already commercialised before the First World War.

Originally, hydrogenation of edible oils was mainly used to improve the oxidative stability of oils that contained multiple polyunsaturated fatty acids, such as fish oil. The high susceptibility of these oils to air oxidation, especially when heated, made them unusable in a lot of food applications (particularly those involving baking and frying). Furthermore, oxidation reduces shelf life as it triggers the formation of off-flavours upon storage. The practice of hydrogenation therefore allowed liquid oils to be stored for much longer than nature would normally permit.

The reduction of unsaturation of the oil also leads to a considerable increase in melting point and thus an increasing transformation of the oil into fat. This technique therefore facilitates the processing of natural oils into tailormade products with specifically desired properties: a selective and partial hydrogenation of edible oil will result in a stable but still pourable liquid oil suitable for cooking, whereas a more complete hydrogenation enables the production of brittle, high-melting fats for use as coatings, spreads and so on. The first successful hydrogenated products on the market were shortenings made from cottonseed oil. This also shows that, certainly in the early days, the introduction of hydrogenation in oil processing made the sourcing of oil with specific physical behaviour a less crucial issue, since this process was able to turn even the most polyunsaturated oil into a fat that could even be as hard as candle wax in a matter of hours. Hydrogenation has steadily developed since, due to increasing understanding of hydrogenation effects, reaction control and, not least, catalyst optimisation. But as the nutritional and health aspects of edible oil consumption began to receive more attention in the 1970s and 1980s, hydrogenation became identified as a major source of not just saturated but also *trans* fatty acids in the human diet. The latter are commonly (but not unanimously) believed to raise low-density lipoprotein (LDL) blood levels and increase the risk of coronary heart disease (Mensink *et al.*, 1992). Despite the controversy around this subject and the sometimes contradictory results from various medical studies, it has led to a general push for reduction, explicit labelling and even elimination of *trans* fatty acids in food. It is not within the scope of this chapter to elaborate upon this multifaceted topic, but the review of the position of *trans* fatty acids and their impact on our food formulation (and hydrogenation) by List *et al.* (2007) is highly recommended. The 'unhealthy' reputation of hydrogenation technology has had an undeniable impact on the global hydrogenation capacity. Where it was estimated in the mid 1990s that about 90% of all edible oils passed through hydrogenation (which for soft oils alone would amount to about 35 million tonnes), 1 decade later only 6 million tonnes of oils were hydrogenated annually (Beers *et al.*, 2008).

6.2.2 Principle

The general hydrogenation reaction consists in a sequence of reactions that allow the net adding of protons (hydrogen atoms) to a fatty acid chain.

Although the basic reaction of adding hydrogen to a double bond is quite straightforward, it is not a pure reaction, and different types of reaction can occur when a fatty acid chain (with double bond) adsorbs to the catalyst surface and dissociated hydrogen is present. The first hydrogen can add to the double bond, and if another hydrogen atom can be added then the result is an irreversible, stable, saturated C-C bond, which desorbs from the catalyst surface (saturation). This 'other' hydrogen atom, however, can also be an H-atom on the neighbouring C-atom, which results in a shifting of the double bond down the fatty acid chain (positional isomerism). This isomerisation has little impact on the physical behaviour of the oil. At the moment of addition of the first hydrogen atom and the existence of the 'half-hydrogenated' intermediate (Beers *et al.*, 2008), the sp3-orbital momentarily allows free rotation over the C-C axis. It is exactly this sort of brief 'randomness' that permits formation of *trans* fatty acids (geometrical isomerisation). These different outcomes of adding hydrogen to a double bond were presented in the 1950s.

It is crucial to understand that there does exist a competition between these reactions: saturation versus isomerisation. Dijkstra (2010) has published various reviews of the reaction mechanism in edible oil hydrogenation, in which the different steps of the complex reaction are discussed, especially in terms of the order and rate of each subreaction. Indeed, in ideal conditions, the hydrogenation of a polyunsaturated fatty acid actually occurs stepwise. Essentially, for each of these subreactions, the competition between the two types of reaction can be reduced to a balance between hydrogen demand

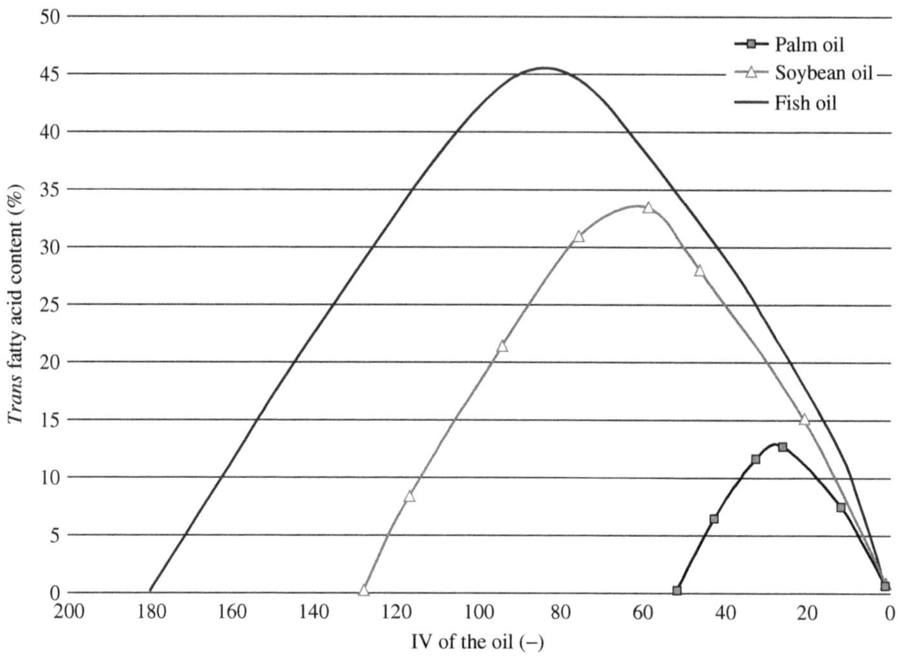

Figure 6.1 *Trans* fatty acid content of different oils as a function of decreasing IV upon hydrogenation.

(depending on the intrinsic reaction kinetics) and hydrogen supply (how fast the hydrogen can be supplied, which is limited by mass transfer). In industry, this balance is commonly expressed through a 'selectivity ratio' for the particular fatty acid and reaction conditions, as a measure of the likelihood of the one reaction taking place over the other (e.g. how much more prone an oleic acid is to form elaidic rather than stearic acid). So, the more isomerisation occurs, the more selective the conditions are considered to be.

Most research on the hydrogenation reaction mainly aims to explore how a variation of process parameters such as temperature and pressure might shift the ratio between isomerisation and saturation.

Generally, the *trans* fatty acid level in the oil reaches a maximum when 50% of the original unsaturation (expressed as IV) remains (Figure 6.1). More selective conditions (temperature, pressure, catalyst) will not so much shift this maximum as flatten these curves. Overall, the highest selectivity can be obtained at high processing temperatures, low hydrogen pressures, low mixing and a high dose of catalyst. These effects will be further discussed in Section 6.2.3.

6.2.3 Process parameters

6.2.3.1 Hydrogen pressure

Overall, a higher pressure of hydrogen gas will shift the reaction towards saturation, since in this case the hydrogen reagent is available to a virtually unlimited extent for the partially hydrogenated intermediate and so a greater chance exists that a second hydrogen atom will be added before the free rotation over the C-C axis makes it possible for geometrical isomerism to take place. However, not only is the *trans* isomerisation reduced by the higher hydrogen concentrations but the selectivity of the reaction also drops (due to the difference in reaction order for different degrees of unsaturation). It should be noted that for a reduction of 50% of the 'normal' *trans* content in a partially hydrogenated rapeseed oil (e.g. from 40 to 20%), hydrogen pressures over 40 barg should be applied in high-pressure reactors. Increased hydrogen pressures are therefore helpful in reducing *trans* fatty acid formation. However, this practice does not result in a full elimination of *trans* fatty acids, and can generate undesired high amounts of fully saturated triglycerides, which limit the possible applications for such fats (List *et al.*, 2007).

6.2.3.2 Temperature

As for many chemical reactions, the rate of the reaction increases substantially with increasing temperature, so hydrogen atoms near the absorbed double bond will react faster with the double bond. Under these conditions, the reaction interface is virtually instantly depleted of hydrogen. Therefore, with increasing temperature, the supply of hydrogen towards the catalyst reaction surface will eventually be the limiting factor of the reaction. As the isomerisation reaction has a lower order dependency on hydrogen than the hydrogenation reaction, increasing temperature will boost the selectivity but also the *trans* formation. In recent years, the trend has been mainly to lower the operation temperatures, resulting in a significant *trans* reduction but also a lower capacity (Figure 6.2).

6.2.3.3 Catalyst

The archetypal catalyst in hydrogenation is reduced nickel (Ni) supported on natural earth, such as kieselguhr. These supports have a high surface–volume ratio, which is an evident asset for adsorption processes. The high porosity additionally permits swift mass transfer from and to the bulk phase. The powder, about 2–12 μm mean particle size, is commonly formulated in solid droplets of hard fat such as palm stearin or fully hydrogenated oil, in order to prevent the Ni from oxidising and to facilitate handling and dosing.

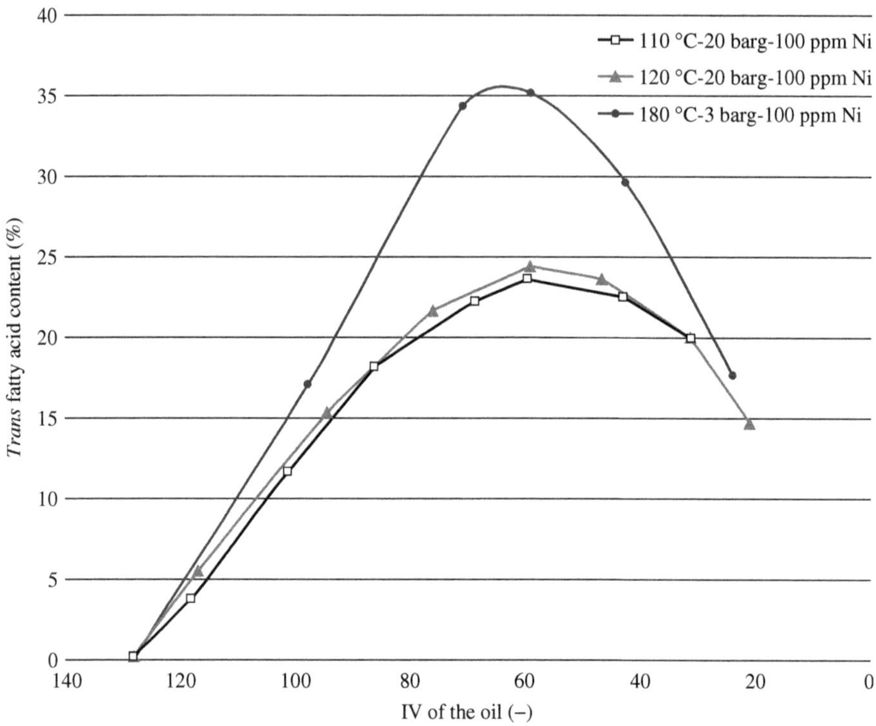

Figure 6.2 *Trans* fatty acid content at different temperatures and hydrogen pressures as a function of decreasing IV upon hydrogenation.

The type and concentration of the catalyst chosen is quite dependent on the purpose of hydrogenation: if an increased oxidative stability of the liquid oil is intended, a high selectivity towards polyunsaturated fatty acids is preferred. A good catalyst selection can help steer a reaction towards more or less *trans* formation, although this is mainly true for nonselective conditions such as high hydrogen pressures. As soon as hydrogen becomes the limiting reagent (through increased temperature, lower pressure or increased catalyst concentration), the catalyst type itself has very little impact on the (large) quantity of *trans* isomers formed.

Using spent catalyst or poisoning it with sulfur will favour *trans*-isomerisation. This is because upon reuse (or with insufficient refining in the preceding bleaching and deodorisation section) the catalyst will lose activity and/or selectivity due to absorbed impurities such as sulfur and phosphorus. Nickel sulfides will affect the dissociation of the hydrogen molecules in such a way that less hydrogen is available for saturation, and therefore the rate of (*trans*) isomerisation will considerably increase. The inverse is also true: some manufacturers aiming for the lowest possible *trans* content prefer to dose the catalyst in multiple stages, in an attempt to use only the freshest and most active of all catalysts and minimise side reactions due to impurities. Alternatively, sometimes spent catalyst is added prior to hydrogenation in order to absorb the impurities, and then filtered from the oil. This is a way of cleaning the oil of sulfur or phosphatides. Such phosphorus compounds, even at concentrations as low as 4 ppm, can physically occupy the inner channels of the catalyst, restricting the adsorption surface.

Many other catalysts have been investigated, among which the noble metals probably represent the most relevant alternative to standard nickel catalyst, due to their superior catalytic activity at – interestingly – lower temperatures. Although the use of alternative catalysts for edible oil hydrogenation is by no means novel, this topic will be discussed in the following sections on recent developments.

6.2.4 Process design

In essence, a hydrogenation plant requires a heating system, a properly designed reactor and a post-process filter system to recover the spent catalyst (Figure 6.3).

Hydrogenation is exothermic, and each drop in iodine value can cause an oil temperature increase of about 1.7 °C. Most industrial designs are therefore provided with cooling/heat recuperation.

The plant usually operates batchwise, although the integration of various buffer tanks can lead to a more continuous form of operation. Continuous reactors have also been designed.

Before entering the reactor and mixing with hydrogen, the fresh oil is deaerated and dried in a buffer tank kept under reduced pressure. The incoming oil is sprayed in a thin film on a coil in which the previously hydrogenated batch of oil circulates, for heat-recovery reasons. The temperature of the incoming oil then rises from 50 to about 150 °C for hydrogenation pressures up to 6 barg and to about 90 °C for hydrogenation pressures >10 barg.

As soon as the reactor is filled with the oil, the catalyst is added, and the mixture is then brought to vacuum conditions. Only then is a controlled flow of hydrogen introduced, according to the hydrogenation pressure requirement. A typical batch reactor is usually equipped with an agitator, not only to keep the catalyst in suspension but also to create a good dispersion of the introduced hydrogen bubbles (which is also a function of the position and type of spray nozzles used) and thus to provide intense mixing of all reagents.

A well designed agitator serves multiples purposes: it should disperse the bubbles properly through the bulk, recirculate the overstoichometric gas excess and homogenise the (more viscous) oil phase to ensure the required mass transfer. The straightforward parameter used to assess the performance

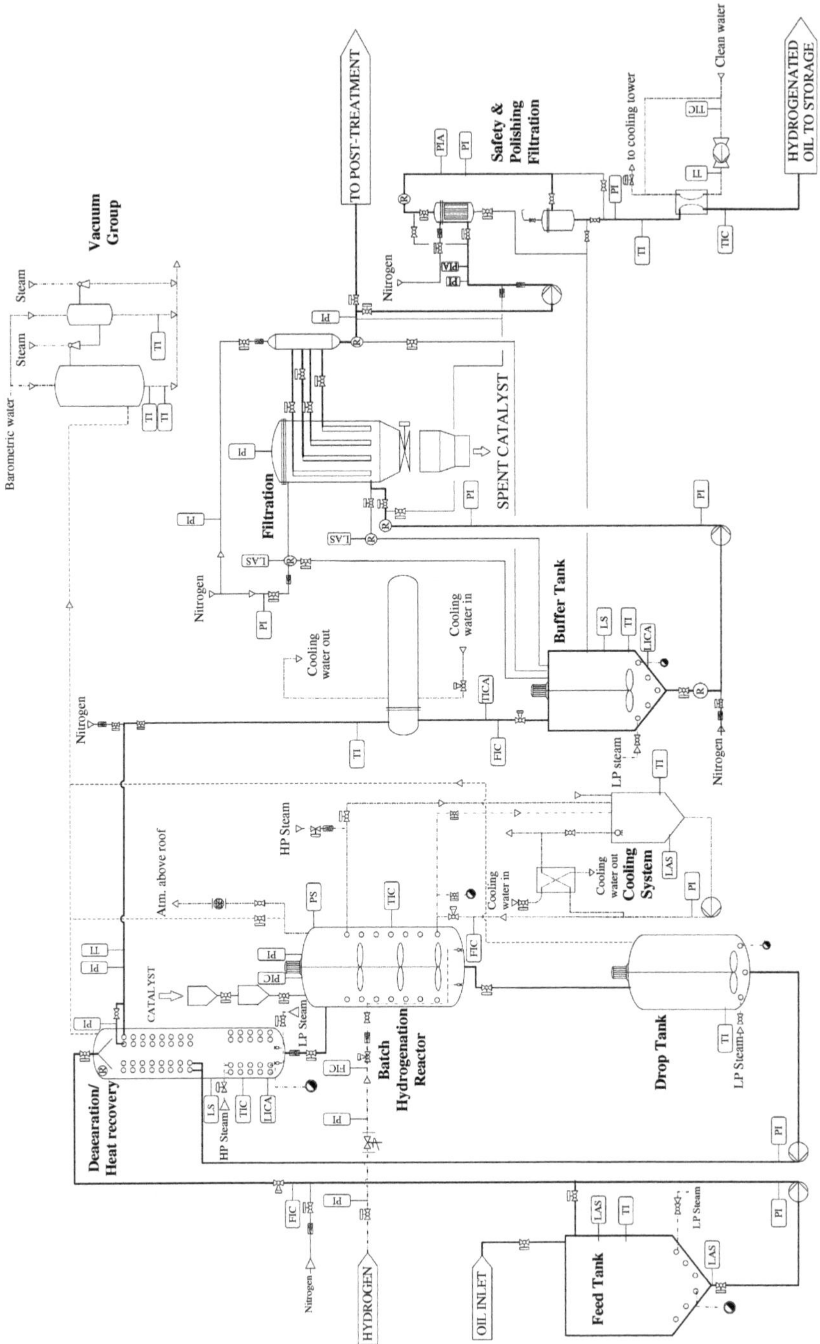

Figure 6.3 Schematic process diagram of hydrogenation.

of a reactor design is the rate of hydrogen uptake, r_H, expressed in mol/m^3.s, which can be calculated thus (Koetsier, 1997):

$$r_H = k_L \cdot a \cdot \Delta C_H \quad (6.1)$$

where k_L is the mass transfer coefficient in quiescent liquid (m/s), a is the specific air–oil interface area (m^2/m^3) and ΔC_H (mol/m^3) is the concentration difference of hydrogen at the interface and the bulk oil, which is the actual driving force of uptake into the oil. It is clear that the specific area parameter, and thus the overall hydrogen uptake in the reactor, will be greatly affected by how well the agitator manages to break up the introduced hydrogen bubbles into microscopic bubbles.

Another reactor type used for industrial hydrogenation is a loop reactor, in which the suspension is recirculated at high velocity through a mixing nozzle while hydrogen is introduced. The two main benefits of this type of reactor are the absence of mechanical agitation and that the hydrogen bubbles act as a stripping agent for moisture and oxygen left in the oil, making a preceding drying step and vacuum conditions redundant. In practice, however, it appears to be difficult to obtain r_H rates comparable to those of a dead-end batch reactor, because of the lower specific interfacial area created. In certain reactions, the higher degree of side reactions such as oxidation and hydrolysis due to the presence of oxygen and moisture can present an additional quality issue, nullifying the original advantage of a loop reactor.

Since the hydrogenation reaction is highly exothermic, the heat generated by the reaction itself can be used to reach the optimal reaction temperature at first (~180 °C for hydrogenation pressures up to 6 barg and ~120 °C for pressures above 10 barg). There are several systems available to subsequently recover the excess heat, such as steam/hot water production through the pumping of soft water through a coil or internal heat exchanger inside the reactor. Once the optimal reaction temperature is reached, the IV should drop by 0.5 to 3 units per minute. This means that the reactor should be able to reject the still-developing heat, as without cooling, the oil temperature would keep on rising at about 5 °C every minute. At elevated hydrogen pressures (20–25 barg), the hydrogenation reaction will proceed considerably faster; up to twice as many batches can be processed on a daily basis using elevated as compared to normal hydrogen pressures (3–5 barg).

In the case of partial hydrogenation, the reaction simply ends when the supply of gas is discontinued. The hydrogen should be dosed with great accuracy in order to obtain consistent production. For full hydrogenation, it is obviously the scarcity of double bonds left in the oil that will finally slow the reaction. The process can be steered sufficiently in terms of time, temperature and so on in the light of the desired end product quality and degree of hardening (Tables 6.1 and 6.2).

Table 6.1 Typical reaction times for hydrogenation pressures up to 6 barg (dead-end reactor, INTERMIG agitator).

Oil type	ΔIV	Final IV	Reaction time (h)	Number of cycles per day
Soybean, sunflower	10–25	>100	<1	~15
Soybean, sunflower	max. 50	>70	1.5	10
Soybean, sunflower	max. 60	70	2.5	8
Palm oil	10–20	>30	2	10
Palm oil	max. 50	<2	4	5

Table 6.2 Typical process parameter variation for different hydrogenated products (from soybean oil).

Characteristic	Brush	Shortening	Margarine	Coating	Stearin
IV	115	75	70	75	<5
Trans isomers (%)	15	30	50	65	-
C18 : 0					
absolute (%)	5	9	7	7	85
increase (%)	1	5	3	3	80
Temperature (°C)	170	150	210	210	220
Time (min)	30	60	60	60	300–150
H_2 pressure (bar)	1	2	1	2	max.
Catalyst type	selective	selective/nonselect	selective	Ni-S	high active
Ni (ppm)	50	100	100	500–1000	500–1500

The hydrogenated oil batch is then dropped into a tank under reduced pressure. This vessel is also equipped with an agitator to prevent settling of the catalyst. Cooling to the required filtration temperature is carried out by means of heat exchange with fresh cold oil circulated through the coils of the reactor. If needed, an additional cooling is performed in the safety oil cooler. The cooled oil is collected in an oil buffer tank, from where it is sent to filtration over – typically – a back-pulse filter (such as leaf or candle filters). For plants that operate many stock changes, a simple plate-and-frame filter might still be preferred, in order to reduce possible contamination in the heel volume. After the filtration, the nickel concentration in the hydrogenated oil should be below 5 ppm and preferably not greater than 1 ppm

The drying of the spent catalyst cake is carried out by nitrogen blowing, and the thus recovered oil and heel volume of the filter can be recirculated, while the dried cake is discharged by nitrogen pulses through a filter cake hopper to the disposal container. The performance of the filter then determines to what extent the catalyst can be reused. Some processors prefer to install an additional safety filter and oil cooler prior to sending the hydrogenated fat to the tank farm.

6.2.5 Future for hydrogenation technology

Most advances and research in recent years have been dedicated to reducing *trans* fatty acids without sacrificing the high selectivity of the reaction.

6.2.5.1 Smarter combinations of the conventional technology

In recent years, the boundaries of *trans* reduction in classic technology have been explored by combining higher pressures (5–15 barg) and extremely low temperatures (such as in the Losatra® process, with operating temperatures between 40 and 50 °C). Effectively, the setup could lower the *trans* content in soybean oil IV 100 to a mere 5%. Typically, such processes require high catalyst dosage (>1500 ppm Ni), however, and are noted for their very slow conversion rates, in the order of 0.2–0.4 IV drop/minute.

In a different set of experiments, similar low temperatures were combined with adapted Ni catalyst. These too have shown a reduced *trans* level: as low as 6%. There are considerable downsides to this practice, mainly due to the reduced catalyst activity and consequently long reaction times (Beers *et al.*, 2008).

6.2.5.2 Alternative catalysts

Much research effort has been invested in catalyst optimisation. Nickel is nevertheless still the archetypal catalyst. In terms of selectivity, the precious metal palladium (Pd) probably has the performance closest to typical nickel catalyst results for partial hydrogenations. Rhodium (Rh)-based catalysts show results shifting to higher saturation over *trans* formation ratios, but the best values in this respect are obtained by platinum (Pt) catalysts (Table 6.3).

This comes at the expense of fairly increased saturate levels, and consequently flat (as in 'unsteep') melting behaviour and excessive waxiness at elevated temperatures are observed in the resulting fat products. Effectively, the presence of *trans* fatty acids is often a means of having steeper melting curves, and these products are, from a functional point of view, generally preferred over the softer and/or flatter low-*trans* alternatives.

Copper catalysts and homogenous metal complexes have also been studied over the years. In some cases, these studies have demonstrated results at the other end of the spectrum: fats with over 60% *trans* fats are no exception.

In terms of the applicability of these alternative catalysts, it is good to realise that many require deviating (often more demanding, and thus more expensive) reaction conditions compared to those usually applied when using nickel catalysts. This severely limits the practicality of these catalysts in the common hydrogenation technology for edible oils. A second and equally important disadvantage is their cost: they require a very large emphasis on catalyst recovery and reuse, and often the value added to the fat products (which are often considered commodity fats with low added value) does not

Table 6.3 Effect of nickel versus platinum catalyst on fatty acid composition and solid fat content (SFC, %) of soybean oil.

Parameter	IV = 105		IV = 70	
Catalyst	Ni	Pt	Ni	Pt
FAC (% w/w)				
C18 : 0	4.7	15.2	10.3	30.1
C18 : 1t	*12.6*	*1.5*	*31.3*	*3.0*
C18 : 1c	36.7	28.8	43.5	35.9
C18 : 2t	*5.3*	*0.8*	*3.0*	*1.0*
C18 : 2c	28.1	38.5	0.5	17.9
C18 : 3t	*0.1*	*0.4*	*0.0*	*0.1*
C18 : 3c	1.8	3.8	0.0	1.0
TFA	***18.0***	***2.6***	***34.3***	***4.2***
SFC (% @ °C)				
10	7.6	19.2	63.6	50.2
20	1.5	14.2	38.0	38.7
30	0.0	9.6	11.8	26.7
35	0.0	7.5	3.0	20.8

offset the extra costs involved, limiting these catalysts' broader industrial implementation.

The catalyst type has not been the only subject of study: so too has the type of formulation. Noteworthy are the use of zeolite as a carrier material and the dosing of nanoparticles of platinum. This formulation comes down to a dispersion of (clustered) tiny particles (nanoscale), about 1000 times smaller than conventional particle sizes (microscale). One aspect of these improvements should be the reduction of mass transfer limitations, but at the other end of the scale the limits of improvement seem to be set by the far more difficult filterability of these superfine dispersions.

6.2.5.3 Advanced process technology

Currently, the standard approach for (commodity) oil hydrogenation is a batch reactor as presented in the previous sections. Over the years, several novel developments in the field of reactor design have been tested for edible oil hydrogenation.

Membrane reactors are believed to offer serious advantages with respect to mass transfer, in this case through the pores of the membrane. The key design is in impregnating a porous polymer with a noble metal catalyst such as Pd or Pt and controlling the catalyst availability. Several pilot test data point towards increased *trans* formation, rather than reduction (Table 6.4).

The feasibility of supercritical hydrogenation has also been evaluated. In this process, the oil is mixed with hydrogen and organic solvent (typically

Table 6.4 Membrane hydrogenation of sunflower oil.

Parameter	Feedstock	Membrane-hydrogenated sunflower oil		
Temperature (°C)	–	**100**	**120**	
Pressure (bar)	–	4	10	
Membrane catalyst	–	Pt	Pd	Pt
Time (min)	–	480	240	240
IV	126	89	53	60
Reaction rate (ΔIV/min)	–	0.08	0.30	0.28
FAC (% w/w)				
C18 : 0	3.0	–	32.2	25.6
C18 : 1	29.8	–	55.7	62.1
C18 : 2	59.0	–	2.5	0.9
TFA	–	***20.0***	***26.0***	***25.0***
SFC (% @ °C)				
20	–	–	86.0	76.4
25	–	–	77.4	65.5
30	–	–	66.5	52.8
35	–	–	53.1	38.4
40	–	–	36.7	23.9

butane or propane) and sent to a multiphase mixing reactor. The normal operation temperatures are rather low, but in order to induce the supercritical state, pressures range up to 200 bar. The main asset is the very fast reaction rate under these conditions: the IV can drop about 40 units in just 1 minute. Thus, reactors could be built in smaller sizes, and catalyst consumption would decrease. However, the industrial scale-up of such a configuration is uncertain due to the expensive high-pressure reactor design required, the huge volumes of solvent needed (about 2 to even 20 times the amount of oil), the safety measures involved and the solvent recovery costs. It is clear that only products with a lot of added value could pay back such vast investment/operational costs.

Packed- or fixed-bed reactors, which have already been applied in many other processing technologies, have been seriously considered for edible oil hydrogenation. Such beds can work fully continuously, and as the catalyst is immobilised in the bed (in fact, the catalyst is the bed), a post-hydrogenation filtration step is redundant. The selectivity of the reaction in packed bed seems to decrease, however (Boger *et al.*, 2004), and other limitations, as with most other packed-bed applications, have to be taken into account: packed beds are less flexible in terms of stock change and product quality and require a very smart design in order to master all temperature, pressure and concentration gradients that occur during hydrogenation.

Applying a concept similar to packed-bed reactors, monolithic reactors are in many cases an attractive alternative to conventional multiphase (slurry) reactors. Such monoliths consist of a single block of solid material (typically ceramic or metallic) but with a very porous structure. The material can additionally be coated with a catalyst. Here the advantages are the low-pressure drop, the absence of a need for a catalyst separation and the large geometrical interface area (Boger *et al.*, 2004). In general, *trans* formation in soybean oil seems to be reduced to some extent, but not yet to a sufficient degree to consider implementation industrially (Zieverink, 2007).

6.2.5.4 Summary

Hydrogenation technology has seen some sensible and important changes in the 21st century. Considerable research and development has been invested in the optimisation of reaction conditions, catalyst types and reactor design. Notwithstanding these efforts, industrial hydrogenation technology does not seem able to recover its position as the main modification technology. There are several reasons for this problem:

(1) The production of *trans* fats during partial hydrogenation has not been fully eliminated, as mentioned previously; nor is there any solution in sight.
(2) Meanwhile, palm oil has grown to become the most widely used oil in the world, and is established as a readily available *trans*-free source of saturated fatty acids, giving structure to the matrix.
(3) Today, hardened fats in general, even if they are 100% saturated, are not favoured by consumers, and therefore by manufacturers. This means a 180° turn from the days when *trans* fatty acids were in fact deliberately introduced through hydrogenation, as the specific physical behaviour of these fats cannot be sufficiently matched by any of the non-*trans* alternatives.

Therefore, despite the technological advancements listed in this section, the impossibility of eliminating all *trans* from all fats, the availability of an economic alternative and increased consumer awareness have definitely put '20th century hydrogenation of edible oils' on the back foot. Or, as Zieverink (2007) stated as a gloomy conclusion to his doctorate on catalytic hydrogenation: 'any catalytic solution to the problem of *trans* formation in the hydrogenation of edible oils will most likely be too late in the offering'.

6.3 Interesterification

6.3.1 Historical perspective

The very first publication mentioning the synthesis of a triglyceride (by esterification of glycerol and butyric acid), by Pelouze, dates from 1844. Soon afterwards it was discovered that if a small proportion of glycerol is used, the migration and interchange of the fatty acid radicals lead to the formation of triglycerides of new composition. It was found that the use of glycerol was not necessary, and that the use of alkaline catalysts is sufficient. With the interesterification principle sketched out that early, it might be a bit surprising that it was another 80 years before the actual invention of interesterification in fats and oils. The development of interesterification is rooted in the search for 'cheaper butter', however, for which the demand rose quickly after the First World War. And so it happened that the Germans Grün and (again) Normann both filed patents on interesterification of fats and oils, involving reactions with typically short-chain fatty acids such as butyric acid, in the 1920s.

Later, real ester interchanges were established using alkali compounds. The focus of the research then shifted to enabling real ester interchange between triglycerides through the use of catalysts such as alkali compounds. By the Second World War, after testing a whole spectrum of catalysts at various temperatures, sodium methylate was found to be the best all-round catalyst for the interesterification of fats and oils. A very important contribution in the field was made by Eckey (1945), who introduced the use of sodium methoxide as catalyst, which permits much milder reaction conditions and which therefore established the classic chemical interesterification process as we know it today.

In the first decade of the 21st century, the green equivalent of chemical interesterification, enzymatic interesterification, found a wider entry into the fats and oils industry. Compared to chemical interesterification, enzymatic interesterification offers the advantages of stereospecific acyl exchange in terms of performance, and of milder process conditions in terms of operation cost. For large-scale randomisation processes, however, the edible oil industry still relies on the robust conventional (chemical) interesterification technology.

6.3.2 Principle

The term 'interesterification' refers to all those reactions that involve fatty acid esters reacting with other acids, alcohols or esters, but in most cases it is used to describe 'a rearrangement of the fatty acyl groups within and between different triglycerides'. This reaction normally requires very high temperatures, but, as has been ntoed, the use of catalysts allows for much milder conditions. Typical catalysts are alkali (m)ethylates, metals and sodium/potassium alloys.

The actual mechanism of the interesterification reaction in triglycerides has been the subject of several research projects, publications and discussions within the realm of fats and oils chemistry. Largely, the discussion in the

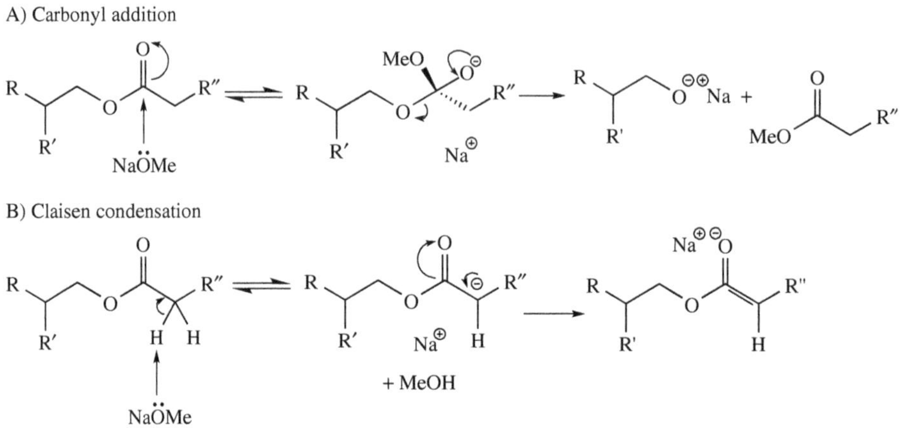

Figure 6.4 Proposed reaction mechanisms for chemical interesterification: carbonyl addition and Claisen condensation.

last few decades has pertained to how the reaction is propagated, or which molecular species really acts as the catalyst. There are various proposed catalytic reaction mechanisms for the interesterification reaction: the carbonyl addition suggests that the reaction begins when the catalyst attacks the α-hydrogen of an acyl group, forming an enolate ion, which then reacts with another ester to form a β-ketoester. Another suggested pathway is that the carbonyl group is directly attacked by the methoxide ion to form a diglyceride anion, which acts as the real catalyst and transfers acyl groups around the glyceride backbones (Figure 6.4).

For people less concerned with organic chemistry, it might seem that the question of whether the catalyst is a glycerolate or an enolate is more cerebral than technological. There seems to be unanimity about the fact that the reaction proceeds with an enolate as active component, since the glycerolate is increasingly formed during the reaction and therefore does not seem to be at the root of the reaction itself, but rather a product of it (Dijkstra, 2009). Also, the temperature sensitivity of the enolate species can explain more convincingly how the reaction is temperature-dependent.

Upon inactivation of the reaction, with citric acid for example, an equivalent amount of soaps (or free fatty acids, FFA) is formed. The typical browning during interesterification is believed to be due to a complex formation involving the active catalyst, possibly associated with oxidation products. The interesterified triglycerides, however, are not involved in the colour formation.

This reaction is not ideal, and the main effect on the oil quality is that the level of mono- and, more importantly, diglyceride content is found to be increased after the interesterification process; doubled concentrations after interesterification are no exception.

6.3.3 Process parameters

As the crux of the process is a catalysed, imperfect chemical reaction in a complex matrix, the main process parameters with which to influence the feasibility of the process are the quality of the oil, the concentration and type of catalyst and a reduction of the oil losses. The latter is achieved by keeping the former two under control.

6.3.3.1 Oil quality

The initial oil quality is important with respect to interesterification in the sense that minor components or contaminants will also consume the catalyst to a considerable degree. The reaction scheme outlined in Section 6.3.2 helps to explain why the presence of water can be a serious drawback for the 'pure reaction': moisture will consume the classic catalysts, such as sodium methylate, in the proportion of 1 : 20. FFA will also deactivate the catalyst, albeit less extremely, and often a bit of caustic soda is added prior to the reaction. For this reason, in practice, the oil to be interesterified is at least neutralised (as in the case of soybean oil) or physically refined (as in the case of palm oil) before it enters into the production line. Also, oxidation parameters such as the peroxide value will lower the catalyst efficiency and thereby increase the cost of achieving full randomisation (Table 6.5).

6.3.3.2 Catalyst

In the light of the interesterification reaction, the active catalyst species needs to be an electron donor. It can be understood that in the early days, pure alkali metals such as sodium or potassium were ideal for such reactions, and in small-scale operations these proved successful. However, for larger operations the most used catalysts for interesterification are the alkoxides, with sodium methoxide (or sodium methylate) as the principal catalyst, because of the high reaction rate at rather low temperatures. This catalyst allows for a relatively clean separation after the reaction. The applied concentrations of catalysts generally range between 0.05 and 1.5%. The relatively broad range

Table 6.5 Feedstock quality for chemical interesterification (Kellens, 2000).

	Required	Preferred
Oil/fat	Refined, bleached, deodorised	–
Water	<0.02%	<0.01%
% FFA	<0.1%	<0.05%
Peroxide value	<3 meq/kg	<1 meq/kg
Phospatides	<0.01%	max. 5 ppm

Table 6.6 Overview of catalysts used in chemical interesterification (Sonntag, 1983).

Type	Example	Required dosage (% oil weight)	Time
High temperature (120–160 °C)			
Metal salts	Acetates, carbonates, chloride, oxides of Zn, Fe	0.1–0.2% 0.2%	0.5–6.0 hours under vacuum
Alkali hydroxides	NaOH, KOH, LiOH or sodium hydroxide + glycerol	0.5–1.0%	45 minutes–1.5 hours under vacuum
Metal soaps	Sodium stearate + glycerol	0.5–1.0%	1 hour under vacuum
Low temperature (25–270 °C)			
Metal alkylates	Sodium methylate	0.1–1.0%	5–120 minutes
Alkali metals	Na, K Na/K alloy	0.2–0.5%	3–120 minutes
Alkali metal hydrides	Sodium hydride	0.2–2.0%	30–120 minutes
Alkali metal amides	Sodium amide	0.15–2.0%	10–60 minutes

of catalyst dosage is due to the large influence of the feed oil purity on the catalyst activity. The data reported by Sonntag in 1983 are still representative for the main catalysts used in the industry today (Table 6.6).

6.3.3.3 Oil losses

Depending on the pH of the water added to inactivate the reaction, the oily material lost will largely consist of fatty acid methyl ester (FAME) and FFA or soaps. Generally, between FFA and soaps, the former is the preferred end form, as in the common concentrations it is good practice to remove FFA in the post-process deodorisation step, whereas soaps generally require a supplementary adsorption or washing step for adequate removal. Apart from the direct loss of oleaginous material, such post-treatment implies an additional 'neutral oil loss' through carry-over in the deodoriser or incomplete separation during the filtration step, respectively. It should also be realised that due to the stoichiometric nature of the interesterification (and deactivation) reaction, the creation of FAME and FFA is proportional to the amount of catalyst used. So as more catalyst generates more side products (direct loss) and more side products will imply higher neutral oil loss (indirect loss) by a factor 1.0–1.5, it is crucial to optimise the catalyst dosing to reduce the losses and keep the process affordable.

Overall, the removal of the minimal added water for the catalyst inactivation itself does not present a lot of problems: it can be easily evaporated after the inactivation. For modest doses of water, this is a much quicker and more effective method than a gravitational separation through a decanter or centrifuge. It is true that when the catalyst can be deactivated without addition of water, the small water-associated oil loss can be largely avoided. Adsorbents such as Trysil® can be used (Kellens, 2000), but it should be noted that adding more solid material implies a higher oil loss through filtration. Taking into account the mandatory bleaching step downstream, it has been proposed that acid-activated bleaching earth be used for catalyst inactivation, but this does not seem to be as effective as water.

Apart from the inactivation-related oil loss, the mass balance of an interesterification process contains a second, rather large 'loss term' due to the mandatory bleaching post-treatment, resulting from the colour change upon catalyst inactivation. The loss here is largely constituted by the actual physical entrainment of oil in the bleaching filter cakes, and not really by an inevitable chemical side reaction as in the deactivation step. Generally, a typical bleaching cake after blowing will still contain about 20–30% residual oil, meaning that for every 1% bleaching earth used, about 0.3% loss of oil is brought about.

As a rule of thumb, a total oil loss of 20 times the concentration of catalyst can be assumed for a chemical interesterification process.

6.3.4 Process design

A customary chemical interesterification batch plant consists of an oil blending tank, a (static) oil–caustic soda mixer, an oil heat exchanger, the interesterification batch reactor, a catalyst dosing device, the products pumps and a vacuum unit (Figure 6.5). Generally, a post-treatment line is designed within the plant for dry catalyst deactivation.

The oil feed pump pumps the neutralised and bleached oil from blending tank into the interesterification batch reactor through an oil heater, where the oil is heated by means of low-pressure steam to about 100–110 °C. The blending tanks can be placed on load cells to measure the exact quantity of oil that is fed to the reactor, but accurate flow meters are also nowadays used for this purpose.

If the initial amount of FFA in the oil exceeds 0.1%, the oil is first neutralised with caustic soda solution in an oil–caustic soda static mixer in order to eliminate the FFA as much as possible. The oil is dried in the reactor at a pressure of about 80 mbara. To enhance the drying, the reactor discharge pump pumps the oil through the spray nozzles into the underpressurised reactor. As the moisture level drops, the pressure is reduced further to about 10–5 mbara. The combination of underpressure and high temperature thus allows the process technologist a minimum residual moisture content in the oil. Only when the oil is sufficiently dried (<0.01%) is the catalyst introduced. In order to increase the life of the catalyst and to minimise undesired side reactions, contact with moisture in the air should be avoided.

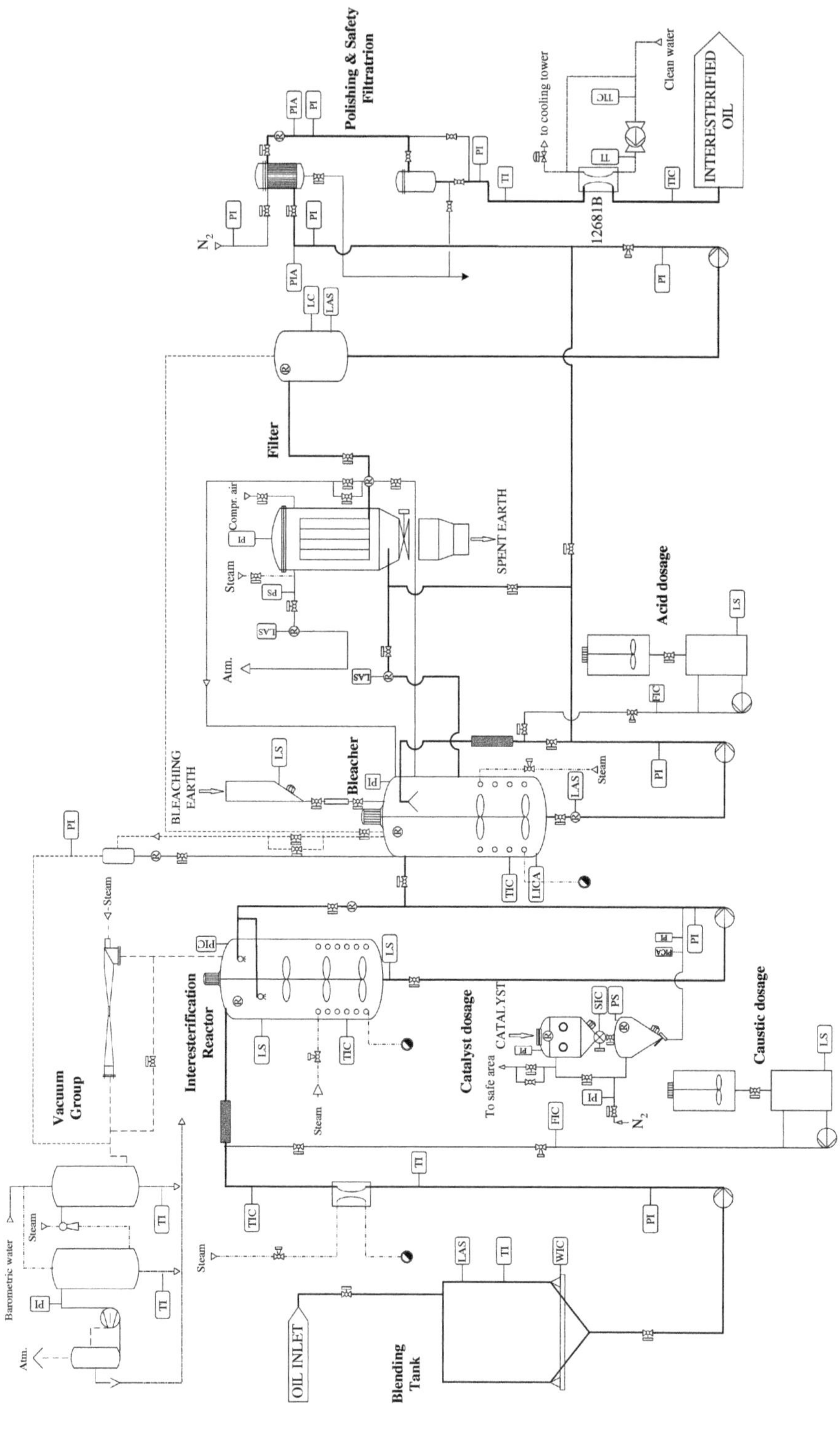

Figure 6.5 Schematic process diagram of chemical interesterification.

A standard randomisation reaction takes about 30–40 minutes, after which the interesterified oil has to be post-treated in order to inactivate the catalyst and eliminate the remaining soaps formed during the reaction. The interesterified oil is pumped to the acid reactor/bleacher. The citric acid dosing pump adds a citric acid solution to the flow of oil pumped in the recirculation by the bleacher oil pump. It is this citric acid that inactivates the catalyst and converts soaps back to FFA. When the acid reaction is over, the residual water that has been introduced in the oil together with the citric acid is evaporated under the low pressure applied in the acid reactor/bleacher.

To counteract the darkening of the oil as a result of the reaction, some bleaching earth is dosed from a bleaching earth hopper. From the acid reactor/bleacher, the oil is sent to the interesterification post-treatment main filter, typically a leaf filter type.

Filtered oil is collected in the bleached oil tank and then pumped over a safety filter (typically back pulse) to the secondary safety bag filter and often to a final oil cooler, as it is good practice to cool the oil before storage.

When the differential pressure in the hermetic leaf filter increases above the threshold value, the filter needs to be emptied and dried. The drying of the cake is done by steam blowing, after which the recovered oil and heel volume of the main filter are sent back to the bleacher.

The dried cake is finally discharged by vibrating filter elements through a spent earth-collecting hopper to the disposal container.

6.3.4.1 Processed product quality

In chemical interesterification, the fatty acids will typically be redistributed over the glycerol backbone in a randomised way, and the outcome of the process can therefore be relatively easily predicted. This also allows for monitoring of the extent of the interesterification reaction as the mixture progresses towards a randomised state, which is especially useful in following up slow reactions and possibly incomplete reactions. At any moment t, the degree of interesterification (DI) can be expressed as:

$$DI(\%) = 100^{*}\left[\frac{\left(\frac{TAG\_A}{TAG\_B}\right)feedstock - \left(\frac{TAG\_A}{TAG\_B}\right)product}{\left(\frac{TAG\_A}{TAG\_B}\right)feedstock - \left(\frac{TAG\_A}{TAG\_B}\right)random}\right] \tag{6.2}$$

where TAG_A represents the triacylglycerol or group of triacylglycerols that most proportionally decreases after interesterification and TAG_B is the triacylglycerol or group of triacylglycerols that most proportionally increases after interesterification. The triacylglycerols (TAGs) are expressed as a relative percentage of the total TAGs after normalisation. This formula is flawed, however, when using very small and/or very large values, which can lead to large proportional differences. This will make the calculated DI insensitive to actual physical changes and thus lose relevance. A more robust calculation would rely on the use of sum of squared residuals (SSR) of all TAG (in which the SSR of the randomised mixture is considered zero):

$$DI(\%) = 100^* \left[1 - \frac{\sum_{i=a}^{z} (TAG_{product,i} - TAG_{random,i})^2}{\sum_{i=a}^{z} (TAG_{feedstock,i} - TAG_{random,i})^2} \right] \quad (6.3)$$

where a and z are the first and the last TAG considered in the TAG distribution. It's important to recognise that most conventional TAG analyses do not differentiate between positional isomers (PPO versus POP), and that this calculation does not therefore give information on how this racemisation proceeds throughout the reaction, nor on its end state.

The possible effect of positional isomers is in fact more easily (though indirectly) detected by taking the physical properties of the mixture as criteria. It is a more pragmatic approach, as generally the physical properties – like the solid fat content (SFC) at a certain temperature – are prime quality parameters by which to trade the product. A common formula based on SFC of the product is therefore:

$$\frac{SFC_t - SFC_\infty}{SFC_0 - SFC_\infty} = e^{-k.t} \quad (6.4)$$

The advantage of this expression is that it contains three parameters that have both a physical and a chemical meaning: the k value is related to the reaction rate of the catalyst and to the SFC_0 initial of the feedstock and its change until equilibrium, which is ΔSFC. When using SFC as a parameter, it has to be taken into account that it is nonadditive, and that crystallisation effects that have little to do with the chemical changes as such can occur.

6.3.5 Future for interesterification technology

In the context of further development of interesterification, possibly the most appealing possibility for process technologists is 'directed' interesterification. The basic idea is to establish a continuously shifting equilibrium within the reaction mixture by letting the formed trisaturated species precipitate and thus become removed from the (liquid) reaction mixture. This would allow a much more controlled and efficient separation of saturated and nonsaturated triglycerides, instead of a series of interesterification, fractionation, interesterification, fractionation and so on. It is perhaps a little amusing that this process is mentioned in a paragraph about the future of the technology, because it is not really new at all; in fact, the principle of directed interesterification was described by one of the best known innovators of fats and oils interesterification, Eckey, in 1948. Such processes, which require (less efficient) low temperatures to invoke the fractional crystallisation of the *in situ* produced saturated triglycerides, found some entry into the industry, but the elegant pathway in theory too often turned out to be a troublesome application in practice. In chemical interesterification, it has for now largely been abandoned as industrial practice.

Nowadays the benefits of enzymatic interesterification (see also chapter 7) over the chemical pathway (less partial glycerides formation, increased tocopherol retention, colour preservation, less post-refining costs) have put the latter in decline for new investments. Indeed, more and more fat processing industries are shifting to enzymatic processing technology. In this light, it should not be forgotten that in the commodity oils and fats industry, the economy of a process is still more important than its novelty. As such, the attraction of an enzymatic randomisation process consists largely in the fact that it generates fewer oil losses than the chemical technology, rather than an enhanced (or faster) randomisation. It also means that enzymatic technology will be adopted by fats and oils manufacturers for as long as the enzyme cost does not exceed the cost of oil losses that can be expected in chemical interesterification. Though the quality and the stability of the enzyme lipases have evolved considerably, the general perception in the industry is that this technology is less robust than its chemical counterpart. In some cases, a properly fine-tuned chemical process is therefore still preferred over the enzymatic pathway.

6.4 Dry fractionation

In edible oils processing, the term 'fractionation' usually refers to any process involving a fractional crystallisation of the oil as the means of separating species based on melting point. The general aim of the process is to extend the applicability of the feedstock, be it through the reduction of an undesired component or the intentional concentration of another.

The actual crux of such fractionation process is a crystallisation process; hence 'fractional crystallisation' is nowadays often used as a more evocative synonym. Also, in fractionation discipline, the edible oil industry is turning away from technically superior solvent or detergent fractionation due to safety and investment issues, as well consumer perception. However, the need for high-quality oil fractions has not disappeared, so the combination of these factors has pushed the boundaries of the most sustainable pathway in this field: the fractional crystallisation of the pure oil, or dry fractionation. Other oil fractionation techniques have been developed and applied in recent decades, among which molecular distillation and supercritical carbon dioxide extraction are certainly noteworthy. Although the said alternative technologies can offer specific advantages with regards to specific separation efficiency (enrichment of high-value minor components, reduction of unwanted minor components), the economics of such processes exclude their broad implementation for bulk edible oil processing. Therefore, in the remainder of this chapter, the focus will be on the fractional crystallisation of edible oil, which is the norm in the fats and oils industry.

It should be remarked that compared to hydrogenation and chemical interesterification, dry fractionation technology has undergone substantial developments in order to achieve the current state of the art. Whereas dry fractionation was often regarded as an unpredictable, tedious and labour-intensive process, and the relatively cost-effective fractionation technique as being without additives, polluting effluents or the necessity for post-refinement, the sustainability and safety of the dry fractionation process are second to none. Because of these features, it is perceived as the modification technology of the 21st century (Timms, 2005), which warrants a more elaborate discussion.

6.4.1 Historical perspective

In most literature on fat fractionation, Hippolyte Mège-Mouriès is credited with the invention of a patented method to produce certain fats of animal origin. In fact, he concocted the production of a sort of margarine fat through the separation of a liquid fraction from ordinary tallow after gentle cooling. With only temperature difference as a driving force, the fractional crystallisation of a fat is a natural, spontaneous phenomenon. It was thus also observed that in palm (kernel) oil harvested in tropical regions, small crystals would appear upon cooling and form a crystal suspension in the wooden barrels during shipping to chillier Western Europe. These slightly denser solids eventually settled, and such fractions could effectively replace hardened fats in margarines. We can therefore consider these wooden shipping drums the very first oil crystallisers, with the ocean waves providing the gentle agitation to keep the developing crystals in suspension. Moreover, the natural fractional crystallisation of fats upon mild cooling is echoed in the term 'winterisation', referring to the habit of leaving large oil tanks quiescent in wintertime to induce some mild crystallisation and obtain a liquid fraction with improved cold stability, in a rather economic fashion.

Despite the apparent spontaneity of the process itself, it took until the 1960s for the fractionation industry (and technology) to boom, when the production of palm oil in South East Asia increased substantially and export taxes on processed palm oil were reduced. At that time, however, the boundaries of the technology were mainly determined by the phase separation. In recent decades, the continuous development of the separation technique in particular – from vacuum belt filtration to centrifuges and membrane press filters – has put fractionation on the map as a versatile and economic modification technique. Although some specific techniques relying on the use of detergents are still applied for very particular productions, only two main fractionation technologies are used in the 21st century's edible oil industry:

- *Dry fractionation, also known as crystallisation from the melt*: this is fractional crystallisation in its most simple form, and the economy of the technology allows it to be used for the production of commodity fats.
- *Solvent fractionation*: already patented in the 1950s, this involves the use of hexane or acetone to allow the high-melting components to crystallise in a very low-viscous organic solvent. This can be helpful with respect to the selectivity of the reaction, but mainly offers advantages in the field of phase separation: much purer solid fractions can be obtained, even with a vacuum filtration. Being a more expensive process, it is far less common than dry fractionation and only comes into the picture when a very high added value of (at least one of) the resulting fractions makes up for the high cost.

6.4.2 Principle

It is not in the scope of this chapter to cover all aspects of crystallisation of oil, such as nucleation, crystal growth and so on; however, the principal concepts need at least to be concisely explained since they do have serious implications for the level of the technology.

Natural oil, even after refining, is a very complex mixture of different triglycerides (and numerous other minor components), which has important repercussions for the melting behaviour of a fat: most oils do not exhibit a sharp melting point, but rather display a steady softening (or increasing liquid content) with increasing temperature, until they are completely liquid. The fact that oil is a mixture allows it to be considered as a heterogeneous system of at least two separable fractions – a solute and a solvent – at a given temperature.

There are three variables that can be used to consider whether a certain triglyceride will remain in the melt (behave like solvent) or crystallise (behave like solute): the absolute melting temperature of the triglyceride, the concentration of the triglyceride and the bulk temperature. Their relationship is expressed by the van 't Hoff equation:

$$\text{In}\,x = \frac{\Delta H_m}{R}\left[\frac{1}{T_m} - \frac{1}{T}\right] \tag{6.5}$$

where x is the solubility of the solute (mol/mol), ΔH_m is the heat of fusion of the pure solute (kJ/mol), T_m is the melting temperature of the solute (K), T is

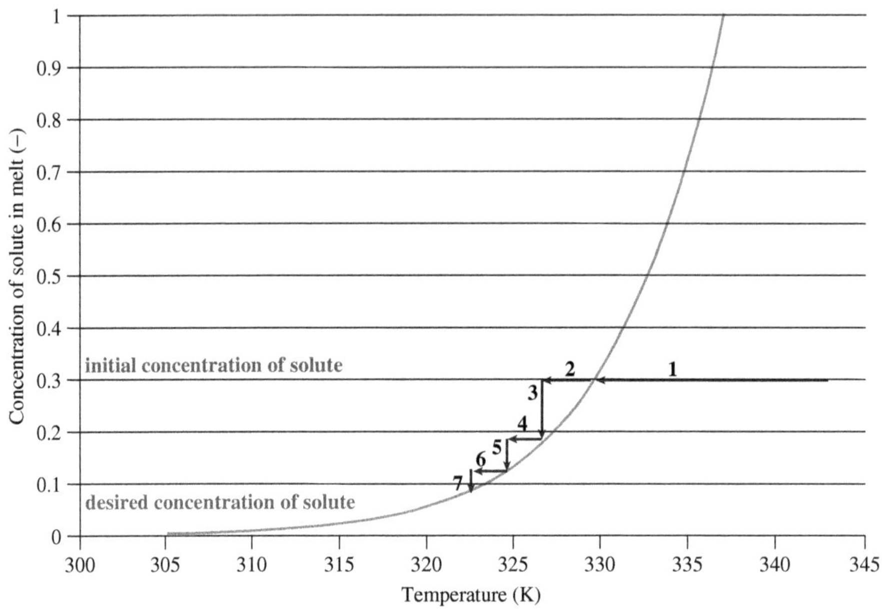

Figure 6.6 Solubility diagram with schematic representation of fractional crystallisation.

the experimental temperature (K) and R is the gas constant (kJ/mol.K). For any triglyceride, a solubility line can be constructed as function of temperature (Figure 6.6).

In this schematic representation, the fractional crystallisation process is thus essentially as follows:

(1) Approach the solubility line (cooling the melt).
(2) Supersaturate the solution to initiate the crystallisation (cooling into the metastable zone).
(3) Equilibrate the solid/solution suspension (the solute triglyceride precipitates till the equilibrium situation on the solubility line is reached again).
(4) Supersaturate the suspension.
(5) Equilibrate the solid/solution suspension.
(6) Supersaturate the suspension.
(7) Equilibrate the solid/solution until the desired final residual concentration of the solute in the melt is reached.

After this process is completed, the crystal suspension is ready for filtration. The solubility line also enables the fact that for every additional decrease in the solubility of a solute, proportionately lower cooling temperatures are needed to be appreciated.

The complex nature of fractional crystallisation in natural oils becomes apparent when we consider that for each type of triglyceride present, such solubility lines can be constructed. More importantly, ideal solubility curves like the one in Figure 6.6 actually ignore the occurrence of interactions between different triglyceride molecules. This indicates that the solution behaviour of a triglyceride is itself influenced by the surrounding triglycerides in the melt. Even in rather simple model fat systems, small but significant deviations will occur (Zhou & Hartel, 2006). In the context of fractionation, 'intersolubility' is the most important kind of interaction: it generally refers to the property of forming a solid 'solution' in which the constituting triglycerides cannot be separately determined, nor divided; it behaves as one phase. The physical chemistry behind this is in fact comparable to that applying to metal alloys. The more structural resemblance two triglycerides have, the higher their miscibility in a solid state.

Consequently, such intersolubility of the TAG often presents the largest fundamental problem in several fractionation processes, as the actual goal of fractionation is to separate different TAGs selectively. Intersolubility, however, can occur to a considerable degree in practice, and can have very negative effects on both the viscosity and the filtration properties of the crystal suspension formed (Calliauw *et al.*, 2010). Although in theory intersolubility between two species decreases with decreasing temperature, the problem in a multicomponent system such as oil is that with decreasing temperature, the number of different triglycerides that can crystallise will increase, making undesired interactions more likely to happen. Therefore, intersolubility will determine to an appreciable extent how many supersaturation/equilibration steps are needed to achieve the desired residual concentration in the melt. The less susceptible the oil system is to intersolubility effects, the deeper the supersaturation can be, and consequently, the fewer 'steps' (or time) needed to reach the final point on the solubility line. Upon stringent cooling, excessive nucleation, crystal growth and intersolubility effects result in the formation of interconnected, homogeneous lumps of fat that are practically unfilterable. This explains why fractional crystallisation in most cases cannot be executed by simply crash-cooling the oil (unlike in margarine production, where fat crystal networks are desired).

Compared to intersolubility, other typical fat crystallisation phenomena such as polymorphism are only of secondary importance; typical fractionation conditions are generally sufficiently restricted in time and temperature range to only allow one type of molecular arrangement to form.

The solubility diagram (or binary phase diagram) also lacks a very important aspect of the crystallisation reaction: the kinetics. For these relatively large organic molecules, crystallisation in the melt is a rather slow reaction, hampered by limited heat and mass transfer between bulk and interface. Intuitively, it can be understood that the deeper the saturation (i.e. the driving force), the more rapidly the crystallisation reaction will occur. The first supersaturation is typically quite severe, in order that it can induce a fast nucleation of the very first crystals. Achieving supersaturated conditions as such is technically not difficult; it just comes down to the removal of the sensible heat of the oil (at a rate of 0.5 kcal/kg.K) and it occurs quickly in most crystallisers. For this reason, the supersaturation trajectories can commonly be drawn parallel to the x-axis: the temperature is decreased well below the solubility line, even before the triglyceride starts to solidify, and thus the concentration remains constant. In the example, the equilibrium stage occurs isothermally, but in practice this is quite difficult to achieve, as it implies zero temperature gradients in the crystalliser during crystallisation. The exothermic character of crystallisation (latent heat: about 50 kcal/kg) and the slow heat transfer in oil evidently make this supposed homogeneity a very tough situation to realise in practice.

Essentially, all of the preceding can be condensed to the statement that a good fractional crystallisation can be achieved when the melt is kept just sufficiently in metastable condition to create a driving force for crystallisation of triglycerides of interest. But at the same time, the conditions need to be sufficiently close to the solubility line in order to prevent formation of solid solutions between desired and undesired triglycerides and/or uncontrolled crystal growth from hampering filtration. If the crystal growth is regular and steady, the crystal aggregates result in sharply discrete and dense spherulitic structures, sometimes measuring up to several millimetres in diameter (generally larger with lower supersaturation conditions), which are fairly uniform in size and shape (Figure 6.7).

To conclude this section, the influence of minor components on oil crystallisation should also be addressed. Overall, 'impurities' such as diglycerides or FFA have a negative effect on the crystallisation rates of the triglycerides of interest at typical fractionation temperatures and on the filterability of the solids formed (Siew & Ng, 1999; Calliauw, 2008).

6.4.3 Process parameters

The basic principles sketched in this section help to explain the key aspects of a crystalliser, the technological heart of the fractionation installation. It should be able to gently cool down a mass of oil (up to 100 tonnes/batch) and keep the resulting crystal suspension as homogeneous as possible.

6.4.3.1 Cooling speed

Fat crystallisation is an exothermic reaction, so the efficiency with which this energy can be evacuated is an important design feature. For most industrial crystallisers, this ranges between 120 and 200 $W/m^2.K$.

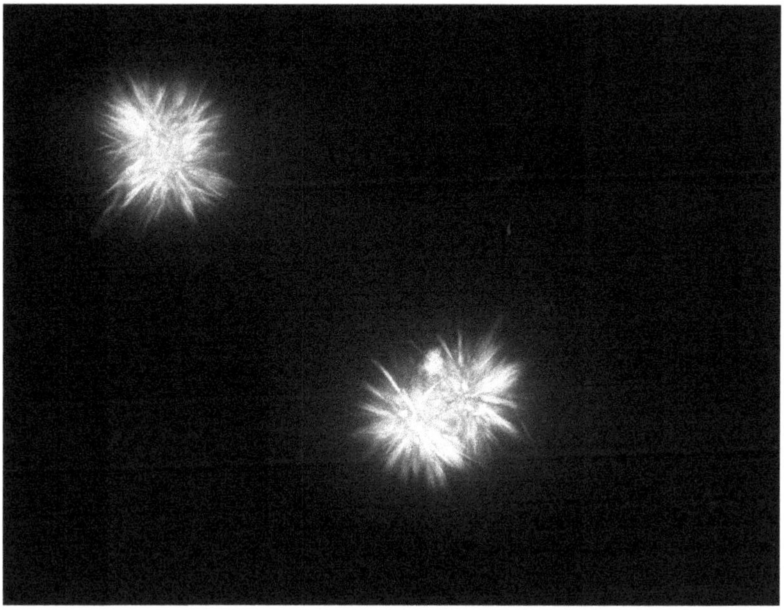

Figure 6.7 Spherulitic crystals formed in palm oil.

In view of this, the most important feature in fractionation technology is ensuring proper and homogeneous heat transfer in the crystalliser. At every stage of the cooling and crystallisation process, the temperature of the oil should be kept in a state that permits crystallisation to the desired degree. Generally, a crystallisation process can be broken down into several temperature stages. In these stages, the manner of oil cooling can be based on:

- selecting a fixed oil temperature (meanwhile allowing the cooling water temperature to vary between set limits); or
- selecting a ΔT, a fixed temperature difference between oil and cooling water; or
- selecting a fixed temperature for the cooling water.

Typically, the first two process control modes have been labelled as the typical cooling strategy in Tirtiaux-designed crystallisers, whereas the water temperature control has characteristically been applied in Desmet Ballestra-designed processes. This distinction has been stated in many fractionation reviews over the last 20 years. However, currently most fractionation plants are built to offer total freedom in terms of process control: in every step of the process, the mode of process control can be selected (Figure 6.8).

In this context, it is a rather widespread misconception that, for example, oil temperature control is slower than water temperature control: the speed of

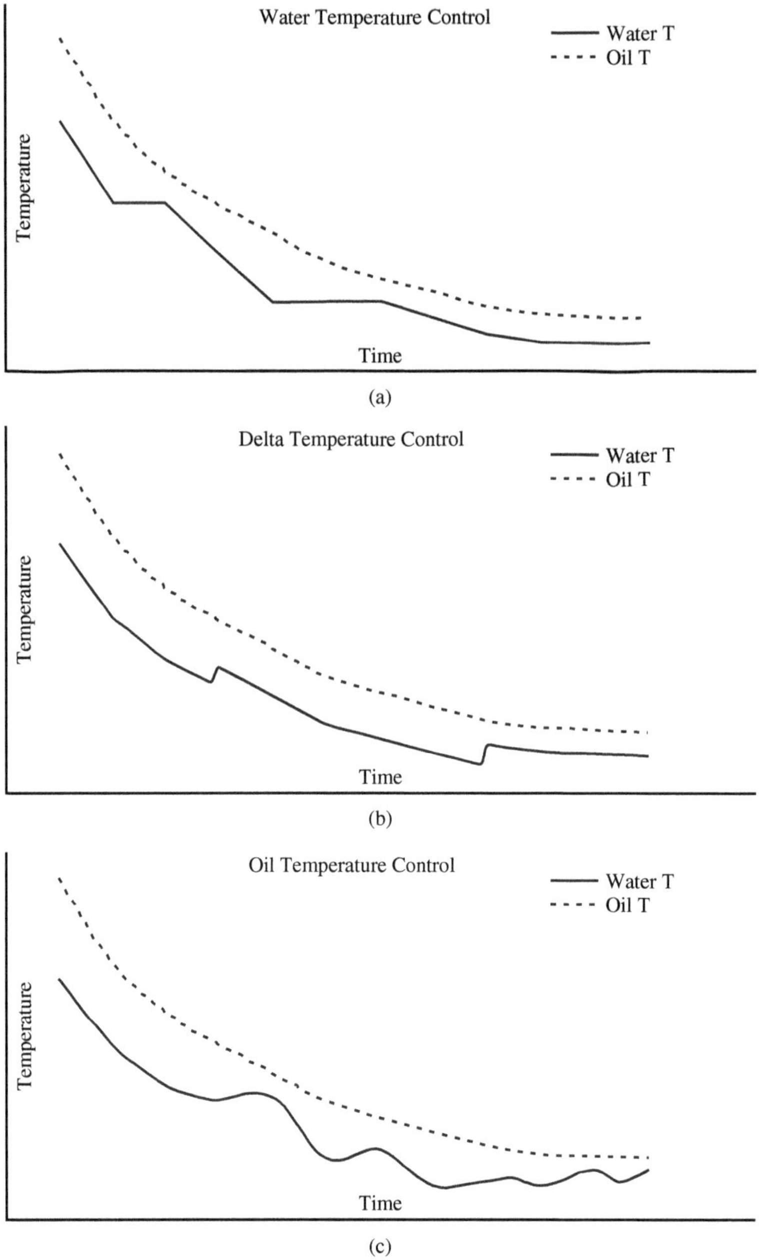

Figure 6.8 Different oil cooling modes used in dry fractionation: (a) water temperature control; (b) delta temperature control; and (c) oil temperature control.

the crystallisation cycle is largely determined by the size of the average cooling driving force imposed. So regardless of the mode of process control, larger temperature gradients between water and oil will lead to shorter processes (so higher throughput), until the point at which filterability is affected (so yield and stearin purity decrease). In practice, the cycle times can vary up to a factor 3, depending on the average imposed ΔT, and the challenge is to optimally balance yield, fraction quality and capacity.

6.4.3.2 Agitation

Since oil is a poor heat conductor, heat transfer should occur mainly through bulk mixing and friction with the cooling surface. This force is supplied by the agitator, which is often positioned quite close to the cooling surface, in order to make sure the oil just cooled can be readily brought back to the bulk.

Agitation is necessary not only for heat exchange but also for mass transfer. In fact, the cooling of oil does not only happen at the cooling surface interface; it's also the result of colder oil getting mixed with the warmer bulk oil. Specifically for mass transfer, the agitation has to prevent the settling of the crystals being formed and distribute them homogeneously over the bulk, so that concentration gradients don't get too high either. Also, as many authors agree that secondary nucleation (i.e. crystal growth on earlier formed crystal surfaces, instead of *de novo* formation in the melt) is highly relevant for industrial crystallisation applications (Mullin, 2001), it is the underlying mechanism of crystal seeding in crystallisation processes. It can explain to some extent why such nucleation is promoted by higher agitation: crystals get shattered when colliding with each other, with the impeller or with the crystalliser wall. Although secondary nucleation is rather inevitable in industrial processes, Timms (2005) stated that this phenomenon is generally undesirable in fractionation, and agitation in an industrial crystalliser is therefore kept to a minimum: just enough to keep the heat transfer and bulk mixing intact. Industrial tip speeds easily range up to 3 m/s at the start of crystallisation, but can drop below 1 m/s in the final (more viscous) stages.

6.4.4 Process design

6.4.4.1 Crystalliser design

Compared to crystalliser designs for oleochemical or pharmaceutical products, for example, edible oil crystallisers are quite basic. In view of the need to reduce concentration and temperature gradients as much as possible, the conservation of the mass and heat transfer properties over the length of the batch process has been the main challenge for edible oil crystallisers. The complete range of viscosity of the oil (from about 50 cP of liquid oil to sometimes really thick slurries with non-Newtonian rheological properties). Oil being a very poor heat conductor, the rate-determining step in heat exchange between the bulk oil and cooling water is the mass transfer from bulk to interface oil. Excessive cooling without sufficient renewal of the interface oil will only result in the precipitation of a vast (unfilterable) mass of crystals on to the cooling surface and consequent complete loss of process temperature control. Only in cases where a very fast crystallisation reaction is used (and thus a great driving force is created) to keep the economy of the process in check, such as in solvent fractionation, can scraped surface heat exchangers be used to counteract this crystal loading, even if this comes with the drawback of crystal fragmentation.

A simple batch crystalliser is the 'tubular crystalliser': a double-jacketed cylindrical vessel fitted with an axial agitator at its centre. Normally a cooling surface of at least 5 m^2 per 1 m^3 oil is needed to ensure proper heat transfer. In practice, this limits the dimensions of tubular crystallisers to the extent that it is only used for low-volume speciality fats and less frequently for commodity edible oil products, such as palm oleins. For the fractionation of palm oil, large crystallisers are needed, and various means of providing additional cooling surface have been developed. A substantial increase in cooling surface per unit volume of oil can be obtained by dividing concentric annular crystallisation compartments that are separated by concentric, annular, double-walled cooling elements (Kellens *et al.*, 2007). The most conventional type of crystalliser used in the last few decades is the tank crystalliser, in which the cooling surface is built inside a tank in the form of (horizontal) cooling coils, (vertical) pipes and/or cooling plates. Generally, such crystallisers are equipped with propeller-type blades, inducing a convection-like flow pattern through the tank.

Notwithstanding the basic character of edible oil crystallisers, several enhancements can be made within or just outside them, such as seeding tanks (to make a kind of pre-mix of crystals and reduce crystal initiation time) and hot oil spray pipes to clean the cooling surface from the residual precipitated crystalline matter after batch draining. Due to concerns over increasing peroxides, this practice is nowadays generally abandoned. Another practice is the implementation of an ultrasonic precrystalliser, facilitating crystal nucleation in the bulk prior to its entry into the crystalliser.

6.4.4.2 Filter design

Although the triglyceride separation is in theory already established during crystallisation, it is clear that the separation stage itself effectively determines the product yields as well as the stearin quality. As more residual olein is expelled from the solids cake, the final stearin will be more concentrated in crystals, will turn out 'purer' and will display higher and steeper melting. The olein quality is determined entirely by the amount and selectivity of crystallisation in the preceding stage. In some applications, the crystals formed are often not sufficiently stress-resistant and get squeezed through

Table 6.7 Separation efficiencies of different filtration technologies in dry fractionation of palm oil.

	Vacuum filtration	Centrifugal nozzles	Membrane press (16 barg)
IV palm oil	52	52	52
IV palm olein	56–57	56–57	56–57
IV palm stearin	40–42	36	30–32
Solids in cake (%)	46	–	65
Olein Yield (%)	72	76	82

the filter medium. Obviously, such contamination of crystals in the olein phase negatively affects the efficiency of the fractionation process and results in a liquid phase with inferior cold stability properties. Overall, the 'permitted' degree of olein dilution in the stearin cake determines the choice for the applied separation technology (Table 6.7).

Membrane press filtration, also used in, for example, sludge dewatering systems, is by far the most widely used separation technology in dry fractionation today. The filters consist of a large steel frame that can easily hold up to 150 filter plates together, each plate counting for up to 7 m^2 of filtration surface and over 100 l filter chamber volume.

Usually, the filter chambers are first filled with the crystal suspension, and a large portion of the liquid olein thus passes through the filter cloths. Watertight membranes (one membrane per chamber) attached to the internals of the plates are then gradually inflated (with water, liquid oil or air) to the desired pressures, reducing the chamber volume and pushing out residual liquid, which is immediately evacuated via internal channels in the plate towards collecting tanks. The volume reduction of the chamber thus compacts and dries the cake. Typical final squeeze pressures are 6 or 15 barg; 30 and even 50 barg membrane press filters are available on the market if higher purity is needed in speciality fat cakes. It is also noteworthy that the mass fraction of solids in the filter cake decays exponentially as a function of the distance to the filter cloth, and consequently thinner filter chambers (from 15 to 50 mm chamber widths) and longer squeezing times can be a helpful (though costly) means of significantly reducing the entrainment.

6.4.4.3 Plant design

A general layout of a present-day dry fractionation process is presented in Figure 6.9. Often multiple crystallisers are used in (overlapping) series. This is not only a matter of capacity; it is also done in order to maximise the use of the filter: through good planning of the crystallisation times, the expensive (batch) filter should be in constant operation. It is important to note that

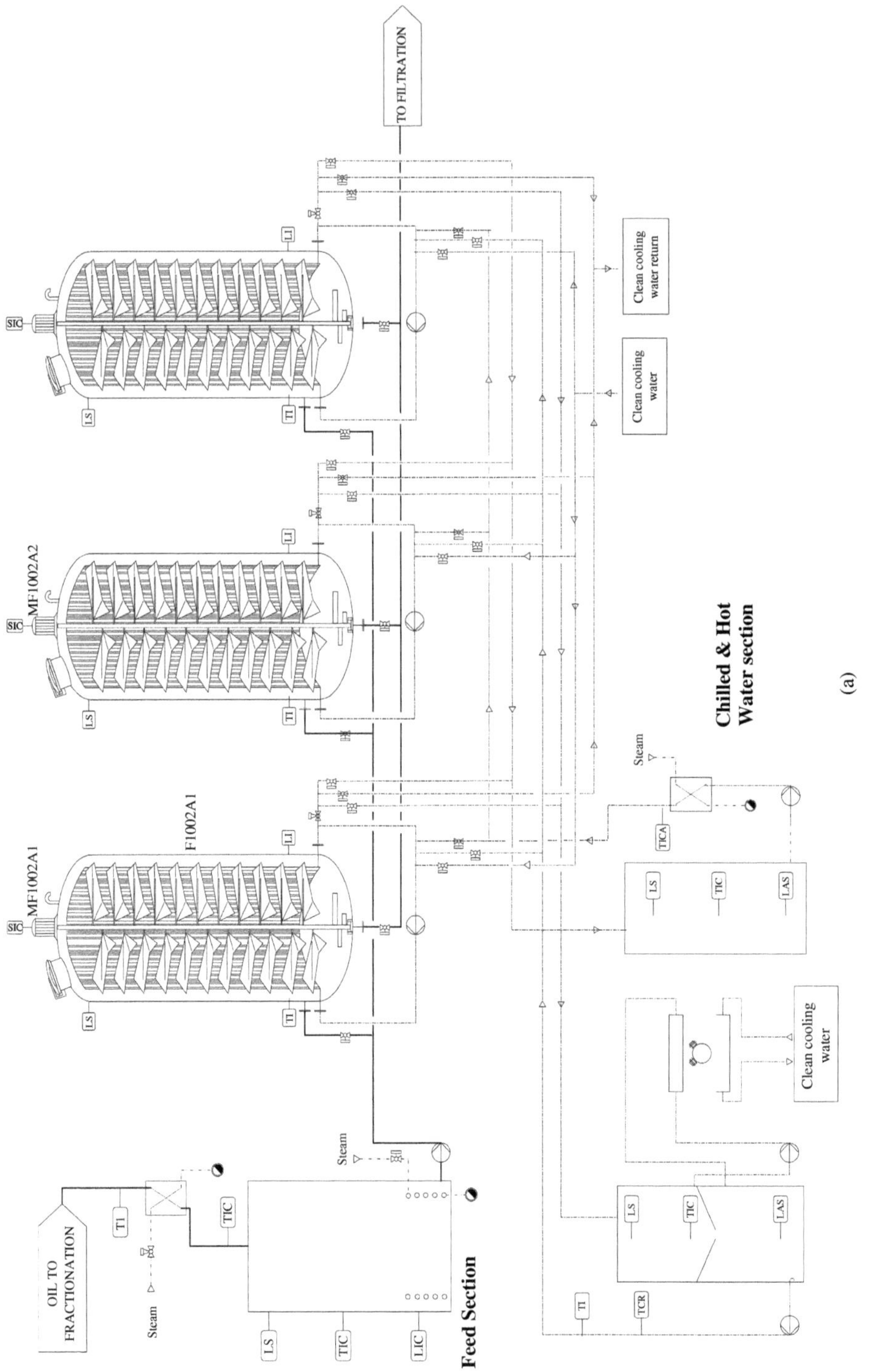
TO FILTRATION
MF1002A1
F1002A1
MF1002A2
SIC
LS
TI
LI
OIL TO FRACTIONATION
Steam
TIC
LIC
Feed Section
Clean cooling water return
Clean cooling water
Chilled & Hot Water section
TICA
LAS
TCR
(a)

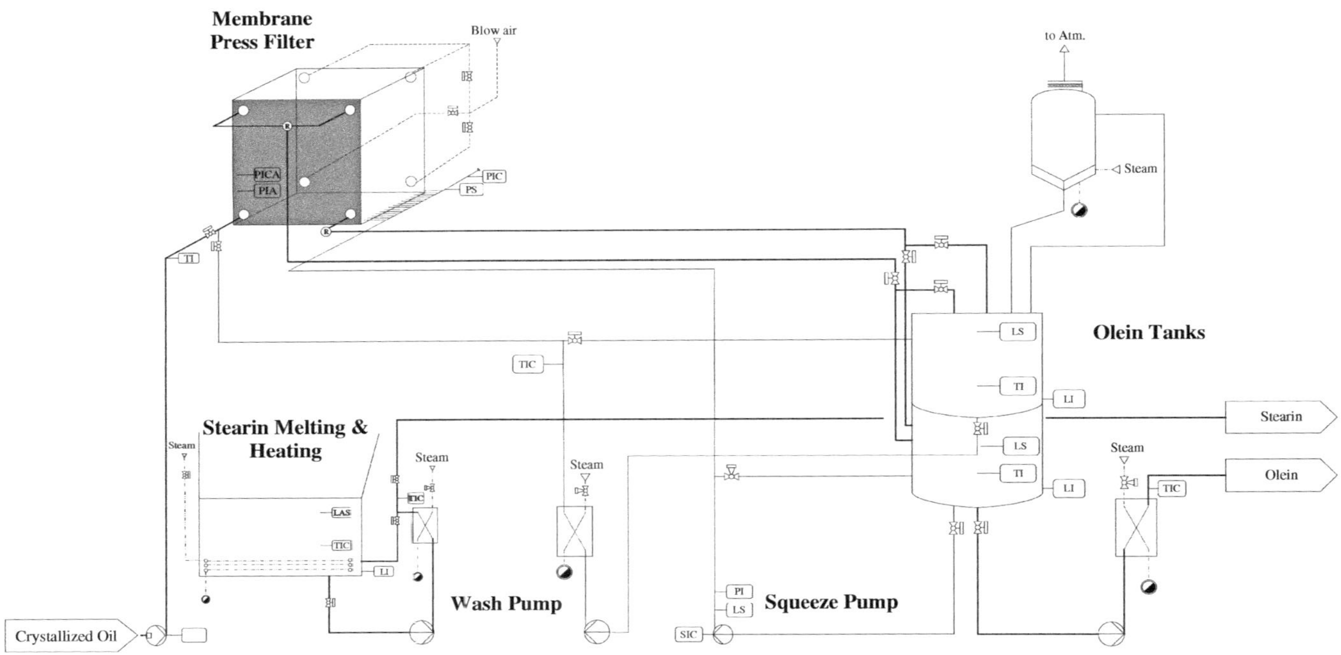

(b)

Figure 6.9 Schematic process diagram of dry fractionation, consisting of (a) crystalliser section and (b) filter section.

fractionation is a quite modular technology and very often results in a tailor-made plant; several water tanks (two to four), all at different temperatures, can be installed in order to maximise heat recuperation while keeping all cooling medium in a closed loop, but this can just as well be done with a single tank feeding chilled water to the crystallisers, while steam is added to supply heat and normal cooling water is fed directly from a clean cooling tower. Similar tailor-made selections can be used for the filter section: the squeezing medium in the membranes can be water, but could equally be the filtrate itself! In some cases, a significant reduction in the dead time of a filter can also be established by means of a crystallised oil buffer tank; each crystalliser can be quickly drained and made ready to receive the next batch of oil. Meanwhile, the cooled buffer tank will send set volumes of crystal slurry to the filter, whenever it is ready.

Continuous filtration systems have also been a very elegant strategy in dry fractionation, although currently the demand for purer solid fractions as obtained by filter chamber compaction has pushed continuous belt filters somewhat out of the dry fractionation market. They are however still the norm in solvent fractionation.

It should be kept in mind that fractional crystallisation of a triglyceride oil is a relatively slow process and is therefore the time-determining stage; some simple fractionations can be established in about 5 hours of crystalliser residence time, whereas more complex oils can require up to 3 days of cooling and crystal maturation before being sent to the filter. The whole filtration plus squeezing operation can take from 30 to 90 minutes.

6.4.5 Future for fractionation technology

6.4.5.1 Optimised crystalliser designs

The innovations in the field of oil crystallisers have been orientated towards improved heat-exchange cooling geometries, in all possible variations: double jackets, cooling plates, cooling coils and their relative positions in the crystalliser body. Additionally, agitator types (propeller paddles, surface scrapers, plate paddles etc.) have also been the subject of study. Industrial crystallisers therefore exist in all shapes and forms. Interesting innovations in this field in recent years use agitation with an integrated cooling system, such as the STAR-crystalliser (Weber *et al.*, 1998), consisting of a collection of rotating cooling bundles, and the MoBulizer™ (Calliauw, 2008). The latter design features a very uniform linear movement of the homogeneously distributed cooling surface (Figure 6.10).

Industrial results indeed point to considerably higher heat-exchange efficiencies compared to slow-stirred crystallisers, and to improved results for shear-sensitive crystals, such as those formed in tallow and lard fat.

Figure 6.10 Detail of MoBulizer™ cooling tubes in vegetable oil. The full colour version can be found on the plates.

Probably the most significant crystalliser design of the last decade is the Statolizer crystalliser (Calliauw *et al.*, 2005), designed specifically to form and deal with very viscous slurries that are practically unstirrable in a conventional crystalliser. It has been introduced in the industry as an automated system to replace panning-and-pressing cold rooms, and over the years has become a standard for palm kernel oil fractionation in the production of cocoa butter substitutes. Further product developments based on Statolizer technology are occurring in the production of cocoa butter equivalents (CBEs) and even plain cocoa butter fractionation (Calliauw *et al.*, 2011).

6.4.5.2 High-pressure filtrations

Spurred on by the demand for speciality (confectionery) fats, the liquid–solid separation technologies are under continuous pressure to push the boundaries of the separation efficiency. Now 30 barg cake squeezing pressure is well established, and there is increasing interest in 40 or 50 barg. It should be realised that at these very high pressures, there is a diminishing marginal return of olein yield per barg of extra pressure applied. However, it could deliver a crucial purity of stearin composition for some high-end applications. There has also been a (renewed) interest in centrifugal separator systems, especially with respect to fish oil fractionation and cross-contamination in sensitive processes. However, separation efficiencies cannot really compete with the high-pressure membrane press filters for most applications.

Other features that are widely integrated in contemporary dry fractionation filter assemblies include antistatic filter plates and cloths, replaceable polypropylene membrane units (instead of the whole plate), cake dropping detectors, facilitated cake discharge shakers or pendulums, 'self-cleaning' inlet ports, membrane leak detectors, light safety barriers and many more, all of which contribute to a safe, fast and consistent functioning of the filter press, but do not touch the core of the technology.

Solvent fractionation technology usually operates with gas-tight vacuum belt filters, an important advantage of which over membrane press filters is the possibility of (multiple) countercurrent washing of the stearin cake, resulting in a physical displacement of remaining olein miscella by pure solvent and therefore the production of purer stearin fractions.

6.4.5.3 Continuous fractional crystallisation

Mainly in the final decade of the 20th century, the possibilities of continuous fat fractionation on anhydrous milk fat were studied, with results that resembled those obtained by traditional batch dry fractionation (Illingworth, 2002). Quite atypically for dry fractionation purposes, a scraped-surface heat exchanger (SSHE) can be used as a plug-flow reactor (PFR). In all such designs, however, a batch membrane press filtration is still recommended, in order to reduce the entrainment of olein within the stearin cake. Therefore it is more correct to define them as continuous crystallisation techniques rather than complete continuous fractionation systems. A vacuum belt filter or hyperbaric filters can be considered to meet this objective of total continuity of the process, although they yield less thorough phase separation compared to membrane press filters.

Continuous fractionation of palm oil is regarded as probably the most promising pathway for future dry fractionation. An important challenge in continuous fractionation is to avoid complete mixing while maintaining adequate heat and mass transfer in a low viscous suspension. In order to achieve this, contemporary batch crystallisers with the desired homogeneous mixing should be redesigned to function as PFRs, allowing minimal back-mixing and ensuring uniform residence time upon continuous product throughput.

A good continuous fractional crystallisation technique has considerable advantages for plant operation: it can increase throughput, it will reduce interbatch variability, it can improve the homogeneity of the particle size distribution in the slurry leaving the crystalliser and it offers a substantial energy saving. This economy exists because the longer a crystalliser can run at a the same temperature, the more equivalent 'batches' it can produce without reheating or recooling of the crystalliser's steel and water. Additionally, substantial energy economisation can be achieved by crossing the hot ingoing

Table 6.8 Comparison of batch and continuous dry fractionation of palm oil (Calliauw *et al.*, 2011).

	Batch		Continuous	
Average residence time (hours)	5		4.5	
Average runtime before draining (hours)	5		150	
SFC of the slurry (%)	8.7		8.4	
	Stearin	**Olein**	**Stearin**	**Olein**
Mettler cloud point (°C) [−3 °C]		9.8		9.4
IV (Wijs)	32.5	56.2	30.1	56.3
Yield (%)	18	82	16.5	83.5

feed with the cold olein leaving the plant for storage. Such a setup can amount to a 20–30% steam (or hot water) and proportional cooling power reduction in the first step of palm oil fractionation. The constant presence of crystals in the crystalliser helps to reduce the degree of supercooling needed to initiate the crystallisation (as in a batch process).

Table 6.8 demonstrates the industrial results of a recently developed continuous crystallisation process with crystallisers operated in a plug-flow mode. It was to be capable of producing various palm (super) olein qualities, at yields superior to those of the batch process, for a duration up to 50 equivalent batch cycles.

6.4.5.4 Alternative multistage processes for specialty fats production

The apparent simplicity, low operation costs and sustainable character have made dry fractionation a 'technology to stay', but, as for all modification technologies, developments in the field are propelled by two driving forces: cost reduction and higher added values of the products. Whereas cost reduction is a relatively universal and direct process-related issue (in the form of lower consumption costs, lower investments, increased yields, higher degree of automation, reduced losses etc.), adding more value to the fat fractions is a more specific market-driven matter, imposed in fact by the consumer. This 'added value' is the premium price a customer is willing to pay for an edible oil product that exhibits specific functional properties such as improved cold stability, low saturates contents, increased docosahexanoic acid (DHA) levels and so on for liquid products; and steeper melting curves and fat structuring properties at minimal saturates content for solid products. The economic driver can also exist in the form of the successful imitation of specific properties of expensive 'natural' fats and oils by relatively cheaper oils such as palm oil fractions, which explains the huge interest in CBEs, replacers and substitutes.

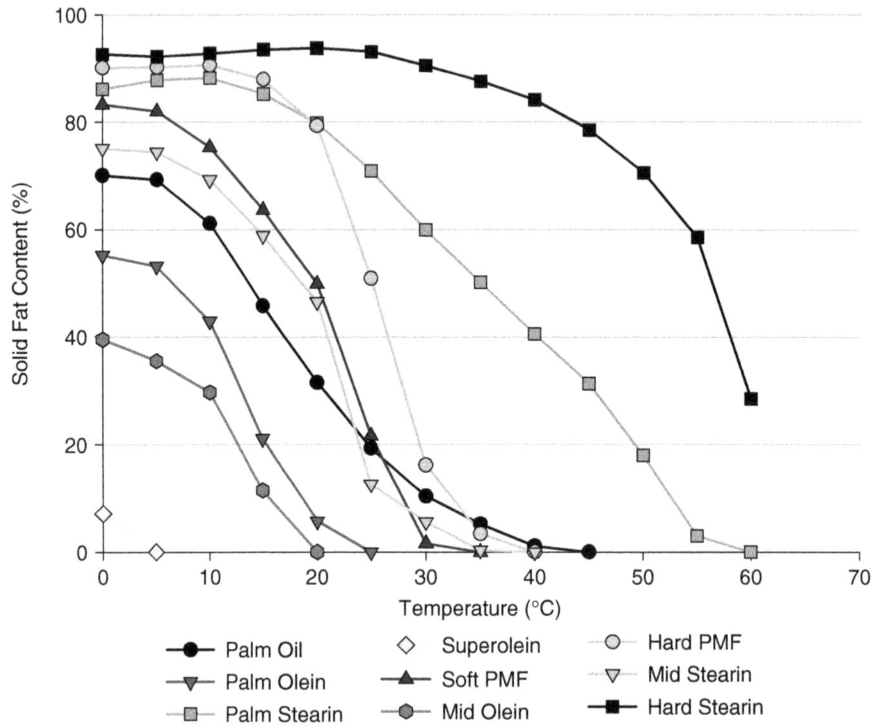

Figure 6.11 Overview of palm oil fraction melting curves in multistage fractionation. The full colour version can be found on the plates.

So the main economics of dry fractionation is determined by how much added value both fractions produced can generate payback of the feedstock cost and operation costs. This explains the ongoing diversification of multistage fractionation processes: as the consumer raises the bar in terms of quality, fractionation processes need more concentration steps to achieve the product properties that can meet these demands.

A good example is how palm superolein (the olein fraction of the olein of palm oil) has to meet ever-increasing cold stability demands: at the end of the 1990s, commodity superolein exhibited IV values around 63, whereas 10 years later a minimum IV of 65 or 66 is considered a standard superolein. The desired quality (specifically the 'cold stability') of these commodity olein products is thus slowly yet steadily increasing (Figure 6.11).

At the other end of the spectrum of multistage palm oil fractionation, superstearin production is receiving increased attention, as superstearin, with IVs ranging from 9 to 13 (depending on the squeezing pressure used), is today considered an optimal (*trans*-free) concentration of saturated triglycerides.

Finally, the mid fractions, particularly the hard palm mid fraction (PMF), can be regarded as those fractions in which the symmetrical monounsaturated

Table 6.9 Basic cost estimation of edible oil modification processes. All costs in US dollars.

			High-pressure hydrogenation (+post-treatment)	Chemical IE (+post-treatment)	Enzymatic IE	Palm oil fractionation	Palm olein fractionation
Plant capacity (tpd)			180	140	100	200	100
Annual capacity (at 340 working days/year)			61 200	47 600	34 000	68 000	34 000
Capital investments		*Equipment and engineering*	$1 500 000	$1 100 000	$1 000 000	$1 600 000	$2 000 000
		Structural works	$600 000	$500 000	$450 000	$800 000	$900 000
		Installation	$750 000	$700 000	$600 000	$855 000	$900 000
ROI			**$2 850 000**	**$2 300 000**	**$2 050 000**	**$3 255 000**	**$3 800 000**
Capital cost/tonne			**7.8**	**8.1**	**10.0**	**8.0**	**18.6**
Annual maintenance costs			$40 000	$40 000	$50 000	$50 000	$60 000
Operation costs		*Manpower*	2	1	1	1	1
Consumption/tonne	Steam	kg/tonne oil	30	150	12	40	63
	Electricity	kWh/tonne oil	10	15	4	10	16
	Ni catalyst	kg/ton oil	2	0	0	0	0
	Hydrogen	m^3/tonne	50	0	0	0	0
	NaOMe catalyst	kg/tonne oil	0	1	0	0	0
	Enzyme	kg/tonne oil	0	0	0.4	0	0
	Citric acid	kg/tonne oil	0.5	2	0	0	0
	Bleaching earth	kg/tonne oil	1.5	5	0	0	0
	Oil losses	kg/tonne oil	−3	18	0.6	0	0
Utility unit costs					**Operation cost/tonne**		
$90 000	Manpower	$/year	$2.9	$1.9	$2.6	$1.3	$2.6

(continued overleaf)

Table 6.9 *(continued)*

			High-pressure hydrogenation (+post-treatment)	Chemical IE (+post-treatment)	Enzymatic IE	Palm oil fractionation	Palm olein fractionation
\$0.03	Steam	\$/kg	\$0.75	\$3.8	\$0.3	\$1.0	\$1.6
\$0.15	Electricity	\$/kWh	\$1.5	\$2.3	\$0.6	\$1.5	\$2.4
\$4.0	Ni catalyst	\$/kg	\$8	–	–	–	–
\$1.0	Hydrogen	\$/m³	\$50	–	–	–	–
\$2.5	NaOMe catalyst	\$/kg	–	\$2.5	–	–	–
\$55.0	Enzyme	\$/kg	–	–	\$22.0	–	–
\$1.78	Citric acid	\$/kg	\$0.9	\$3.6	–	–	–
\$0.65	Bleaching earth	\$/kg	\$1.0	\$3.3	–	–	–
\$0.85	Oil losses	\$/kg	\$(3)	\$12.2	\$1	–	–
		Miscellaneous	\$2	\$2	\$2	\$2	\$2
Operation cost/tonne			**\$65**	**\$32.3**	**\$29.5**	**\$6.1**	**\$10.4**
Total cost/tonne			**\$73**	**\$40**	**\$40**	**\$14**	**\$29**

triglycerides (mainly POP) are concentrated. These generally generate the most added value, but are not obtained in high yields. In general, 100 tonnes of palm oil entering a multistage fractionation will yield only about 8–12 tonnes of good-quality hard PMF.

There is no reason to limit multistage processing within the field of fractionation. Indeed, a logical step is to actively combine the technological novelties of all available modification technologies to add value to the final product. This is already done on the industrial scale, where many oil manufacturers combine hydrogenation, interesterification and fractionation to customise specific fat properties. Another way of approaching possible combinations is not from a product point of view, but rather from a process point of view. One of the avenues of investigation is to conduct a partial (enzymatic) interesterification before dry fractionation, in order to improve the reaction kinetics in the latter technology, rather than improve the oil product quality itself.

6.4.6 Summary

In the fractionation discipline, the edible oil industry has turned away from technically superior solvent or detergent fractionation due to safety and investment issues, as well as consumer perceptions. However, the need for high-quality oil fractions has not ceased, so the combination of these factors has led to a pushing of the boundaries of the most sustainable pathway in this field: the fractional crystallisation of the pure oil, or dry fractionation (Table 6.9). Crystalliser designs are evolving towards more sophisticated tanks, with a primary focus on reducing temperature gradients in the bulk.

Contemporary filter designs exhibit greater ease of use, a higher degree of automation and increased safety. Further, the overall separation efficiency is definitely superior to that of the standard design of 10 years ago. The most recent innovations are to be found in the expansion of the possible edible oil applications spectrum by means of multistage fractionation, as well as in economic improvement of the process through integration of continuous crystallisation (plug flow) reactors and heat-recovery systems in the plant.

References

Beers, A., Ariaansz, R. & Okonek, D. (2008) *Trans* isomer control in hydrogenation of edible oils. In: A.J. Dijkstra, J. Richard, R.J. Hamilton & W. Hamm (ed.) *Trans Fatty Acids*. doi: 10.1002/9780470697658.ch6.

Boger, T., Zieverink, M.M.P. Kreutzer, M.T., Kapteijn, F., Moulijn, J.A. & Addiego, W.P. (2004) Monolithic catalysts as an alternative to slurry systems: hydrogenation of edible oil. *Ind. Eng. Chem. Res.*, **43**, 2337–2344.

Calliauw, G. (2008) Molecular interactions affecting the phase composition during dry fractionation of palm olein. PhD thesis, Ghent University, Belgium.

Calliauw, G., De Greyt, W., Dijckmans, P., Foubert, I., Dewettinck K. & Kellens, M. (2005). Production of cocoa butter substitutes via two-stage static dry fractionation of palm kernel oil. *J. Am. Oil Chem. Soc.*, **81**, 783–789.

Calliauw, G., Fredrick, E., Gibon, V., De Greyt, W., Wouters, J., Foubert, I. & Dewettinck, K. (2010) On the fractional crystallization of palm olein: solid solutions and eutectic solidification. *Food Research International*, **43**, 972–981.

Calliauw, G., Hendrix, M. & Kellens, M. (2011). Dry fractionation of speciality fats: a decade of statolizer fractionation technology. Paper presented at the MPOB International Palm Oil Conference, Kuala Lumpur, Malaysia.

Dijkstra, A.J. (2009) Interesterification revisited. *Alton E. Bailey Award Address presented at the 100th AOCS Annual Meeting & Expo*, Orlando, FL, USA.

Dijkstra, A.J. (2010) Selectivities in partial hydrogenation. *J. Am. Oil Chem. Soc.*, **87**, 115–117, doi: 10.1007/s11746-009-1507-z.

Eckey, E.W. (1945) Process for treating fats and fatty oils. US patent 2 378 005 (to The Procter & Gamble Company).

Eckey, E.W. (1948) Directed interesterification in glycerides. *Industrial & Engineering Chemistry*, **July**, 1183–1190.

Illingworth, D. (2002). Fractionation of fats. In: A.G. Marangoni & S.S. Narine (eds) *Physical Properties of Lipids*. New York, NY, USA: Marcel Dekker, pp. 411–477.

Kellens, M. (2000). Oil modification processes. In: W. Hamm & R.J. Hamilton (eds.) *Edible Oil Processing*. Sheffield, UK: Sheffield Academic Press, pp. 129–173.

Kellens, M., Gibon, V., Hendrix, M. & De Greyt, W. (2007). Palm oil fractionation. *European Journal of Lipid Science and Technology*, **109**, 336–349.

Koetsier, W. (1997) Hydrogenation of edible oils. In: F.D. Gunstone & F.D. Padley (eds) *Lipid Technology and Applications*. New York, Basel, Hong Kong: Marcel Dekker, pp. 265–303.

List, G.R, Kritchevsky, D. & Ratnayake, W.M. (2007) *Trans Fats in Foods*. Champaign, IL, USA: AOCS Press.

Mensink, R.P., Zock, P.L., Katan, M.B. & Hornstra, G. (1992). Effect of dietary *cis*- and *trans*-fatty-acids on serum lipoprotein (A) levels in humans. *Journal of Lipid Research*, **33**, 1493–1501.

Mullin, J.W. (2001). *Crystallization*, 4th edn. New York, NY, USA: Butterworth-Heinemann.

Normann, W. (Herforder Maschinenfett-und Ölfabrik Leprince und Siveke) (1903) Process for converting unsaturated fatty acids or their glycerides into saturated compounds. British patent 1515.

Siew, W.L. & Ng, W.L. (1999). Diglycerides in palm oil products: composition and effects in oil properties. In: N. Widlak (ed.) Physical properties of fats, oils and emulsifiers. Champaign, IL, USA: AOCS Press, pp. 129–139.

Sonntag, N.O.V. (1983) Fat splitting, esterification, and interesterification. In: D. Swern (ed.) *Bailey's Industrial Oil and Fat Products*, 4th edn. New York, NY, USA: Wiley Interscience, pp. 97–173.

Timms, R.E. (2005). Fractional crystallization- the fat modification process for the 21st century. *Eur. J. Lipid Sci. Tech.*, **105**, 48–57.

Weber, K., Homman, T. & Willner, T. (1998). Fat crystallizers with stirring surfaces: theory and practices. *Oléagineux Corps gras Lipides*, **5**, 381–384.

Zieverink, M.M.P. (2007) A catalysis engineering approach to selective hydrogenation. PhD thesis, Delft University of Technology, Netherlands.

Zhou, Y. & Hartel, R.W. (2006) Phase behavior of model lipid systems: solubility of high-melting fats in low-melting fats. *J. Am. Oil. Chem. Soc.*, **83**, 505–511.

7

Enzyme Processing

David Cowan

CS Application Scientist/Global Coordinator, Novozymes, Chesham, UK

7.1 Introduction

The application of lipases in edible oil processing is a relatively recent development compared to their more general use within food processing. For example, there is almost no mention of enzyme processes in *Introduction to Fat and Oils Technology* published in 2000 by O'Brien *et al.* (2000). In the first edition of *Edible Oil Processing* (published 2000), the possibility of using enzymes for interesterification was discussed from a largely theoretical point of view, signifying how little entry enzymes had found in the fat processing industry. Enzyme reactions were still considered to operate in aqueous systems, and as such their use in oil processing was not anticipated. However, when microaqueous environments were studied, it was revealed that enzymatic reactions could take place under these conditions. One of the first potential applications was in the enzymatic synthesis of cocoa butter equivalents (CBEs) by Macrae (1985). Although the proof of principle was delivered, the nature of the product being produced limited the potential for this application and the process of production of the enzyme and its immobilisation was complex. It was costly to produce and hence required that the product itself was of a high enough value to justify this, and the result was almost no immediate industrial application. Further developments refined this somewhat and Eigtved (1992) described a simpler process, but also one of limited application.

Enzymatic degumming also entered the picture in 1992, utilising a porcine pancreatic phospholipase. The high cost of the enzyme required recirculation of part of the gums to allow the enzyme's reuse, and its source limited the areas in which the refined oil could be used.

Edible Oil Processing, Second Edition. Edited by Wolf Hamm, Richard J. Hamilton and Gijs Calliauw.

Lipases are an enzyme group that can catalyse the hydrolysis and synthesis of ester bonds involving fatty acids, but this is not the only group of bonds that can be involved. In addition, they can form and/or hydrolyse amide, carbonate and thio-ester bonds, as described by Hayes (2004). Because of this synthetic capability in nonaqueous environments, lipases are frequently used for biotransformation. Their properties of regio-, stereo- and substrate selectivities, while useful in these areas, have focussed attention on synthesis reactions, and this has resulted in more research being carried out in this area than in general oil processing. Overall, their reactions can be summarised as follows:

(1) hydrolysis: RCOOR′ + H2O ↔ RCOOH + R′OH;
(2) synthesis:
  (i) esterification: RCOOH + R′OH ↔ RCOOR′ + H2O;
  (ii) interesterification: RCOOR′ + R″COOR*↔ RCOOR*+ R″COOR′;
  (iii) alcoholysis: RCOOR′ + R″OH ↔RCOOR″ + R′OH;
  (iv) acidolysis: RCOOR′ + R″COOH ↔ R″COOR′ + RCOOH.

In addition to chain length preferences, lipases also show regioselectivity. Most regioselective lipases act preferentially on ester bonds at the *sn*-1 and *sn*-3 position of the triglyceride structure, whereas few lipases are active at the *sn*-2 position. Lipases without regioselectivity also exist and are particularly useful where complete hydrolysis of a fat is required. Finally, lipases may differentiate between saturated and unsaturated fatty acids, showing preferential hydrolysis or modification of one type or another.

Two scientific developments changed this situation and resulted in enzymes being suitable for use in oil processing for nonspeciality products. The first was the application of gene transfer within microorganisms. The first degumming enzyme, being of porcine origin, could not be used for kosher and halal oils. Also, supplies of this type of enzyme are limited, so the ability to make a microbial phospholipase opened up this application. Second, in a related area (interesterification), development had been hindered by the high cost of the enzyme process. The creation of a low-cost immobilisation system coupled with gene transfer moved this application from speciality to standard modified fat products (Cowan & Holm, 2008).

7.1.1 Objectives of enzyme processing

There are two main objectives to enzymatic processing of edible oils. These are to improve an existing process or to allow for the production of unique products that cannot be produced by chemical or other modifications (Figure 7.1). In improving an existing process, the main focus is upon improving yields,

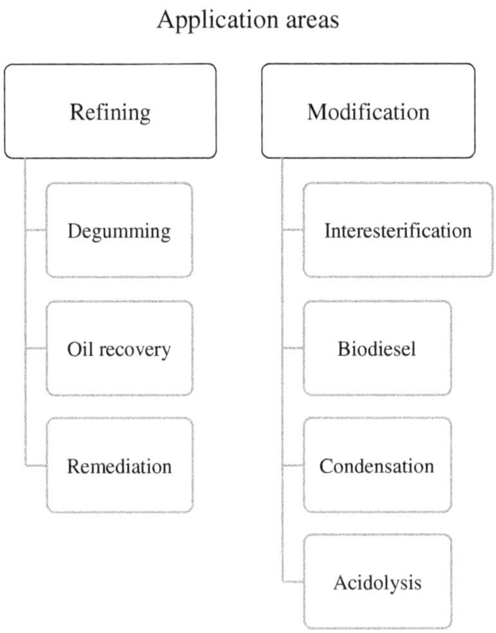

Figure 7.1 Application areas for enzyme processing in edible oils.

reducing byproduct formation and improving the sustainability of the overall process.

To produce unique products, the ability of lipase enzymes to work in either synthetic or hydrolytic modes is applied. Enzymatic condensation, in which fatty acid ethyl esters of eicosapentanoic acid (EPA) and docosahexanoic acid (DHA) are reacted with glycerol to produce high-omega-3 triglycerides, is an example of this synthetic approach. Enzymatic degumming and oil recovery from gums use predominantly the hydrolytic capacity of phospholipases. However, in both types of application, yields, low operational temperatures, lack of byproducts and sustainability are equally important.

7.2 Enzyme applications before oil refining

7.2.1 Enzyme-assisted pressing

Concerns about the yield from pressing and the danger of using hexane have prompted the search for alternative processes that are both high yielding and hexane free. Two enzymatic approaches have been used: enzyme-assisted pressing and total solubilisation of the oil-bearing seed or plant material.

In enzyme-assisted pressing, a combination of cellulases and hemicellulases is applied to a milled or flaked material under conditions of low moisture addition and a limited degradation of the plant cell material achieved. Subsequent drying and pressing of the treated material will result in a higher yield than can be obtained by pressing alone (Cowan, 2010a). However, on the industrial scale, the difficulty of applying the enzyme solution evenly and providing the correct incubation conditions and the subsequent drying costs has resulted in this application remaining at pilot plant level. The increased reluctance of the authorities in a number of countries to sanction the use of hexane may result in more research being applied in this area.

The second approach is to solubilise all the plant materials by hydrolysis of the non-oil fractions. In principle, if the plant cell wall structure is totally degraded and all large-molecular-weight structures are hydrolysed then the oil should be freed and be able to be separated from the water phase by centrifugation. De Moura *et al.* (2009) describe one version of such a process in which proteolytic enzymes are used to degrade soybean and to free the entrained oil. The main disadvantage of this approach is that in order to achieve high yields of oil, solid to water ratios of 1 : 10 have to be applied. This results in the production of large amounts of an aqueous side stream, which requires concentration if the protein fraction is to be recovered. These issues need to be addressed if such an approach is to succeed on a large scale.

7.2.2 Enzymatic degumming

Enzymatic degumming was one of the first industrial applications of enzymes in edible oil refining. During the oil extraction process, the amount of phospholipid (see Figure 7.2 for the most common types) extracted with the oil is dependent on the extraction temperature. Both the oil extraction rate in hexane and the content of phospholipids in the oil increase as the extraction

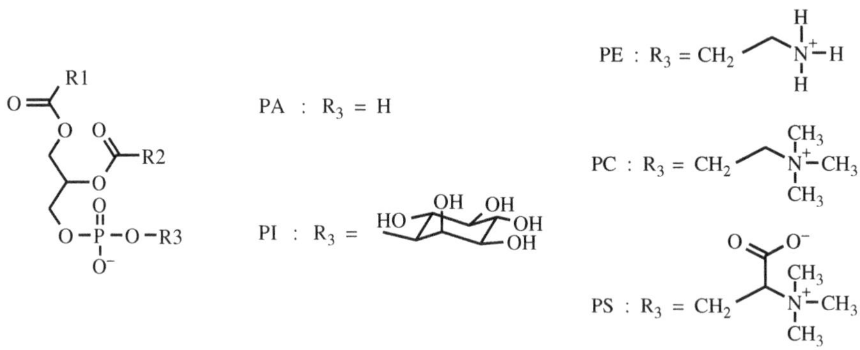

Figure 7.2 Structures of the most common phospholipids. R_1, R_2, fatty acid residues; PA, phosphatidic acid; PI, phosphatidyl inositol; PE, phosphatidyl ethanolamine; PC, phosphatidyl choline; PS, phosphatidyl serine.

temperature rises. A temperature increase from 55 to 57 °C can result in a change of phospholipid content in soybean oil from 0.70 to 0.85% (Bockisch, 1998). Maximising extraction yields in processing will also result in increased levels of phospholipid in oils.

The removal of these phospholipids is typically the first step of the oil refining process. Removing phospholipids by water washing requires that they are hydratable and that they will transfer from the oil to the water phase. The order of the hydration rate of the different phospholipids is phosphatidyl choline (PC) > phosphatidyl inositol (PI) > phosphatidyl ethanolamine (PE) > phosphatidic acid (PA). It is in this respect that enzymes, or more specifically phospholipases, can be used: to increase the hydratability of the phospholipids in the oil (List & Mounts, 1993).

From a biochemical perspective, there are four main types of phospholipase. Their common mode of action is depicted in Figure 7.3. Phospholipases (A_1 and A_2) that remove one of the fatty acids from the glycerol backbone to produce a lysophospholipid are the most commonly used in edible oil refining. The hydrolysis of the phospholipid converts nonhydratable phospholipids into a more hydratable form, which can then be removed by water added to the oil as part of the physical degumming process.

Recently, an enzymatic degumming process utilising phospholipase C has been introduced, in which hydrolysis of the bond between the glycerol backbone and the phosphate group takes place. This results in the formation of a diglyceride and a water-soluble portion containing the phosphorus and head group portions of the phospholipid (Dijkstra, 2010).

The fourth type is phospholipase D which removes the phospholipid head group to produce a glycerophosphate. These enzymes can be applied where total hydrolysis of the phospholipid is required to maximise release of free fatty acids (FFA) and reduce gum volumes as much as possible. They can be used in conjunction with the A_1 and A_2 phospholipases, but an A_1

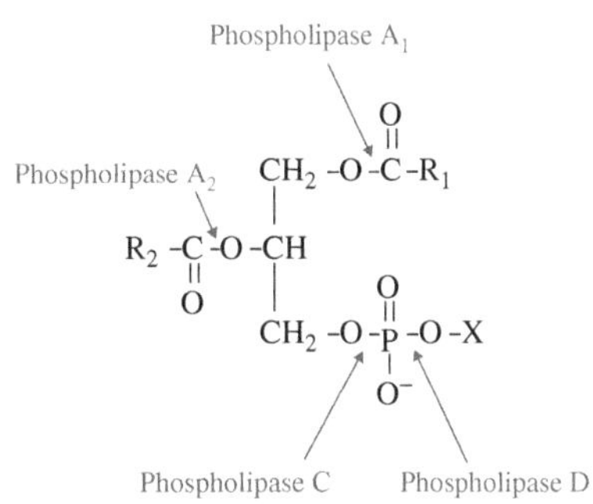

X = H, choline, ethanolamine, serine, inositol, etc.

Figure 7.3 Mode of action of phospholipases.

phospholipase requires an A_2 lysophospholipase and vice versa if complete hydrolysis is required.

7.2.3 Enzymatic degumming process (phospholipase A_1)

As all degumming processes have the aim of reducing phosphorus, the main advantage of the enzymatic approach to physical degumming (Figure 7.4) is that it provides a higher yield of oil than other methods, such as phosphoric acid-based degumming ('acid degumming'). In chemical neutralisation of oils, sodium hydroxide reacts with fatty acids to produce soaps, which are then removed by centrifugation. Phospholipids are removed as part of the overall process but the soap carries with it entrained neutral oil, resulting in a yield loss here and in the soaps produced. In non-enzymatic physical processes, while the phospholipids are converted into hydratable forms, principally by conversion of calcium and magnesium salts of phosphatidic acid, their oil-binding capacity is not removed and there is a resulting yield loss.

Enzymatic degumming with an A_1 phospholipase is a combination of a mild acid treatment to convert the nonhydratable phospholipids and a limited enzymatic hydrolysis to eliminate their oil-binding capacity. The end result is a low level of phosphorus, calcium and magnesium in the oil, coupled with an overall increase in yield of 1.0–1.5%, depending on the type of oil

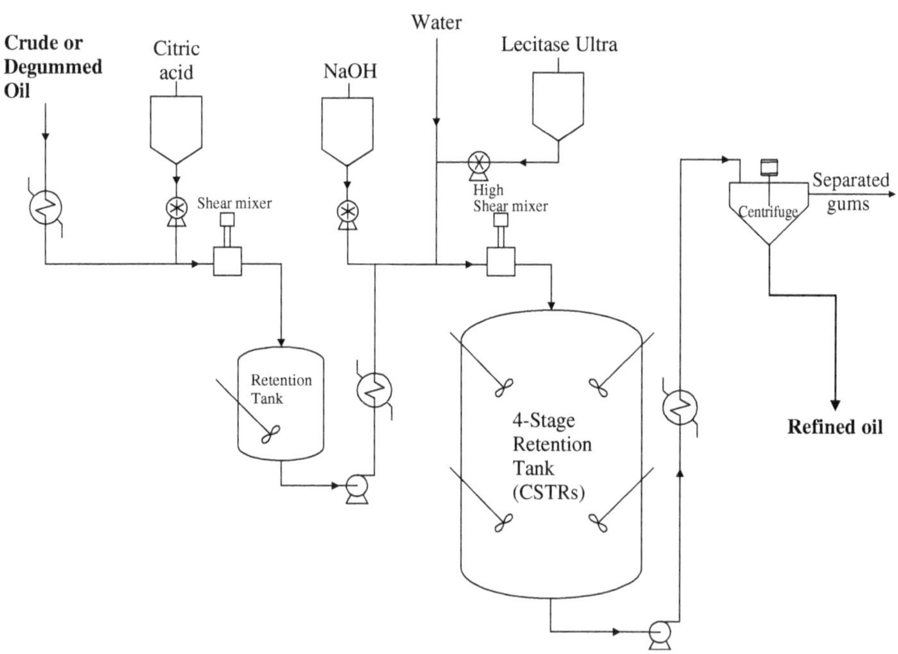

Figure 7.4 Enzymatic degumming flow sheet.

Table 7.1 Expected yields of different degumming methods for 1000 kg soybean oil.

Degumming method	Oil obtained (kg)	Oil loss (%)
Water degumming + chemical refining	942.7	5.73
Water degumming + semiphysical refining	944.3	5.57
Water degumming + acid degumming	945.9	5.41
Water degumming + enzyme degumming	948.2	5.18
Full enzyme degumming	952.1	4.79

being processed. These yield increases are experienced in the case of oils with high levels of phosphatides, such as are found in seed oils. Palm oils and other sources which have lower levels of phosphatides can be successfully degummed, but the increase in yield will not be as high. In essence, the yield increase is proportional to the level of phosphatides in the oil.

A comparison of the expected yields for soybean oil is given in Table 7.1.

Crude or water degummed oil is passed through a heat exchanger to raise its temperature to 70–80 °C. Citric acid is added at 0.065%, based on oil and normally in the form of a 50% w/w solution. The oil then passes through a high-shear mixer to ensure the acid is finely distributed throughout it. The aim is to convert the calcium and magnesium salts of PA to the free acid; a contact time of ~20 minutes is required to complete this process. Some oils with lower levels of NHP may require a shorter incubation time, but it is not usually feasible to adjust this on a batch-by-batch basis. Following this, the oil temperature is reduced to 55 °C, which is the optimum temperature for the enzymatic stage of the process. Sodium hydroxide is then added to neutralise the remaining citric acid, which raises the pH in the water phase to the optimum for the activity of the phospholipase (pH 4.8–5.5).

In total, 3.0–3.5% water is added to provide a route for the removal of the hydratable gums produced by the combined enzyme and acid treatments. This is the total water addition and includes any that is used to dilute the citric acid and sodium hydroxide additions. The optimum process is for most of this water to be added together with the enzyme in order to aid the dispersion brought about by the second high-shear mixer, which produces a large number of very small water droplets distributed throughout the oil. The process of conversion of PA and the hydrolysis of phospholipid takes place at the interface between oil and water, so maximising the number of droplets and maintaining them is key to a good degumming process.

Originally, a contact time of 6 hours was recommended, but more recent research by Cowan & Horsholm (2009) has shown that this can be reduced to below 90 minutes (Figure 7.5). This has the double advantage of reducing the tank volumes required and limiting the generation of FFA in the process, which would need to be removed later.

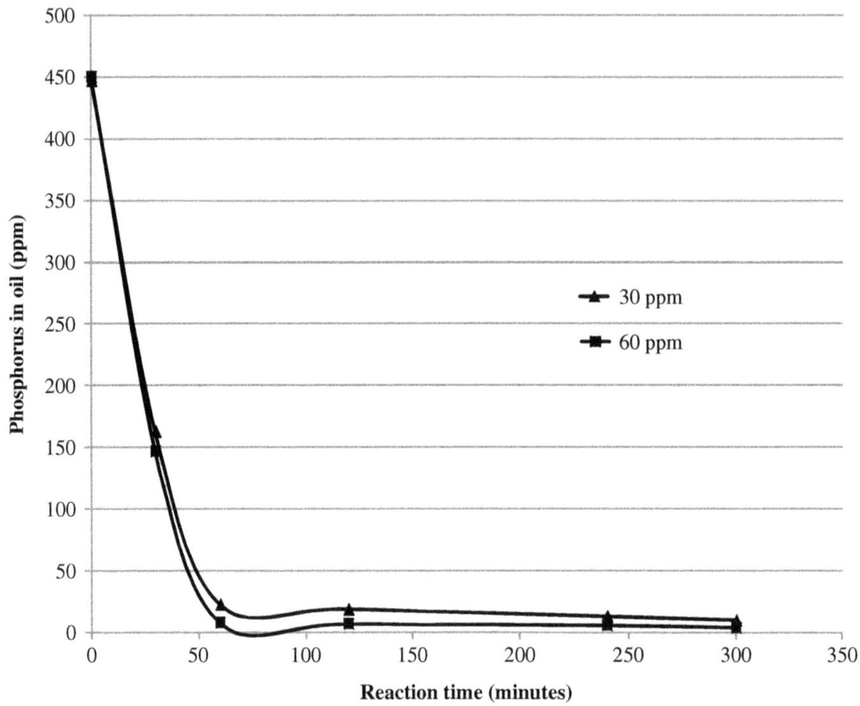

Figure 7.5 Influence of time and enzyme dose on phosphorus in oil.

The final stage of the process is a heating step to help coagulate the gums, squeeze out entrained oil and promote good gum separation in the centrifugation step. While in many cases a single centrifuge is sufficient, some oils, such as rice bran oil, will often benefit from the use of a second water wash centrifuge, which will help reduce the phosphorus level to below 10 ppm. In the total processing chain of edible oils, the additional loss for this second wash is generally compensated for by the reduced consumption of bleaching earth (and thus reduced oil loss in that stage) in the case of difficult oils.

In order for enzymatic degumming to be successful, the enzyme must be able to operate under the temperature and pH conditions found in oil processing. Microbial phospholipases are heat stable, but if temperatures are not well controlled, enzyme inactivation can take place. Although oil as such cannot have a pH, the water phase will, and it is here that the enzyme is located. Although direct measurement is not accurate, the water-soluble acidity can easily be extracted and the pH measured, either with a pH electrode or a narrow-range pH paper. Online adjustments can then be made to ensure optimal performance, but the system is robust and few changes are normally required.

In cases where the incoming nonhydratable phospholipid content is very high, which can occur if seed damage has occurred and more PA has been produced (by phospholipase D), the addition of an increased amount of citric acid may be beneficial.

7.2.4 Other phospholipases

Along with phospholipase A_1, two other enzymes are available for use in the edible oil industry. An A_2 phospholipase is available, which according to the published information functions by transferring the fatty acid to a receptor molecule, usually a sterol in the edible oil (Soe, 2006). The claimed advantage of this approach is that there will be no release of fatty acids into the oil as long as there are sufficient sterols or other suitable receptors present.

Phospholipase C offers an interesting alternative approach to enzymatic degumming by removing the phosphorus and head group from the phospholipid (Gramatikova *et al.*, 2005). The diglyceride produced is retained in the oil and so is normally included in the overall yield (but if it is later removed, the final yield may be lower than anticipated). Currently, the available phospholipase C is also limited in that it cannot degrade all the phosphatides in oil, showing little or no activity against phosphatidic acid. One solution to this is to combine phospholipase C and A_1 in such a way that the second enzyme completes the hydrolysis of the phosphatidic acid (Dayton *et al.*, 2008). As phospholipase A_1 hydrolyses all phosphatides, dosing and treatment conditions need to be carefully controlled in order to maximise the effect of both enzymes. A phospholipase C capable of attacking all the phosphatides in oil would therefore offer some advantages compared to the enzymes currently in use, which are already providing substantial benefits in edible oil processing.

7.2.5 Oil recovery from gums

When water washing is applied to edible oils ('water degumming'), the phospholipids that are removed are predominantly those which are classified as hydratable (see Figure 7.6).

The gums themselves however do entrain oil despite being classified as hydratable; as a rule of thumb, 1 kg of gums removed carries with it approximately 1 kg of neutral oil. These gums are normally added back to the meal, which is incorporated into animal feed. The entrained oil is lost to the process and has a much lower value as an animal feed component.

In enzymatic degumming, the ability of this gum fraction to bind oil is destroyed and the oil remains in the main lipid fraction, with the gums left virtually oil free. The process of removing the oil-binding power of gums can be applied either in enzymatic degumming or once the gums have been

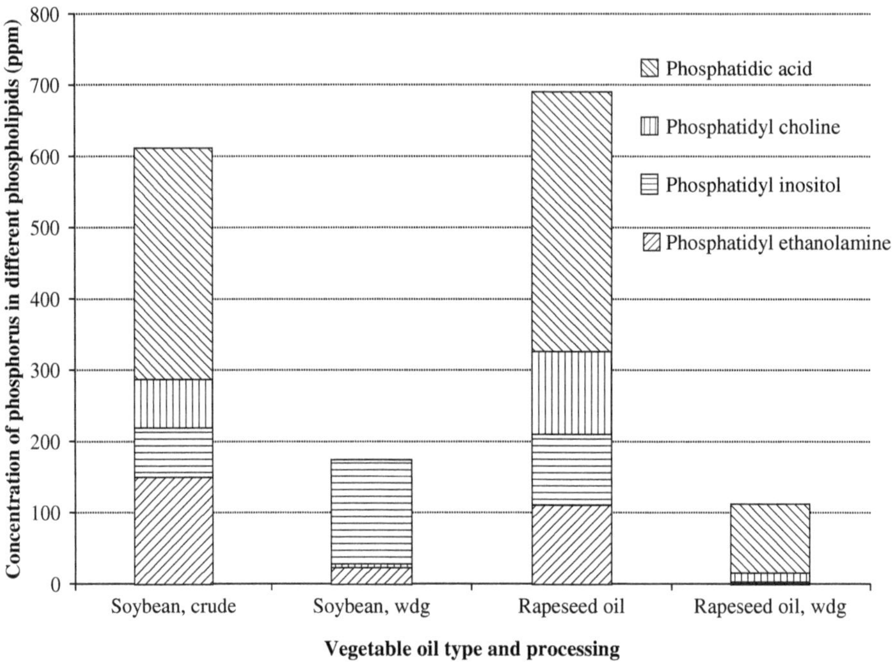

Figure 7.6 Composition of phospholipids in oil before and after water degumming. wdg, water degummed.

washed out of the oil. The advantage to treating these gums separately is that any existing water degumming process line does not require modification. Also, because the wet gum volume is ~5% of the total oil volume, the plant size for such treatment will be proportionally smaller.

To treat gums after water degumming, their pH is adjusted to 4.8–5.5 and their temperature to 55 °C. Phospholipase A_1 is added at 300–400 ppm, based on the dry weight of the gums, and the enzyme reaction is allowed to proceed for 2–3 hours. Following this, the gums are heated to 80 °C to promote the breaking of the emulsion and the oil released is separated by centrifugation. The extracted oil is normally recirculated within the–much larger–main oil flow from extraction and goes on to further degumming and processing as normal. In this way, approximately 80% of the oil that would otherwise be lost in the meal can be recovered from the gum fraction. The resulting deoiled gums are of lower viscosity than before treatment, which promotes ease of handling; due to their lower oil content, protein dilution of the meal is reduced.

7.2.6 Oil remediation

The oil contained within oilseeds and palm fruits can be damaged by lipase activity if the materials are not dried and stored correctly after harvest.

Palm fruits and rice bran are particularly susceptible to damage because of their inherently high moisture levels. Lipases derived either from the plant materials themselves or following microbial growth can hydrolyse a portion of the lipids present, resulting in the production of fatty acids and partial glycerides. As these fatty acids will need to be removed for the sake of stability and oganoleptic quality, this removal will result in a reduction in yield and an increase in processing costs.

Oil remediation is defined as the use of specific lipases (esterases) to reform triglycerides by catalysing the condensation of glycerides with FFA, with the water generated being removed by operation under reduced pressure. When there is an excess of fatty acids, the esterase can convert a portion back to triglycerides, but equilibrium between FFA and released water is reached before complete conversion (Figure 7.7). By applying vacuum to the reaction, water is removed from the system and the condensation reaction continues. In some cases there will be insufficient mono- or diglycerides to absorb all of the FFA if they have been removed as part of another process. In such cases, addition of glycerol can be used to provide the acceptors required for the fatty acids.

The physical form of the enzyme is also of importance here. Typically an immobilised catalyst will be used, as this allows for recovery of the enzyme and its application in subsequent batches. However, as has already been shown in enzymatic degumming, liquid water-based enzymes can also function at the oil–water interface, and a water-based esterase can also be used to carry out the condensation reaction. Recycling of the enzyme can also be carried out by adding an excess of glycerol to the reaction. The water-soluble enzyme will locate in this phase, and once the reaction is completed, separation of the glycerol phase allows for the reuse of the dissolved enzyme in a subsequent batch of oil.

The nature of the material to be treated will influence the choice of physical form of the enzyme catalyst. In cases where filtration has already occurred or

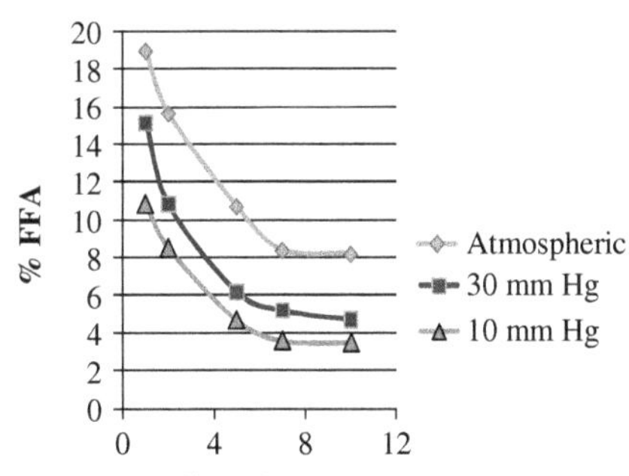

Figure 7.7 Influence of vacuum on the condensation of FFAs in rice bran oil.

where the material has been recovered from another process, the immobilised form can be preferred. However, for materials that have not been filtered or fully degummed, such as crude palm oil, an immobilised enzyme may rapidly clog up due to deposition of particulates and impurities, including phosphatides, on the granule's surface. In these cases, a liquid form of the enzyme coupled with a recycling procedure can be the preferred route.

Also, fatty acid distillates can be converted into a mixture of mono-, di- and triglycerides by reacting them with glycerol in the presence of an esterase. The resynthesised oils can then be further used for biodiesel production, utilising waste glycerol coming from the normal production, which uses methanol and neutral oil. The conversion of distillates in this way masks the fatty acids and allows the resulting material to be added to oil that will be converted in a conventional (oleochemical) biodiesel process.

7.3 Applications within edible oil modification

Enzyme applications post-refining are concerned with modification of the physical properties of the oils, often to change their melting profiles. These techniques have been described as physical (fractionation) and chemical (hydrogenation or interesterification), and in some instances have been combined, such as with the hydrogenation of palm olein or palm kernel olein derived from the fractionation of palm oil; this topic is covered in more detail in other chapters.

In the chemical interesterification processes, a number of side reactions occur. Typically, this results in colour generation, which is seen as a sign that the reaction has gone to completion. In addition, the catalyst promotes the formation of diglycerides and reduces the level of natural antioxidants (tocopherols), as they can also enter into the reaction.

Although enzymatic interesterification was considered at an academic level for many years, the high cost of the catalyst restricted its use to the production of high-value fats such as CBEs. The development of low-cost enzyme immobilisation methods, coupled with the ability to markedly increase fermentation yields, resulted in the development of an enzymatic interesterification process that could compete with the earlier chemical process.

The aim of enzymatic interesterification is identical to that of its chemical alternative: to change the melting properties of a fat blend so as to render it suitable for use in a final fat product. One example of such a modification is shown in Figure 7.8. In this example the raw material is a blend of 75% soybean oil and 25% fully hydrogenated soybean oil. Modification of this mixture by chemical (CIE) or enzymatic (EIE) interesterification results in a fat which is then suitable for margarine production. As fully hydrogenated soybean oil is used as one component, the level of *trans* fats is very low:

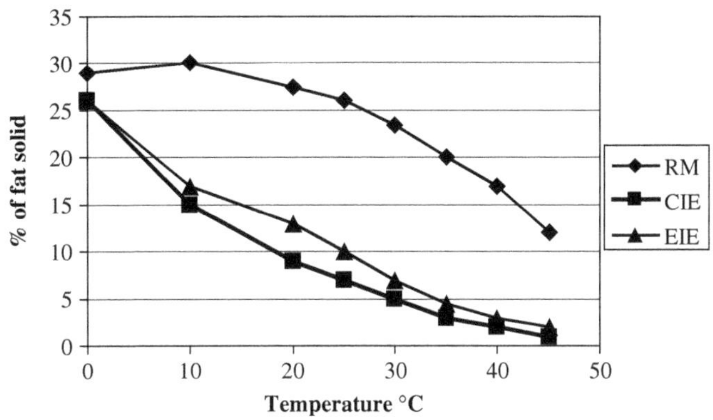

Figure 7.8 Change in solid fat content (SFC) following interesterification. RM, raw material; CIE, chemical interesterification; EIE, enzymatic interesterification.

basically that coming from the associated processes used in the production of the raw materials.

7.3.1 Industrial-scale enzymatic interesterification

For large-scale use, an enzymatic interesterification process needs to be both robust and economic. The key to this is to have a catalyst which is simple and robust and to have a good understanding of the factors which influence enzyme activity and working life. The main reason enzymatic interesterification did not progress beyond speciality fats was that these two key areas were not well enough understood.

The development of a low-cost enzyme immobilisation system utilising an inert carrier as an enzyme support made it possible to carry out enzymatic interesterification for bulk rather than speciality fats. Christensen *et al.* (2003) stated that for successful implementation:

- The immobilisation should increase the thermal stability and working life of the enzyme.
- The enzyme should maintain its activity throughout the immobilisation process.
- The process of immobilisation should be robust and reproducible.
- The immobilisation should be cost-efficient and not occupy too many production resources.
- The materials and production equipment should be suitable for the production of food-grade enzymes.
- From an application standpoint, the enzyme should be physically robust.

These key points were used in the design of an immobilised lipase for interesterification of food fats. The thermal stability of the enzyme is increased by operating in an environment of low water activity. In this case, the heat stability of the immobilised enzyme is approximately 20 °C higher than in water. Operation and product development have proven that a robust and reproducible product can be created and operated on a large scale to produce modified food fats.

For enzymatic interesterification, an inorganic support material, a lipase and an organic binder are combined to produce a final enzyme particle. Typical particle sizes are in the range of 300–1000 nm, with a mean of 500–600 nm. The final particles are hygroscopic, which allows for the small amount of water required to promote the reaction to be retained within the granule. Thus the problem of having constantly to add water, which was a key part of the earlier immobilised enzymes, is avoided. Also, because of the thermal stability of the enzyme, the need for organic solvents is avoided as the raw material can be applied as liquids above their melting points.

Normally the enzyme is packed into several fixed-bed column reactors with a capacity of 250–1000 kg and operating in a series configuration. The rationale behind the use of a series configuration is to allow for constant output from the interesterification plant. Figure 7.9 demonstrates how enzyme activity within a single reactor falls as a function of time of operation. If a single enzyme column is used then in order to produce the same degree of fat modification as the enzyme ages, a reduction in flow is required, which is not generally desirable in a production facility.

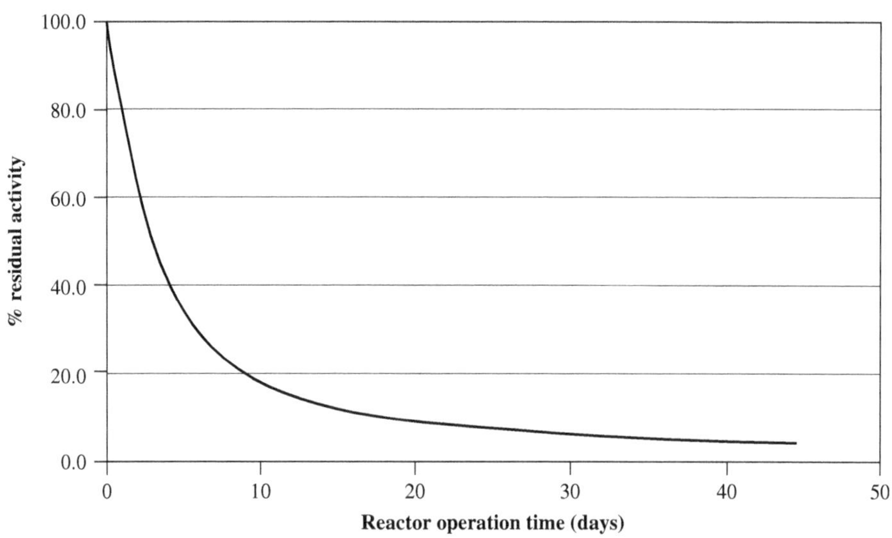

Figure 7.9 Reduction in activity of an immobilised enzyme with time.

To overcome this, a number of reactors are operated in series, with the oil to be interesterified passing from one reactor to the next in the chain. At the initial startup of the system, all reactors (normally four) will contain fresh enzyme, but as the oil passes through the reactors, the enzyme in the first in the series gradually loses activity. In a single reactor setup, this would require the flow to be reduced, but as there other reactors following, they can complete the conversion and allow constant flow to be maintained.

A typical configuration for an enzymatic interesterification unit is shown in Figure 7.10. Oil enters at the top of reactor 1, with the enzyme being supported on a splitscreen sieve. This has an inverted triangle section (base uppermost) and the gap between its elements is 150 nm, which is smaller than the minimum particle size of the enzyme granule (250 nm). Oil enters at the top of reactor 1, exits at the base and then travels to reactor 2, before finally exiting the system at the base of reactor 4. In operation, the desired control parameter (e.g. solid fat content (SFC) or melting/dropping point) is monitored at the exit of each reactor. When the value no longer changes at the exit of reactor 1, or if reactor 4 does not show full conversion, the enzyme in reactor 1 must be replaced. Oil flow to the first reactor is stopped and instead routed directly to reactor 2 and the process of interesterification continues. The spent enzyme is removed from reactor 1 by a vacuum system and is discharged together with spent bleaching earth or other solid wastes. Fresh enzyme is added to reactor 1, which is then reconnected to the oil flow, but now as the last reactor in the series, so that the process keeps working in a countercurrent fashion. As interesterification proceeds, activity in the old reactor 2, which is now number 1, will likewise become insufficient, and so this too will have to be replaced and the process of removal and replacement of the enzyme will continue. In normal operation, one reactor will require replacing approximately every 21 days, and the complete chain will cycle back in approximately 3 months.

The optimum temperature for conversion and long enzyme working life is 70 °C. This is maintained by adjusting the incoming oil to this temperature and by having jacketing and tracing of the reactors and associated pipework. While the process is designed to be continuous, there will be occasions when the process will need to be halted. This might be due to a lack of raw materials, insufficient downstream storage capacity or other problems. In these cases, it is critical to maintain the heating on the reactors in order to avoid oil solidification. Under practical conditions, even a shutdown of 1 or 2 weeks will not produce a significant decrease in enzyme activity.

7.3.2 Factors influencing enzyme working life

A key factor in the overall economy of an application utilising immobilised enzymes is that the productivity (kg oil converted/kg enzyme) should be as

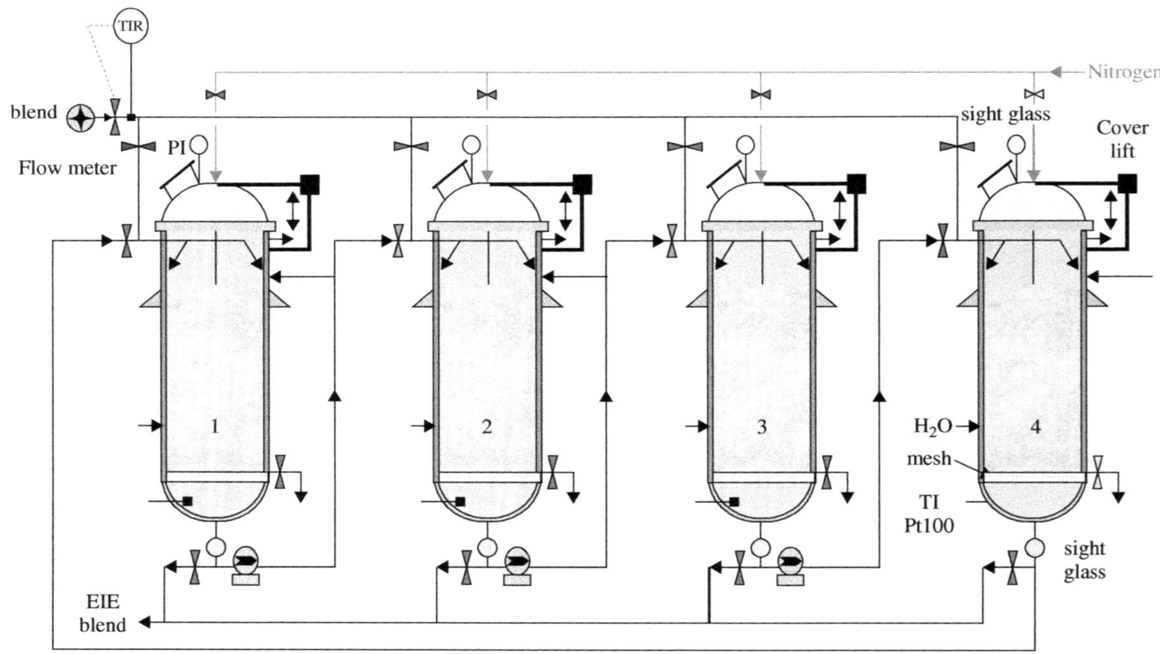

Figure 7.10 Typical packed-bed reactor configuration and connections for enzymatic interesterification.

high as possible. Enzymes share with chemical catalysts the ability to be recovered from the reaction media and reused. However, in some situations, such as degumming, the cost of the recovery is higher than the value of the recovered enzyme and so this is not carried out. In enzymatic interesterification, the activity within the enzyme product is gradually lost. Understanding the mechanisms behind this is critical to the process control.

For an immobilised enzyme product, an apparent loss of activity can occur for three main reasons:

(1) Enzyme protein is washed off the carrier as oil passes down the column.
(2) Particulates and/or soaps, phosphatides and so on coat the particles, preventing the oil from reacting with the enzyme.
(3) Contaminants or components within the oil directly react with the enzyme protein, resulting in modification of the active site and loss of enzyme activity.

In addition, operating outside of the recommended temperature range can also result in an activity loss. Above the recommended operation temperature, thermal denaturation of the enzyme protein will occur. And although reaction rates normally increase with temperature, so does thermal inactivation, and this second effect will eventually result in a decrease in conversion.

Laboratory studies have demonstrated that in continuous fixed-bed operation there is no loss of enzyme protein from the enzyme itself (Cowan, 2010b). The protein level in a fixed-bed operation was monitored and it remained relatively constant while enzyme activity decreased with time of operation. If activity loss is not due to physical loss of enzyme protein then other factors must be involved. The presence of particulates and/or phosphatides will result in coating of the enzyme articles and prevent access of the oil. This can occur if the upstream processing is not sufficient or if there is contamination from tanks used for other oils or other products. Generally this will result in an increase in the pressure drop across reactor 1 from the normal 0.5–0.8 bar up to 3–4 bar, depending on the scale of the contamination. However, as this enzyme is regarded as being of largely sacrificial function, as most of its activity is exhausted, the simplest solution is to replace the enzyme and 'move' this reactor to the last position in the series. It is always recommended to have a 15 nm polishing filter before and after the reactor chain. Should such pressure drops occur, finding the cause to avoid its reoccurrence is critical.

Two groups of components have been identified as being capable of interacting with the enzyme and causing activity loss. Oxidation compounds, measured as peroxide value (PV), will react with the amino acids in the enzyme protein, particularly lysine. Their resulting oxidation changes the configuration of the active site of the enzyme, resulting in an irreversible loss of enzyme activity. Nitrogen blanketing and a PV < 2.0 will reduce the impact of oxidation on enzyme performance.

The second group is mineral acids and citric acid present in the oil, as a result of either upstream processing or deliberate addition to improve oil stability. Bleaching earths are often used in an acid-activated form as this enhances colour removal and general performance. Sulfuric and phosphoric acid residues are found in these earths and can dissolve in the small amounts of water present in vegetable oils. When these oils are then used in enzymatic interesterification, the water can be absorbed by the enzyme granule and will carry the acidity into the enzyme particle. This acidity then lowers the pH of the bound water in the granule and results in the enzyme not being at the correct pH for optimum activity. This is typically seen as a need to reduce flow rate, as overall enzyme activity is lower. However, because the enzyme is no longer operating at its optimum pH, thermal inactivation occurs at a lower temperature. The end result is that not only is activity reduced but it can also be lost, and overall enzyme productivity is irreversibly reduced.

If both of these parameters are controlled then enzyme productivity will be high. The PV can be controlled by the correct combination of bleaching and deodorisation of the feed stock. Residual acid in oils can be avoided by the use of neutral bleaching earths or by neutralisation of the feed stock by alkali. The critical factor in the use of alkali is to limit the addition rate so that only the strong acids are neutralised and the FFA in the oil are left untouched. If excessive alkali is used, soaps will be formed, and these too will have an impact on performance.

Acidity in oil can be monitored by extracting the oil with a 1 % w/w solution of potassium chloride and then measuring the pH of the resulting water extract. When the pH of this extract is above 5.5, good enzyme performance will be obtained. It should also be noted that both parameters can interact and that if the PV is very high (>6), enzyme activity will be significantly reduced, irrespective of the oil's acid content. The oil quality values required for overall good enzyme productivity are summarised in Table 7.2.

Table 7.2 Oil quality parameters for maximum enzyme productivity.

Type	Identity	Level required
Particle occluding	Soaps	<1 ppm
	Phosphorus	<3 ppm
	Nickel	<0.2 ppm
Oxidation compounds	Peroxide value (PV)	<2 meqO_2/kg
	Anisidine	<5
Mineral acids	Acid extract value	pH 6–9

7.3.3 Formulating with interesterified oils and fats

As has already been described, enzymatic interesterification is used to change fat melting properties. In formulating recipes for products, two broad approaches can be used. If there is an existing product produced by chemical interesterification, the blend ratios used for this will form the starting point for the new formulation. Figure 7.8 demonstrates that the melting curves for products of the same fat ratios but made by CIE versus EIE are similar. The difference is mainly due to the lower level of diglycerides formed during EIE. To obtain an identical melting curve, the proportions of the two components must be modified slightly to (in this case) soften the resulting product. Figure 7.11 illustrates the alteration in melting properties obtained when different proportions of the same fat are interesterified. The y-axis indicates the change in SFC for the interesterified blend compared to the same blend without treatment.

The second approach starts with a specification that requires matching. Batch interesterification trials are carried out using the available raw materials to find combinations that are close to the desired end point. Figure 7.12 shows one example of this approach, where the SFC curves of three blends are compared to that desired. The 30 : 70 blend is closest, but matching will require a second round of tests, focussing on the intervals between 30 : 70 and

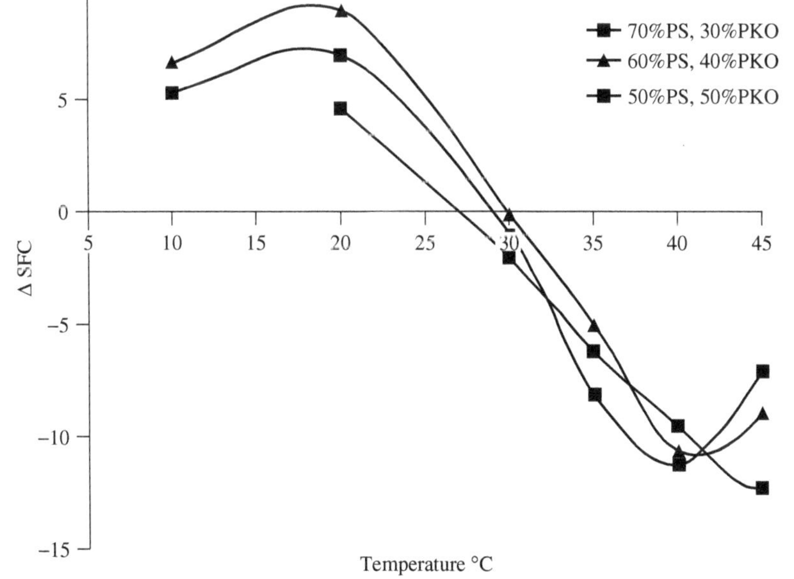

Figure 7.11 Diagram of SFC change of an interesterified versus noninteresterified blend.

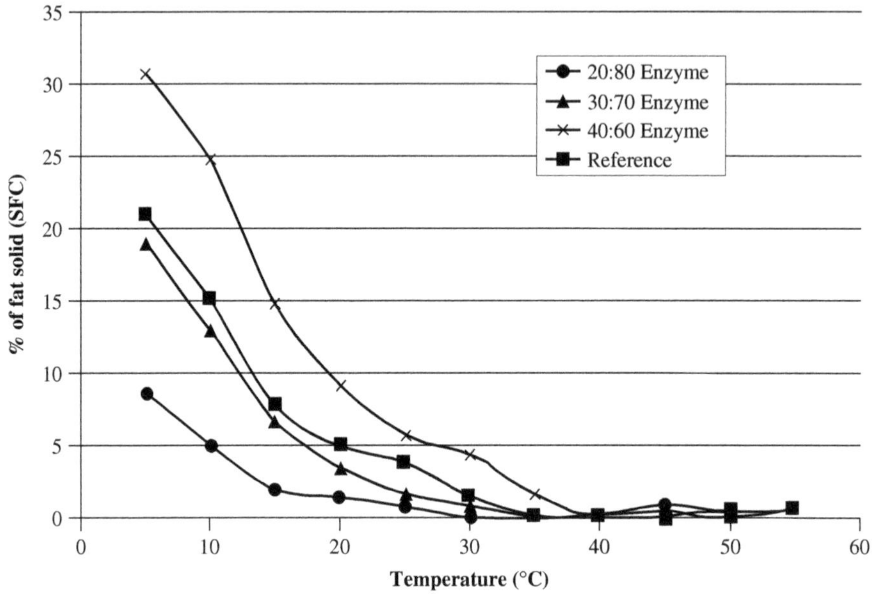

Figure 7.12 Comparison of interesterified blends with desired values.

40:60. Once established, this blend ratio can be used in the plant, and the performance of fat seen in the figure will be obtained.

By combining these approaches, modified fats of high quality are produced by EIE, with none of the issues–such as *trans* fats and byproduct generation–seen in the two chemical modification technologies, leading to a more sustainable process for the production of hard stocks for margarine and other products.

7.3.4 Enzyme reactions for speciality fats

Lipases are also used in the speciality fat sector for the production of analogues to existing fats, such as human milk fat or CBE, or to increase the content of a particular fat in a mixture of fats, such as high omega-3 fatty acid-containing products.

Acidolysis, in which a lipase can be used to increase the content of a particular fatty acid, is used for products such as CBE in which an increase in the amount of a specific fatty acid is required. Due to small batch sizes and the need to have somewhat longer reaction times than are used in the margarine and shortening fats system, these reactions may be carried out in either batch or continuous reactors. An example of a batch reaction for the production of a modified fat is one in which oleic acid is used to replace palmitic acid, as shown in Table 7.3.

Table 7.3 Acidolysis using *sn*-1,3-specific lipase for 48 hours at 60 °C.

Triglyceride type	Palm stearin	After acidolysis with enzyme
PPP	62.7%	1.8%
POP	13.2%	17.2%
POO/OPO	4.4%	37.8%
OOO	0.6%	25.0%

PPP, tripalmitin; POP, 2-oleo-dipalmitin; POO, 1-palmitin-di-olein; OPO, 2-palmitin-do-olein; OOO, triolein.

The process reaches an equilibrium between the level of fatty acid in solution and that incorporated into the triglyceride, so that the end product of the enzyme reaction still requires fractionation to remove some of the unwanted fats, which can then be recycled for a further reaction together with fresh oleic acid. In this application, an *sn*-1.3-specific enzyme is used, which catalyses exchange on the outer part of the fat molecule but largely leaves the fatty acid occupying the *sn*-2 position unchanged.

Larger-scale operations can utilise the same type of reactor design used for continuous interesterification, and the same concerns about oil quality apply. It should be noted that in this application there will be a need for downstream processing to remove the fatty acids liberated, and for some degree of separation to concentrate the desired triglycerides. Typically, short-path distillation and fractionation are employed to remove the fatty acids and concentrate the triglycerides.

7.3.5 Production of fats high in omega-3 fatty acids

Fish oils contain DHA and EPA, both of which are associated with enhanced cognitive function, a reduction in cardiovascular problems and a range of other benefits. These fatty acids cannot be synthesised by mammalian biochemistry and hence need to be consumed as dietary components. The main source is fish oils, derived from oily fish caught predominantly off the west coast of South America and from menhaden in the Atlantic Ocean. Fish caught in Northern European waters tend to contain elevated levels of contaminants (dioxin, heavy metals etc.), so there are recommended limits on their consumption.

For the existing fish oils, the concentration of EPA and DHA varies according to species and the general conditions encountered, but is normally in the region of 20% of the fatty acids in the oil being of the desired type. Consumption of fish oils as a source of EPA and DHA therefore requires approximately a fivefold increase in the amount of oil consumed; for example, 5 g of oil will deliver 1 g of EPA/DHA. Masking of 'fishy' odours and taints in products used as supplements is difficult, and for this reason processors try to separate the desired from the undesired fatty acids, in the process hopefully losing these negative organoleptic taints. The most common process utilises a sodium hydroxide degumming stage followed by chemical esterification to yield the fatty acids as ethyl esters. Short-path (molecular) distillation then serves to make a partial separation, producing a fraction enriched in DHA and EPA esters. Recent investigations have revealed that these esters are not as well digested and incorporated as the same fatty acids as are triglycerides (Cowan, 2010b), and an enzymatic condensation reaction is now used to convert them to this form (Figure 7.13).

The reaction utilises a nonspecific esterase to convert the ester or FFA into a triglyceride by a condensation reaction with glycerol as the acceptor. Typically the reaction is carried out in a stirred batch reactor, under reduced pressure in order to remove the generated water or ethanol (see Figure 7.14). Although either fatty acids or esters can be used, as the distillation process yields ethyl esters, these are the normal raw materials employed in the process.

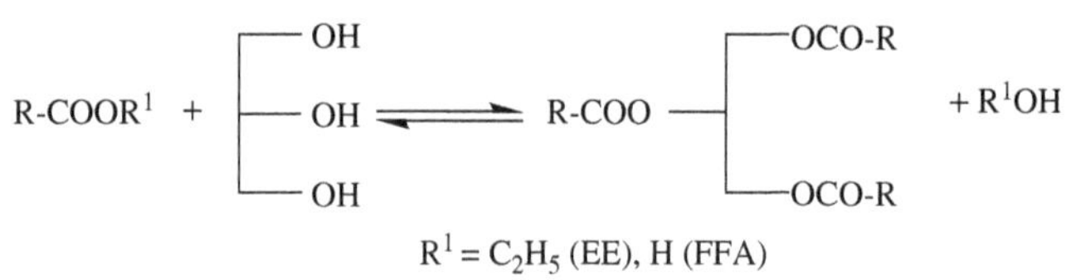

Figure 7.13 Condensation between glycerol and fatty acid or fatty acid ethyl ester to yield triglycerides.

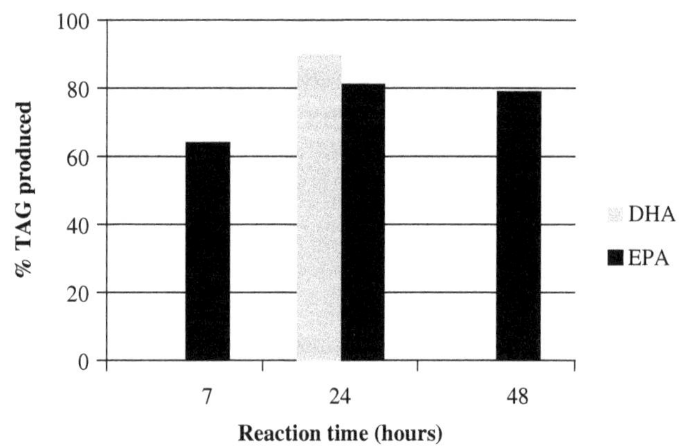

Figure 7.14 Batch synthesis of EPA or DHA TAG (adapted from Kosugi & Azuma, 1994).

Batch reactions do however suffer from some drawbacks. First, their speed is dependent on the rate at which ethanol can be removed from the reaction. Thus a large evaporation surface is required and glycerol may need to be added in small amounts to avoid too much ethanol being liberated, which would poison the enzyme. Second, a stirred reactor will cause more enzyme damage than a packed-bed reactor, and this may limit productivity and/or cause filtration issues.

An alternative approach is to use a packed-bed reactor like that developed for enzymatic interesterification and circulate the reactants through it via an external flash tank to remove the generated ethanol. The eventual decrease in enzyme activity can be compensated for by increasing the number of cycles through the column and by introducing a second column that acts as a finishing reactor. When enzyme activity in reactor 1 is too low, it is refilled and the polishing reactor takes over to provide the main conversion and a 'new' finishing reactor is brought on line.

7.4 Improving processing sustainability through enzyme usage

Edible oil processing uses a large amount of energy and produces byproducts that range from bleaching earth to soapstock to fatty acid distillates. With the current high cost of both raw materials and energy, the opportunities for yield saving through enzymatic processing are considerable. Further, enzymatic processes allow a reduction in environmental impact.

A system of life cycle assessment (LCA) has been used to analyse a number of enzymatic processes that can be applied in edible oil refining and compare their environmental impact with an alternative chemically based process. In this case, LCA has been applied to four processes used within the oils and fats industry (biodiesel, degumming, oil recovery and interesterification) to compare the environmental impact of their conventional and enzymatic alternatives. In all cases, inputs and outputs have been quantified and the potential savings in terms of different environmental indicators have been calculated.

For two of the studies, a simple LCA was carried out; in the others, a full, in-depth analysis was performed. In one of the simple studies, the production of vegetable oil via sodium hydroxide was considered and the influence of oil recovery from the gum fraction examined.

In these studies, a functional unit is chosen to serve as basis of comparison, which in this case was the production of 1 tonne of refined soybean oil. The main environmental factor considered was CO_2 production. The main effect of introducing an enzymatic treatment of the water degumming waste fraction was found to be that less crude soybean oil is needed for the production of 1 tonne of refined soybean oil. As a consequence, less waste for use as animal feed is formed. There is also a slight reduction in the use of processing chemicals for degumming and soapstock treatment, but as including these will favour the enzymatic treatment, they have not been taken into consideration. Overall, the CO_2 production is reduced by 3.4 kg/tonne oil produced, which if 1 million tonnes of oil is processed is approximately equal to the load of 300 000 people.

A more detailed analysis of enzymatic degumming and enzymatic interesterification has also demonstrated how these processes can contribute to reducing environmental impacts. In a full LCA, the environmental impacts of production of the chemicals used are also considered. For example, in enzymatic interesterification the environmental impacts of the production of both the enzyme and the sodium methoxide used as catalysts are analysed and the results are entered into the overall calculation. For the analysis to be accurate, full details of all stages are required, as is a large database of information on the environmental costs of producing energy, chemicals and so on. The efficiency of a process also has a considerable impact on the end result. For example, poor oil quality might result in an increased consumption of enzyme, which increases the environmental impact. However, if this were achieved by reducing the processing costs, this could potentially offset this increased impact. Thus it is critical in all these investigations that a full set of data is collected and used in the analysis.

The three in-depth studies reveal that significant reductions in environmental impact can be achieved by replacing a chemically driven process with one that uses enzyme catalysis (Table 7.4).

Enzymatic processing provides both new and more environmentally friendly routes to high-quality fats and oils, which may explain the rapid growth in the number and scope of its applications since its first mainstream introduction in 2002.

Table 7.4 Reduction in environmental impact from enzymatic processing per 1000 kg unit.

	CO_2 (global warming)	Energy consumption	Acid rain (SO2)	Smog (ethylene)
EIE (European conditions)	−22 kg	−270 MJ	−62 g	−4 g
EIE (US conditions)	−136 kg	−1280 MJ	−960 g	−26 g
Enzymatic degumming	−44 kg	−400 MJ	−527 g	−18 g

EIE, enzymatic interesterification.

References

Bockisch, M. (1998) Extraction of vegetable oil. In: M. Bockisch (ed.) *Fats and Oils Handbook*. Champaign, IL, USA: AOCS Press, 345–445.

Christensen, M.W., Andersen, L., Husum, T.L., Kirk, O. (2003) Industrial lipase immobilization. *European Journal of Lipid Science and Technology*, **105**, 318–321.

Cowan, W.D. (2010a) Application of second generation enzyme technologies in oil processing. New Technologies in Oilseed Extraction and Edible Oil Refining Short Course. Available from http://www.aocs.org/archives/am2010/index.cfm?page=101st-AOCS-Annual-Meeting-Program-Short-Course-new-technologies-oilseed-extraction, last accessed 8 January 2013.

Cowan, W.D. (2010b) Concentration of omega-3 fatty acids using enzymes. *AgroFOOD Industry Hi-tech*, **21**(4).

Cowan, W.D. & Holm, H.C. (2008) The evolution of enzymatic interesterification in the oils and fats industry. *European Journal of Lipid Science*, **110**, 679–691.

Cowan, W.D. & Horsholm, N. (2009) Improved oil recovery in an adapted enzymatic degumming process. *Proceedings of the 100th AOCS Conference Orlando*.

Dayton, C., Rosswurm, E.M. & Galhardo, F.D.S (2008) Enzymatic degumming utilizing a mixture of PLA and PLC phospholipiases. Patent no. WO/2008/094847 A1.

De Moura, J.M.L.N, de Almeida, N.M. & Johnson, A. (2009) Scale-up of enzyme-assisted aqueous extraction processing of soybeans. *J. Am. Oil Chem. Soc.*, **86**, 809–815.

Dijkstra, A.J. (2010) Enzymatic degumming. *Eur. J. Lipid Sci. Technol.*, **112**, 1178–1189.

Eigtved, P. (1992) Enzymes and lipid modification. In: F.B. Padley (ed.) *Advances in Applied Lipid Research*, Volume **1**. London, UK: JAI Press, pp. 1–64.

Gramatikova, S., Hazlewood, G., Lam, D. & Barton, N.R. (2005) Phosholipases, nucleic acids encoding them and methods for making and using them. US patent 2005108789 (A1).

Hayes, D.G. (2004) Enzyme-catalyzed modification of oilseed materials to produce eco-friendly products. *J. Am. Oil Chem. Soc.*, **81**, 1077–1103.

Kosugi, Y. & Azuma, N. (1994) Synthesis of triacylglycerol from polyunsaturated fatty acid by immobilized lipase. *J. Am. Oil Chem. Soc.*, **71**(12), 1397–1403.

List, G.R. & Mounts, T.L. (1993) Origin of the non-hydratable soybean phosphatides–whole beans or extraction. *J. Am. Oil Chem. Soc.*, **70**, 639–641.

Macrae, A.R. (1985) Microbial lipases as catalysts for the interesterification of oils and fats. In C. Ratledge, P. Dawson & J. Rattray (eds) *Biotechnology for the Oils and Fats Industry*. Champaign, IL, USA: AOCS Press, pp. 189–198.

O'Brien, R.D., Farr, W.E. & Wan, P.J. (2000) *Introduction to Fats and Oil Technology*, 2nd edn. Champaign, IL, USA: AOCS Press.

Soe, J.B. (2006) Enzymatic oil degumming method. Patent no. WO 2006/008508 A1.

8 Application of Edible Oils

Arjen Bot[1] and Eckhard Flöter[2]

[1]*Unilever R&D Vlaardingen, Vlaardingen, The Netherlands*
[2]*Technical University Berlin, Berlin, Germany*

8.1 Introduction

Fats and oils play a major role in our diet and food preparation habits. Their properties are manifold and, depending on your point of view, can be regarded as either a blessing or a curse. Fats and oils have the highest energy density of any macronutrient. They typically deliver 9 kcal/g of energy, compared approximately 4 kcal/g for proteins and carbohydrates. This high energy density easily stigmatises fats and oils as a key contributor to the obesity problem; on the other hand, with an ever growing global population, they might be the cornerstone of our future food needs. In the past several decades, fats and oils have been in the spotlight of consumer attention over their nutritional properties. Besides their energy density, the issues relate mainly to their fatty acid profiles. *Trans* fatty acids and subsequently saturated fatty acids have been intensively negatively discussed regarding their nutritional contribution (e.g. see Micha & Mozaffarian, 2010; WHO, 2003), while the *n*-3 fatty acid family, and in particular eicosapentanoic acid (EPA, 20 : 5) and docosahexanoic acid (DHA, 22 : 6), have received a lot of positive attention (e.g. Zevenbergen *et al.*, 2009). Lately, the sourcing of fats and oils in terms of sustainability and natural occurrence have been considered, though with less intensity than the nutritional aspects.

Against this background, the application of fats and oils in food products has been subject to great change. In the mid 1990s, emerging evidence on the effects of *trans* fatty acids in nutrition (e.g. Judd *et al.*, 1994; Mensink & Katan, 1990; Willet *et al.*, 1993) caused the practical ban of partially hydrogenated fats in many parts of the world, either by legislation (Denmark, USA) or

Edible Oil Processing, Second Edition. Edited by Wolf Hamm, Richard J. Hamilton and Gijs Calliauw.

through public opinion and consumer awareness. This eliminated a major tool for the development of fat-based products. Consequently, a great wave of reformulation swept in and many applications were reviewed concerning the actual role fats and oils play in their respective products (Korver & Katan, 2006). For obvious reasons, the drive for this reformulation effort, and thus the speed of implementation, depends on the importance of fats and oils for the functionality of a particular product and their inclusion level. With this in mind, this chapter discusses the characteristics of successful fat compositions for various applications. The ongoing effort to formulate products that are considered more beneficial for health with respect to their fat and oil content and composition makes the challenge of designing successful products for 'fat technologists' even greater. Delivering the desired organoleptic sensation with reduced levels of fat and reduced levels of saturated fatty acids within an oil phase is as much a challenge as implementing the desired increased levels of highly unsaturated fatty acids – EPA and DHA – which are prone to oxidation and hence off-flavour development.

The designed functionality of fats and oils in food products initially splits into two categories: nutritional and physicochemical. Beyond those attributes already mentioned, the nutritional quality of fats and oils will not be the subject of this chapter. Liquid oils primarily account for the delivery of essential fatty acids, according to their respective fatty acid compositions. The various liquid oils also deliver a certain amount of nutritionally less desirable saturated fatty acids and other unsaturated fatty acids. Finally, the presence of a fatty phase is necessary for the successful delivery of the fat-soluble vitamins A, D, E and K.

The physicochemical functionality of fats and oils in products can arbitrarily be divided into three categories: transport, structuring and organoleptic properties. In applications like shallow or deep frying, the liquid lipid phase primarily serves as a heat transfer medium. Since chemical stability and flavour delivery are dominant parameters for these applications, the fatty acid composition of the lipid phase is a primary design parameter. High-stability oils are typically characterised by high levels of oleic acid and low levels of highly unsaturated fatty acids. The ability of lipid films and phases to limit the transport of water is used for both preservation purposes and as a simple humidity barrier.

The two other functionalities of fats in products, namely structuring and organoleptic properties, are not completely independent, as both relate to structure. Fat crystals are, in many fat-based products, a major contributor to the product's macroscopic structure. The fat phase may function as a sort of glue to bind numerous solid particles into a solid mass, as for example in bouillon cubes. In fat-continuous shortenings and spreads, the fat crystal network supplies the soft-solid nature to the product and prevents oil leakage as the porous structure, analogous to a sponge, binds liquid oil. However, in water-continuous emulsions and aerated products, such as whipping creams, the macroscopic properties of the product structure might also be influenced by the physical state of the lipid phase inside the dispersed droplets. The ability of fat crystals to stabilise fat-continuous emulsions via the so-called Pickering stabilisation (Pickering, 1907) deserves explicit mention here. In spreads, the physical properties of the fat crystals that form a kind of shell around the oil–water interface of the droplets – the Pickering stabilisation – strongly influence the characteristics of the emulsion (Walstra, 2001). Features affected are temperature stability, initial and long-term droplet size and the coalescence/inversion kinetics of the emulsions during storage and use.

Finally, the presence of the lipid phases and their physical properties can have a profound effect on the organoleptic properties of a product. On one hand, the lubrication effect of fat during mastication and swallowing is important to the perceived quality of food products (de Bruijne *et al.*, 1997; de Hoog *et al.*, 2006; Lillford, 2000; Prinz & Lucas, 1997). On the other, the transition of fat from the solid to the liquid state is accompanied by a melting sensation. In emulsion products the characteristic dissolution of the fat crystals is typically coupled with emulsion breakup and thus flavour release.

8.2 Physical chemistry of triacylglycerides

For the purposes of this chapter, the physical chemistry of fats and oils is reduced practically to the solidification behaviour of triacylglycerides (TAGs) and the aggregation of crystals in a network structure. This follows from the previously described basic functionality of fats and oils in food products. To understand the solidification behaviour of fats and oils, and in particular the process of designing functional fat phases, it is necessary to have a look at individual TAGs and their physical properties. Depending on the point of view of the technologist, one should account for 10–15 different fatty acids as constituents of the TAGs in vegetable fats and oils. This number leads, under the assumption of random arrangement, to more than 1000 different TAG species. It is obvious that it is beyond the scope of the food technologist to take all these species into account. However, it remains important to understand how the key physical properties – melting point and heat of fusion – change with the triplet of fatty acids that constitutes a specific TAG. The systematic of increasing melting point and heat of fusion for fully saturated TAGs, as depicted for example in the *Lipid Handbook* (Foubert *et al.*, 2007), is quite intuitive. A more comprehensive overview, taking other fatty acid configurations into account, is given in *Crystallization and Polymorphism of Fats and Fatty Acids*, edited by Garti & Sato (1988). Some correlations for these properties have been suggested by Wesdorp *et al.* (2004) and

Zéberg-Mikkelsen & Stenby (1999). For practical reasons, the description of fat compositions is often simplified. Saturated fatty acids are classified into three length categories: short-chain (Sh) (caprylic (8:0) and capric (10:0) fatty acid), medium-chain (M) (lauric (12:0) and myristic (14:0) fatty acid) and long-chain (H) (palmitic (16:0), stearic (18:0), arachidic (20:0) and behenic (22:0) fatty acid). With regards to the physical properties, it appears to be sufficient to distinguish further between *trans*-containing fatty acids (E) (mainly elaidic acid (18:1 t)), monounsaturated fatty acids (O) (mainly oleic acid (18:1)) and polyunsaturated fatty acids (U) (linoleic (18:2), linolenic (18:3) and EPA (20:5) plus DHA (22:6)). Here the code in brackets indicates first the number of carbon atoms in the aliphatic chain and second the number of unsaturated bonds. Combination of these fatty acid classes into TAG classes results in a much less overwhelming compositional space. For the purposes of describing the solidification and hence structural behaviour of fats, only a short list of most relevant TAG classes remains. In detail, when omitting *trans* fatty acids, these are first combinations of saturated fatty acids (HHH, HMH, HHM, HMM, MHM, MMM), second combinations of saturated and monounsaturated fatty acids (HOH, HHO, HOO) and finally combinations of saturated and polyunsaturated fatty acids (HUH, HHU). In broad terms, the sequence of structure-relevant melting points shows that HHH-TAGs melt at temperatures above 58 °C, combinations of two H and one M in the range around 45 °C and combinations of two H and one O in the range around 40 °C. With the exception of *trans* fatty acid (E)-containing TAGs, the remaining classes melt at lower temperatures, which implicitly makes them less interesting for structuring purposes, because of their high solubility. The *trans* fatty acid (E)-containing TAGs, clustered in TAG classes such as HHE and HEE, are characterised by a melting point around 50 °C. Compared to the fully saturated long-chain TAGs (HHH), this makes them particularly suited for structuring, because they combine good structuring behaviour with good organoleptic properties, due to their intermediate melting point.

Fat mixtures are mostly characterised by their solid fat content (SFC) over a temperature line, which is also referred to as an N-line. This is routinely measured using nuclear magnetic resonance (NMR) techniques (Gribnau, 1992). A few typical N-lines and their contributing TAG classes are depicted in Figure 8.1. The relation of TAG species to a temperature range over which they contribute to the SFC is not straightforward, as this contribution is governed by the solubility of the TAG in the liquid lipid phase. The solubility of a TAG in the liquid lipid phase depends on the physical properties of the pure TAG, the molecular interactions in the solid and the liquid phase, and the temperature. Applying simple mass balances and ignoring co-crystallisation and mixing effects, the SFC contribution of TAGs with melting points higher than the system temperature can be estimated as the

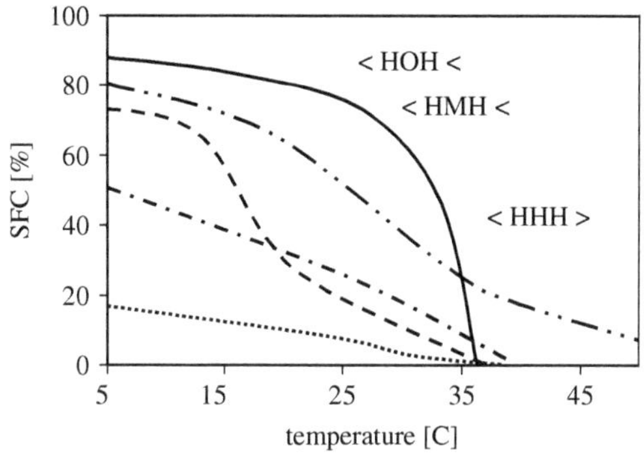

Figure 8.1 Solid fat content (SFC) over temperature line for various fats. Full line, cocoa butter; dashed double dotted line, palm stearin; dashed dotted line, wrapper fat blend; dashed line, milk fat; dotted line, tub-spread fat blend.

concentration of the respective TAG in the overall mixture reduced by the solubility of that TAG times the fraction of the liquid phase. Looking beyond the simple characteristics of fat compositions by SFC, one has to realise that among the crystallisation processes in foods, fat crystallisation assumes a special position (Hartel, 2001; Garti & Sato, 1988). At first the crystallisation process involves a wide range of TAG species and can result in the formation of multiple mixed crystals. The solid phase behaviour of TAGs and thus of fats is characterised by monotropic polymorphism (Garti & Sato, 1988, 2001; Hartel, 2001; Marangoni & Narine, 2002). 'Polymorphism' describes the ability of a material to assume different crystal structures. In the case of monotropic polymorphism, only one of the possible structures is considered stable (Hagemann, 1988; Sato, 1999; Turner, 1971). Many polymorphs or subpolymorphs have been identified for TAG mixtures (e.g. Sato, 2001a; Sato *et al.*, 1999). However, the three structures mainly referred to are the so-called α state, a relatively loose crystal packing with a hexagonal crystal structure, the orthorhombic β' state and the stable tricline crystal structure, β (e.g. Larsson, 1966; Sato, 2001b; van de Streek, 2001). The appearance of the metastable α and β' states is in accordance with the 'rule of stages' formulated by Ostwald (1897), which states in short that 'it is easier to convert to an energetically similar state than to the energetically most favourable'. There is, however, an energy gain through increased packing density related to the transition from the less stable to the more stable polymorph which drives these transitions to proceed in time ranges varying from seconds to weeks. The kinetics of the polymorphic transition depends on the composition of the crystal, its structural quality and the circumstances in terms of temperature, pressure, shear and available solvent. The basic rules for nucleation and growth certainly do apply to fat crystallisation. However, depending on the type of product under consideration, the complication introduced by the polymorphic behaviour of fats must be accounted for in the process design. Subtle differences in the applied cooling profiles, such as whether undercooling is applied to a metastable polymorph or not, can significantly change the properties of the final product.

8.3 Fat crystal networks

The macroscopic properties of a fat phase, such as consistency, are defined by the fat crystal network present. Kloek (1998) has shown that a space-filling network within a lipid phase can be achieved for SFC levels as low as 1% w/w. Above this threshold, when establishing a space-filling network, the structure of the pure lipid phase depends on the number of crystals, their size and shape (morphology) and the strength of the interactions between them. Practically all aforementioned properties are a function of the TAG mixture composition, processing, temperature history and consequently also polymorphic form. Obviously, crystal morphologies with a high surface-to-volume ratio are better able to obtain effective networks: platelets and needles are more effective than cubes or dense spherulites. The interaction between the crystals is traditionally described as consisting of either primary or secondary bonds. Here, 'primary bonds' mean solid bridges between adjacent crystals, which are formed due to stepwise crystallisation (Haighton, 1965; Johansson, 1995; Johansson & Bergenståhl, 1995b; Johansson *et al.*, 1995b). These result in a brittle, nonplastic structure. On deformation, this structure is significantly softened and will only reappear, if at all, over long periods of time (Johansson & Bergenståhl, 1995a; Johansson *et al.*, 1995a). For some product applications, such as spreads, such brittle structures are not acceptable. The much weaker link between the crystals due to London or van der Waals forces is called a secondary bond. A structure based on secondary bonds reveals a plastic behaviour and its hardness is practically maintained after deformation (Haighton, 1965; Kloek, 1998). The destruction of a network based on primary bonds by shear results in a reduction in hardness, and the remaining product structure is based on secondary bonds. To document the significant effort spent in recent years to better understand the nature of fat crystal networks is beyond the scope of this chapter; the book *Fat Crystal Networks* (Marangoni, 2004) and a recent review by Marangoni and coworkers (2012) give a comprehensive overview of the field.

Due to the fact that the process of polymorphic transitions is sometimes slow, manufacturing processes are not necessarily designed to produce fat structures in their final equilibrium state. Consequently, packed products are often characterised by metastable states. Moreover, for some applications the metastable state is produced deliberately, since it is the desired product state, reflecting the best product quality; for chocolate, for example, this is the metastable polymorphic form V (β polymorph). On the other hand, it would be convenient if products represented an equilibrium state, since the risk for changes over shelf life would be small. There are two major processes relating to the fat structure that cause deterioration of product quality: recrystallisation and Ostwald ripening. Ostwald ripening (Ostwald, 1897) simply describes the growth of larger domains at the expense of small domains due to the energetically favourable surface-to-volume ratio. In general, this process is slow and cannot be controlled. However, a narrow particle size distribution reduces the driving force for particle coarsening, while temperature fluctuations speed up the process. Recrystallisation phenomena, in contrast, are a function of the composition of the solid fat phase and can be influenced by processing. In particular, compositions rich in palm oil, containing TAGs with the fatty acid configuration palmitic-oleic-palmitic (POP), are known for their tendency to develop large POP-rich spherulitic aggregates (Tanaka *et al.*, 2009; Watanabe *et al.*, 1992). This phenomenon is referred to as POP or tropical graininess. The growth of these aggregates, of up to 3 mm in diameter, is stimulated by temperature fluctuations, which allow for mobility of the POP TAGs within the oil phase and thus separation from the original metastable solid fat matrix.

8.4 Design of functional TAG compositions

In the design of functional fat compositions, the previously mentioned functionalities must be satisfied simultaneously. For nutritional purposes, it is often desired to maximise the level of liquid oils. For structural purposes, as a first approximation a certain N-line profile has to be met (Bot *et al.*, 2003). From the previous section it should be obvious that just meeting an N-line specification will not necessarily result in a successful product application. Designing compositions with the same N-line based on either *trans* fatty acid (E)-containing TAGs or TAGs of the HOH type will deliver very different crystallisation behaviours and hence different products. It is possible in general to deliver equal or at least similar product functionalities with quite different TAG mixtures, however. To meet the product requirements within the same or a similar window of the manufacturing process is in such a case quite difficult, because the different TAG mixtures show different crystallisation behaviours. It is much easier to ensure the functionality of a new fat formulation if one can start from a proven solution with a defined TAG profile. Alternatively, expert knowledge on both the contribution of different TAGs to the N-line and the compositional effects on miscibility and crystallisation kinetics allows for *a priori* TAG mixture design. Once target

TAG mixtures can be formulated, it is rather straightforward to identify different routes, combinations of raw materials and oil processing steps in order to manufacture these TAG profiles. In the following section this process will be outlined.

Since it is not very productive to characterise a desired fat mixture at the level of individual TAGs, a classification as previously described is more suitable. However, depending on the product application and, for example, the apparent risk of POP graininess, it might be necessary to maintain a more detailed characterisation scheme. TAG mixtures can be expressed in levels of HHH, HHM plus HMH, HOH plus HHO, and POP. The modifications available to the fat technologist are described in detail in Chapters 6 and 7 and will only be discussed here in terms of their ability to manipulate the TAG composition. Beyond the application of mixing, chemical interesterification (IEC), enzymatic interesterification (IEE) and wet and dry fractionation, the choice of different starting materials must also be considered as a tool. Even though full hydrogenation should technically still be considered part of the toolbox, consumers are most likely not capable of distinguishing between full and partial hydrogenation. This means that products containing fully hydrogenated fats are at risk of being falsely related to *trans* fatty acid. Depending on the local legislation with respect to hydrogenation labelling, this risk of discrediting full hydrogenation may be either latent or real.

Of the processes mentioned, mixing or blending is by far the simplest, as it merely delivers a linear combination of components. However, in order to manipulate the crystallisation behaviour of fat compositions, it should be noted here that significant changes in crystallisation behaviour and consequently product properties are sometimes induced by small additions to a fat composition.

In simple terms, fractionation is the formation of a lipid suspension by specific cooling procedures, with the subsequent separation of one part of the liquid phase, the so-called olein, from the mother suspension. In contrast to a lot of fat applications in products where small crystals are desired, fractionation aims at large crystals with a low oil binding capacity. The crystallisation process can be influenced by the choice of adequate cooling profiles. However, the characteristic crystallisation behaviour of the starting material with respect to its tendency to form small crystals is difficult to overcome. With a lesser oil binding capacity of the solid material, better separation efficiencies can be obtained. This means that the level of solid material in the final suspension phase, the so-called stearin, is higher. Obviously, more advanced processes or higher filtration pressures improve the separation efficiencies. The best separation efficiencies can be obtained by either washing of the filter cake or execution of the fractionation process as solvent-supported fractionation. In the wet fractionation process, the lipid phase is diluted with an organic solvent. In this process, crystallisation, for solubility reasons, takes place at lower temperatures. Wet fractionation is obviously more costly due to the larger volumes and lower temperatures involved and the necessity of regenerating the solvent. It is, however, beneficial with respect to the quality of the separation, since the lipid liquid residue level in the stearin can be very low compared to the typical value of around 50% when using a vacuum filter in dry fractionation.

Dry fractionation is extensively applied to palm oil, with several standard fractions available from all major suppliers. The dry fractionation of palm oil is described elsewhere (e.g. chapter 6; Deffense, 1985, Gibon 2006, Kellens *et al.* 2007). Beside the standard commodities of palm olein and palm stearin, there are three speciality directions. When subjecting the olein to further fractionation, one gets a multiple olein or super olein, with a small amount of TAGs, which tend to crystallise at ambient temperatures. Most TAGs with more than one palmitic acid have been separated out of this stream. This super olein is practically a liquid oil, with a high concentration of oleic acid and palmitic acid. The stearin of the first olein is referred to as palm mid fraction and is an intermediate for the production of cocoa butter equivalents (CBEs) and hence needs to be highly concentrated in HOH TAGs. In contrast to this, the further fractionation of the first stearin yields a multiple fractionated stearin with levels of saturated fatty acids of around 90%. The olein from this fractionation (soft stearin) is also a palm mid fraction. However, due to its molecular composition, this fraction does not have a specific application. The fractionation tree of palm oil is shown in Figure 8.2. The different fractions are indicated by a typical value for their yield compared to the starting amount of palm oil and by their iodine value (IV).

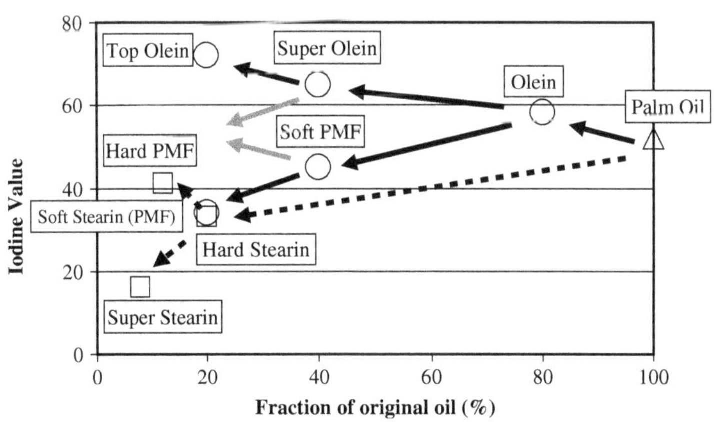

Figure 8.2 Palm oil fractionation 'tree'. Iodine value (IV) over mass fraction of starting material. Circles, fractions out of the first olein; squares, fractions out of the first stearin; grey arrows indicate fractions that are recirculated.

Wet (solvent) fractionation, in contrast, is exclusively applied for high-value applications such as the production of CBEs. Here, wet fractionation is used to concentrate the HOH (with H predominantly stearic acid) TAGs of, for example, shea butter from approximately 30 to 90% in one processing step. This is documented mainly in the patent literature.

Apart from blending and fractionation, other oil modification techniques change the composition of an oil phase by actually changing the molecules of the starting material. Full hydrogenation simply converts all unsaturated fatty acids into saturated fatty acids. Depending on the configuration of the fatty acids in the starting material, only TAGs constituted from medium-chain saturated fatty acids (M) and/or long-chain saturated fatty acids (H) are obtained. The hydrogenation of lauric fats (coconut or palm kernel fat) predominantly yields TAGs with high levels of M and thus a relatively low melting point. Slip melting points can vary between 31 and 41 °C, depending on the starting material, which might also be a fraction of a lauric fat, with consequently varying H/M fatty acid ratios in the hydrogenated fat. Hydrogenated liquid oils and palm oil or its fractions yield straight concentrations of HHH TAGs after full hydrogenation, with slip melting points ranging from 55 to 70 °C. It should be noted that the crystallisation behaviour of HHH-rich compositions is influenced by the nature of the long-chain saturated fatty acids, whether all H fatty acids are identical or whether there is a mixture of different H fatty acids. An exception to the trend that HHH TAGs tend to form relatively large crystals is given by fully hydrogenated high erucic rapeseed oil or fish oil. Because of the disparity of the saturated fatty acid chains, varying from 18 to 22 carbons length in fully hydrogenated high erucic rapeseed oil, this fat crystallises similarly to HHM TAGs in small crystals but has a high slip melting point of 72 °C. This makes it particularly suited for effective high-temperature structuring.

Partial hydrogenation offers the ability to change physical properties gradually through control of the degree of hydrogenation. It has however fallen out of favour because of the accumulating evidence of the negative health implications of *trans* fatty acids. Consequently, this process, which deliberately produces *trans* fatty acids, is not discussed here.

Interesterification offers practically the only alternative means of systematically changing the physical properties of a fat composition. The actual technical process of interesterification is described in many textbooks (e.g. Bockisch, 1998) and in Chapter 6 of this book and will not be discussed here. However, the effect chemically catalysed interesterification has on the TAG composition and hence the physical properties is profound. In simple terms, the process redistributes the fatty acids present in the starting mixture randomly over all triglycerides present in the mixture. Due to this, the TAG profile resulting from a starting mixture is dependent only on the fatty acid composition of the feedstock, and is hence independent of the particular raw

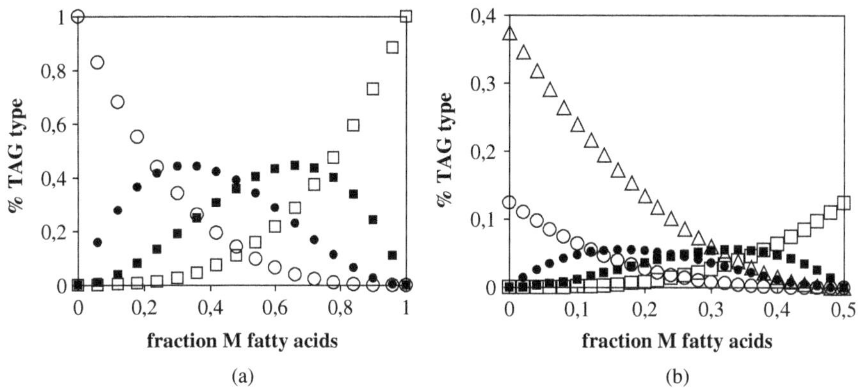

Figure 8.3 Fraction of TAGs in chemically interesterified fat as a function of medium-chain fatty acids in the fatty acid feed mix. Full circles, H_2M TAGs; full squares, HM_2 TAGs; empty circles, HHH TAGs; empty squares, MMM TAGs; empty triangles, H_2O TAGs. (a) Only H and M fatty acids in feedstock. (b) H and M fatty acids and 50% oleic acid in feedstock.

materials constituting the fatty acid mixture. Figure 8.3 shows the translation of the fatty acid composition into the composition of the relevant TAG classes after IEC. This translation can be calculated by means of simple statistics. Figure 8.3a illustrates for mixtures of solely long- and medium-chain saturated fatty acids, H and M, the relationship between the HHH, H_2M, HM_2 and MMM TAG concentrations. In essence, Figure 8.3b reveals the same relationships, but here the effect of the presence of 50% monounsaturated fatty acids is illustrated. For reasons of clarity, numerous TAGs with lower melting points are not shown. A comparison of Figures 8.3a and 8.3b reveals that for any given H/M ratio, interesterification with 50% unsaturated fatty acids only yields one-eighth of the fully saturated TAGs. The relative concentration of the different fully saturated TAGs remains unchanged, however. For the application of different interesterified structuring fats with a constant H/M ratio, this defines a very simple rule for inclusion levels in the final fat composition. When delivering the same amount of fully saturated TAGs, the inclusion levels of different structuring fats scale with the inverse ratio of the fractions of saturated fatty acids in the different interesterification mixtures to the power three.

In recent years, substantial capacity to perform interesterification catalysed by enzymes instead of by metal catalysts has been installed worldwide. The enzyme-catalysed process (see also Chapter 7), although much slower than the chemical one, also generates a random distribution of the fatty acids of the feedstock over all triglycerides. Due to the lower reaction rate, it is possible to control the process to a much greater degree than one can the chemical variant. It differs from IEC in that the commonly used *sn*-1,3-specific enzymes primarily target fatty acids moieties at the terminal positions *sn*-1 and *sn*-3 of a triglyceride. The randomisation of the middle position of the TAGs, which certainly occurs in current industrial processes, is a function of process conditions and the choice of enzyme. In any case, it appears that this *sn*-2 randomisation proceeds at a lower rate than, and independently of, the rate of randomisation of the *sn*-1 and 3 positions. Due to this fact, it is practically impossible to determine *a priori* the exact TAG profile of an enzymatically interesterified fat for a given feedstock. However, good estimates can be made by those skilled in the art (e.g. Xu *et al.*, 2006). Furthermore, it should be noted that due to the nature of the enzymatic process, the properties of the interesterified fat are not solely a function of the fatty acid composition of the feedstock, as in the chemical process, but depend strongly on its specific TAG composition. For example, a mixture of half HOH and half MOM triglycerides will yield different IEE products than a mixture of equal parts OOO, MMM and HHH triglycerides, even though the overall fatty acid compositions are identical. After IEC, however, these mixtures are indistinguishable.

Depending on the TAG profile required for a certain application, different combinations of the described techniques for manipulating the TAG profile can be combined. Sources of long-chain saturated fatty acids (H) for interesterification purposes are either palm oil stearins (as a fractionation product) or hydrogenated fats. Blending is applied for most applications. The general problem of fractionation, that the byproduct stream must be utilised in such a way that the economical burden on the target stream does not become prohibitive, makes application of fractionation subsequent to interesterification (for example) rare.

The framework outlined in this section basically allows for identification of different routes by which to fabricate desired triglyceride profiles once they can be defined for the application at hand. However, the crystallisation behaviour of fat mixtures is not solely dependent on the TAGs themselves and is also strongly influenced by other lipids that might be present (either deliberately or accidently) in any fat phase (e.g. Smith *et al.*, 2011).

8.5 Application in fat-continuous emulsions (spreads)

Rather than the narrow definitions of margarine and halvarine, it is nowadays more appropriate to categorise spread products arbitrarily as high-fat spreads (70–82% fat), medium-fat spreads (48–60%), low-fat spreads (35–42%) and very low-fat spreads (<30%). These categories evolved from uses and consumer preferences in different parts of the world. Products that are used for cooking purposes are best formulated with high fat levels. Due to the progress that has been made in the product quality of low-fat spreads, one finds that products predominantly used for spreading on bread typically have intermediate (medium and low) fat levels. Very low-fat spreads, in contrast, aim much more at the delivery of a spread with lower caloric value. Other fat levels can be found as well, and their justification can be manifold, including product quality, raw material mix and consumer perception. Spread products are an emulsion suspension system. In the final product, the fat crystals have two major functions: they have to stabilise the emulsion by Pickering stabilisation (see Section 8.1 and Pickering, 1907) and supply the bulk structure to the product. The relative importance of these two functions changes depending on the packing density of the water droplets.

Spreads production is typically realised in a votator unit. This is essentially a series of scraped-surface heat exchangers (SSHEs; so-called 'A-units') combined with pin-stirrer crystallisers (so-called 'C-units'). There are different ways to install the process setup prior to the first SSHE. One option is to work with a premix. This means that all ingredients are mixed to prepare a coarse emulsion that is precooled in static heat exchangers to temperatures above any crystallisation point prior to its entry into the first SSHE. Alternatively, the so-called proportioning system can mix the lipid and an aqueous phase in the right proportions by pumping action into the cooling system. For products with fat levels above 50%, the two-phase system is usually fat-continuous throughout the process. For significantly lower fat levels, the system typically maintains a water-continuous state until the viscosity of the lipid phase significantly increases due to progressing crystallisation in the lipid phase. Hence one finds for higher fat levels that the heat transfer is from wall to oil, whereas for lower fat levels it is initially from wall to aqueous phase. This difference has some consequences for the heat transfer itself and for the fat crystallisation. The water-continuous processing route is more effective in terms of heat transfer, due to the high specific heat capacity of water. In contrast, the nucleation of crystallisation at a cold wall in fat-continuous processing is easier than the nucleation inside the small fat droplets that are initially present in the water-continuous processing route. The transition from water-continuous to fat-continuous emulsion is controlled through the application of an inverter: a high-speed pin stirrer. Figure 8.4 shows the SFC over temperature development for the different processing line configurations. On cooling a spread composition, the so-called α-point is reached when the first substantial amount, approximately 2% of solid fat, starts to form. This is, as has been noted, sufficient material to form a space-filling network. However, in the high-fat emulsion products the final droplet size tends to be determined at this point. Once temperatures fall below the α-point, the amount of solid fat in the α polymorph further increases. In a traditional A-A-A-C line configuration (Figure 8.4a) the cooling section is followed by the crystalliser. The polymorphic transition must

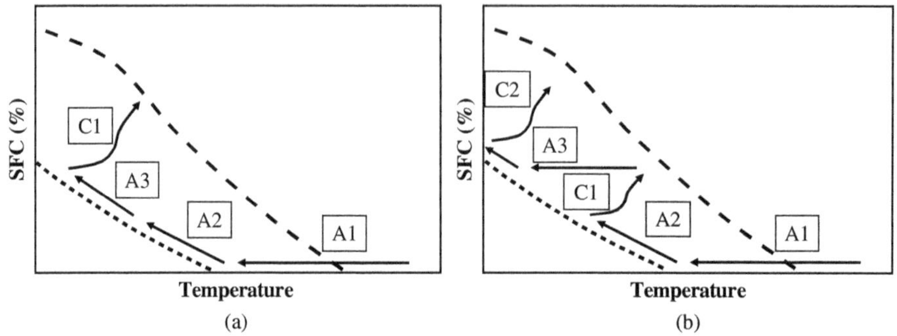

Figure 8.4 Solid-fat-over-temperature diagram. Dotted line, α solids; dashed line, β′ solids; arrows indicate process path within the respective unit. A, scraped-surface heat exchanger (SSHE); C, pin stirrer. (a) Straight through-process, A-A-A-C. (b) Process with an intermediate crystalliser, A-A-C-A-C.

take place in this unit. Depending on the completeness of this transition process within the processing line, this means including a piping and packaging machine, and the structure of the final product will be more or less brittle. Consequently, the speed of the polymorphic transition directly influences the setup of the processing line. Either slower blends need to be produced at reduced manufacturing capacities or lines need to be reconfigured to allow for longer residence times under agitation, in order to prevent the formation of brittle product structures due to excessive crystallisation in the pack. A typical temperature-measurement-over-time curve is S-shaped and allows the characteristic polymorphic transition time to be determined. Since the polymorphic transition is exothermic, the highest increases of temperature indicate high polymorphic transition rates. As an alternative to a straight through-process (A-A-A-C), intermediate crystallisers can be incorporated into the processing line (e.g. A-A-C-A-C) (Figure 8.4b) in order to optimise the product structure or line capacity. However, the first pin stirrer in such a setup only makes sense if prior to its entry a minimum solid level or sufficient supersaturation is achieved. The kinetics of the polymorphic transition is a function of the temperature and composition of the fat. Traditional recipes rich in *trans* fatty acid (E)-containing TAGs have short transition times. Fat compositions rich in triglycerides of the HHM, HMH type also crystallise fairly rapidly. In contrast, TAGs of the HUH type are characterised by their slow polymorphic transition.

Since *trans* fatty acids have practically disappeared from spreads products, interesterification has become the key technology in the production of triglyceride compositions for superior spreads functionality. The desired HHM and HMH TAGs have evolved from interesterifications of palm oil or its fractions with a lauric fat. The feedstock for the interesterification will depend on the specific product requirements. According to Figure 8.3a, a high-performance hardstock will have a higher concentration of saturated fatty acids and hence more functional TAGs. This means that the feedstock has to be more concentrated in saturated fatty acids. This is typically accompanied by higher costs.

8.6 Application in water-continuous emulsions

8.6.1 Mayonnaise and dressings

Mayonnaise and dressings are condiments that can be characterised in physicochemical terms as acidified oil-in-water emulsions. Mayonnaises typically contain 80% vegetable liquid oil, and the emulsion is stabilised with egg yolk. Due to the high amount of emulsified dispersed phase, the volume of the tightly packed oil droplets exceeds the close packing fraction. This proximity of neighbours deforms the droplets. The energy required to achieve a certain deformation depends on the droplet size, as this defines the Laplace pressure of the droplet (e.g. Mason *et al.*, 1995; Princen, 1979). This intense contact of the droplets causes the rather firm texture of these emulsions, which is somewhat surprising considering the liquid nature of the dispersed phase (de Bruijne & Bot, 1999).

Because of this relationship between droplet size and texture, full control of the emulsification process is a must in mayonnaise manufacturing. This is usually achieved by applying a two-step process. An initial coarse pre-emulsion is turned in a second step into a fine emulsion using a colloid mill or similar stator–rotor device. The very high shear rates (10^4–10^5 s^{-1}) in a colloid mill reduce the droplet size very efficiently to values in the range between 1 and 5 μm, depending on the width of the gap between stator and rotor.

With respect to the composition of the lipid phase, there are not too many physicochemical restrictions. The origin of the liquid oil may vary. Typically, common oils such as sunflower, soybean, canola and maize oil are found in commercial formulations, but other oils can be used as well. The primary restriction applicable to the lipid phase is the absence of fat crystals, as these would disturb the product structure prohibitively. Even with the use of only the liquid oils listed here, waxes and small fractions of high-melting triglycerides can promote the coalescence of droplets in the emulsion and destabilise the product. To avoid these problems, oils with a good cold stability (as can be obtained through winterisation, for example; see Chapter 6) are recommended. A key challenge with respect to the application of fats and oils in mayonnaise and dressings is the relatively high sensitivity to oxidation of liquid oil in combination with a long (and usually ambient) product shelf life. In practice, a set of actions is taken to manage product quality throughout shelf life. Careful handling of the oil (low concentration of oxidation precursors, avoidance of high temperatures, reduced exposure to oxygen), addition of antioxidants and optimal packaging barrier properties (e.g. glass) are applied to prevent excessive oxidation.

As with other product categories, the trend is for mayonnaises to move from full-fat to lower-fat variants. In traditional mayonnaise, the high-volume fraction of the dispersed fat phase is key to the characteristic product properties. Consequently, reduction of the oil level is practically impossible without affecting the texture, because the contact between the droplets is lost. To resolve this issue, additional ingredients are added, which replace the removed oil in terms of its displaced volume. The presence of starches – emerging alternatives are polysaccharides and proteins – ensures that the remaining oil droplets in lower-fat products are packed in a similar way to that in full-fat mayonnaise (e.g. de Bruijne & Bot, 1999).

8.6.2 Nondairy (fat) creams and spreads

Most dairy emulsions are water-continuous products based on milk protein and dairy fat. These products include sweet and sour cream, fresh cheese and hard cheese.

Nondairy (fat) creams have typically been developed as application-specific alternatives to dairy creams. The most common application is cooking cream: a neutral cream which should provide taste and cohesiveness to products to which it is added. This may be at high temperatures and low pH. The properties of the oil phase in a cooking cream are not extremely critical, so long as the organoleptic properties (mouthfeel) of the full product remain pleasant. Avoiding off-flavour during storage is key to a successful application. Creams are usually stored chilled, and hence storage conditions should be controlled to limit oxidation – although care in the handling of the oil blends is required too. The fat phase is usually composed of a mixture of a structuring fat and liquid oil. But unlike with spreads, for example, the structuring fat can also be left out without too much deterioration of the final product quality.

The requirements for fat blends in whipping creams are much stricter. Whipping creams are neutral oil-in-water emulsions that are destabilised during the mechanical agitation that is applied during whipping. Here the droplets partially coalesce and form a network at the air–water interface, and thus help to stabilise the air bubbles. The success of this process depends on the delicate balance between emulsifier, protein and fat crystal functionality at the water–oil interface. This emulsion destabilisation process is very critical. On the one hand, this process should not happen in the pack. On the other, a limited amount of whipping – a couple of minutes – should be sufficient to cause destabilisation. Due to these conflicting targets, whipping cream formulations aim to be at the edge of stability. To further aggravate this product development dilemma, nondairy cream formulations tend to be low-fat variants of their full-fat dairy counterparts. Typical fats that allow successful product formulations are tropical fats. Palm oil and coconut-based fats can be used either straight or sometimes blended with liquid oils. Compared to other water-continuous emulsions products, these high-saturated fatty acid products are much less prone to oxidation.

8.6.3 Ice cream

Ice cream is in essence a frozen whipped cream. Depending on local legislation, the fat phase can be either dairy fat or vegetable fat. The main difference between ice cream and cream is the sugar content of the water phase, which is much higher in ice cream. The addition of sugar to the water phase protects the emulsion against destabilisation during freezing, due to the existence of a part of the water phase in which the sugars and polysaccharides concentrate, which consequently does not freeze (e.g. Eisner *et al.*, 2005). Vegetable-fat ice cream blends have similar compositions to vegetable-fat whipping creams. Formulations are typically based on lauric fats or palm oil and its fractions, with slip melting points in the range of 25–34 °C. This allows for a quick melt of the fat in the mouth, although most of the loss of firmness during eating is due to melting of the ice crystals. The type of fat and the (low-temperature) storage conditions ensure that fat oxidation tends to be a less imminent problem than in other products. The need for an absence of off-flavour is however nonnegotiable, because ice cream is eaten for pure enjoyment and high-quality taste is a must.

In contrast to whipping cream, which has to be whippable throughout product shelf life, ice cream is aerated during product manufacture. Hence the manufacturing process delivers a dispersion of air bubbles, ice crystals and fat in a continuous freeze-concentrated solution of sugars, proteins and minerals. Ice cream is typically produced with 100% overrun, indicating that 50% of the product volume is gas. Variation of the overrun, as found in premium ice creams, has a direct influence on the product structure. Reduced overrun yields harder products with improved organoleptic properties, as they tend to melt more rapidly (e.g. Sofjan & Hartel, 2004). A typical composition of ice cream is given in Table 8.1, with approximate volumetric and weight fractions.

Vegetable-fat ice cream is processed in a similar way to a vegetable-fat cream, except that sterilisation of the premix is usually not required and pasteurisation suffices, and that the product is aerated and frozen in, for example, SSHEs. A typical processing sequence is shown in Figure 8.5. The process starts with blending of all ingredients, except particulates that are

Table 8.1 Typical ice cream composition by mass and volume fraction of ingredients and fractions.

Phase or ingredient	Volume fraction (%)	Weight fraction (%)
Air	50	–
Ice crystals	26	–
Unfrozen phase	20	–
Fat	4	8
Water	–	64
Added sugar	–	16
Protein source (SMP)	–	11.4
Emulsifiers and stabilisers	–	0.5
Flavouring	–	0.1

SMP, sweet milk powder.

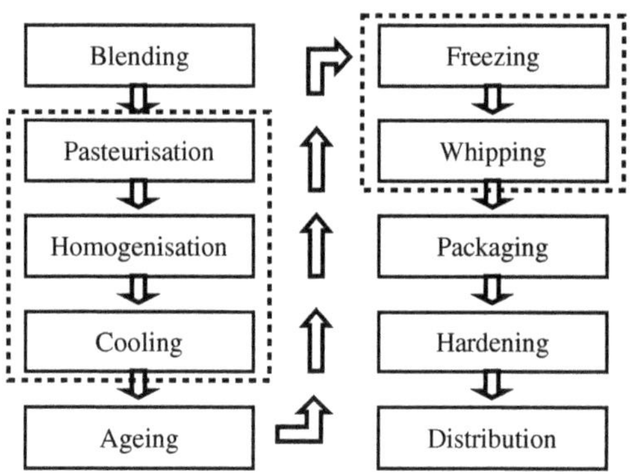

Figure 8.5 Flow scheme of the ice cream manufacturing process. Dotted frames indicate unit operations that are combined in continuous processing.

added later. The steps of pasteurisation, homogenisation and cooling can be integrated in a continuous process, as indicated by the dashed box in the figure. During homogenisation, which takes place in one or two steps, the emulsion is formed. After these steps, the liquid mix, with a temperature of 4 °C, is aged for a period of between 4 and 24 hours. During this time, the viscosity of the mix increases as a macroscopic change. In detail, proteins and stabilisers are hydrated, proteins and emulsifiers rearrange at the interface and fat crystallises in the globules. The crystallisation at this point should however not be complete, but rather in the range of 2/3 SFC, in order to allow for partial coalescence in subsequent processing (Berger, 1997). The aged semifrozen slurry is whipped and frozen under high shear to end up with an approximately 50/50 volume ratio gas/slurry at −5 °C. During this step, partial coalescence of the fat globules and ice formation take place. After the possible addition of flavouring and particulates, the product is packed and subjected to further cooling under quiescent conditions in the hardening tunnel at −30 to −40 °C. Finally, distribution should safeguard the product quality with constant low temperatures of −18 °C.

The domain sizes one typically finds in end products are approximately 1 μm for fat droplets, 10–150 μm for air bubbles and 10–150 μm for ice crystals. During manufacturing, emulsification (beating), aeration and freezing take place simultaneously. The role of the fat phase is primarily to stabilise the air cells after partial coalescence of the fat globules has taken place. Apart from the air bubble stabilisation, the fat also supplies structure to the product and hence influences the melting resistance (e.g. Goff, 2006). Also, organoleptic properties are strongly influenced by the relation of fat to creaminess (Golding & Pelan, 2008). Hence, the key features of a fat that are important to ice cream are similar to those for other applications, namely crystallisation behaviour and crystal structure/size, along with the absence of any flavour disturbance.

8.7 Application in other fat-continuous products

8.7.1 Baking fats

A lot of bakery-related fat products are essentially spreads with a different SFC or N-line, which depends on the product's application. In some cases the structure of the baking fat has a profound effect on the dough structure (e.g. cookies and biscuits), while in other cases the final baked product imparts only limited requirements to the fat (e.g. cakes). In any case, full-fat products – so-called shortenings – are often used for baking purposes (Vaclavik & Christian, 2008). In general, there is no fundamental difference in the rules for the successful formulation of fat compositions between spreads and shortenings. With the previously mentioned move away from partially hydrogenated fats, which were also popular in baking applications, palm oil and its fractions have become the key ingredient in baking fats.

However, even though liquid shortenings have become popular, a lot of applications require fats with relatively high solid fat levels at ambient temperatures. The format of choice for products having elevated ambient solid fat levels is the wrapper, also referred to as a stick or brick. Such products are not supported by a tub or other solid packaging and have to form stable shapes by themselves. This change in fat characteristics is accompanied by a change in the processing setup. The basic processing line, a combination of SSHEs and pin stirrers, remains. However, the high solid level of these products necessitates that post-crystallisation is limited and the polymorphic transition proceeds prior to packaging. On the one hand, the existence of brittle structures at high solid fat levels will lead to nonacceptable hard product structures. On the other, the heat generated by excessive crystallisation after packing can result in critically high temperatures in the stack of wrappers, which negatively influence the product quality. In order to prevent excessive crystallisation after packing and simultaneously ensure the form stability of these products – necessary for packing – so-called resting tubes are placed between the crystalliser and the packaging machine. The detailed design of these resting tubes, their length and the possible presence of sieve plates to avoid channelling all depend on the specific product and its crystallisation behaviour.

Of the various different baking applications, only puff pastry – products based on laminated dough – deserves explicit mention here (Reddy & Jeyaran, 2001). The preparation of laminated dough is basically the repeated folding and rolling of a sheet of dough and a sheet of rolling fat. While creating the typical high number of layers, both the dough layer and the fat layer must stay intact as the fat layer not only separates the different dough layers from each other but also has a fundamental function during the baking process. When the water evaporates from the dough, the water vapour must be trapped between the layers in order to ensure the required puffing. The fat layer must thus function as a vapour barrier. It is easily understood that successful puff pastry applications require fat compositions with high plasticity and adequate melting behaviour.

8.7.2 Chocolate

Chocolate can simply be regarded as a dispersion of ground roasted cocoa seed and sugar in a cocoa butter fat matrix. Vanilla and emulsifiers are also found in the mixture. The expected organoleptic and physical properties of chocolate are well defined in terms of melting behaviour, coolness and creaminess, shiny surface appearance and snap. Cocoa butter has a relatively narrow triglyceride composition, being composed mainly of 2-monounsaturated triglycerides, based on oleic (O), palmitic (P) and stearic(S) fatty acid, such as POP, POS and SOS. As a consequence, the fat has a relatively narrow melting profile, which implies a flat N-line with a steep decent just below the oral temperature. In combination with the lower-than-mouth-temperature melting point of cocoa butter, melting of chocolate is experienced as a 'cooling' sensation. Cocoa butter is quite stable against oxidation, due to its fatty acid composition. Cocoa butter is an important ingredient in chocolate and other confectionary products, as it binds ingredients into a matrix, resulting in a desirable disintegration, melt characteristic. Apart from fat melting, the sensory attributes of chocolate depend strongly on the size and distribution of the solid particles in the product matrix. Particles have to be smaller than 30 μm in order not to be perceived as such and instead to contribute to

Table 8.2 Typical compositions of types of chocolate.

Ingredient	Milk chocolate (w/w)	Dark chocolate (w/w)	White chocolate (w/w)
Cocoa liquor	0.12	0.40	–
Cocoa butter	0.19	0.12	0.23
Milk powder	0.20	–	0.30
Sugar	0.485	0.475	0.465
Lecithin	0.005	0.005	0.005

the creaminess. Either lecithin or polyglycerol polyricinoleate is used as an emulsifier to support the particle distribution and ensure the correct structure.

There are three main types of chocolate: dark (plain), milk and white, all of which have very distinct formulations. In Table 8.2, three typical compositions are given (Timms, 2003). Cocoa liquor is the mass of ground cocoa beans (nibs) and contains approximately equal parts of cocoa solids and cocoa butter. In all three types of chocolate, at least 20% cocoa butter is present. In milk and white chocolate, milk powder substitutes for part (milk chocolate) or all (white chocolate) of the cocoa solids (Lonchampt & Hartel, 2004). However, the cocoa butter fraction given in Table 8.2 is not necessarily 100% cocoa butter. Part of the fat may be replaced by milkfat, CBEs, cocoa butter improvers (CBIs) or cocoa butter substitutes (CBSs). CBEs and CBIs are fat compositions containing TAGs that are present in cocoa butter. These are primarily SOS, POP and, to a much lesser extent, POS, which is less likely to be found in other fats than cocoa or illipe. Typically, fats are rich in either palmitic acid (palm oil) or stearic acid (e.g. kokum, shea, mango seed oil, sal). The level of inclusion of these cheaper fats in chocolate is often regulated. When the designation 'chocolate' is not necessary, fat compositions with similar melting characteristics to cocoa butter can be formulated based on CBEs, CBIs or other fats, typically based on lauric fats (CBSs). If fats other than CBEs or CBIs are mixed with cocoa butter, the mixing behaviour within the solid fat phase requires careful monitoring in order to maintain the desired product characteristics (Smith, 2001; Talbot, 2009).

In line with Angelo Gavazotti's (2007) commentary on crystal polymorphism, in which he cites Walter McCrone's statement 'that the number of polymorphs is proportional to the time and effort in their search', the general approach of considering three polymorphic forms is obviously arbitrary. For cocoa butter it has turned out to be most useful to consider six forms instead. These are simply numbered, and Table 8.3 relates the number to a crystal structure and typical melting point (van Malssen, 1994; Wille & Lutton, 1966). As in the framework outlined earlier, crystallisation of the more stable polymorphs directly from the melt is not common. The polymorphic form V

Table 8.3 Melting temperatures and polymorphic forms of cocoa butter according to Wille & Lutton (1966) and van Malssen (1994).

Polymorph identification	Polymorphic form	Melting point (K) (Wille & Lutton, 1966)	Melting range (K) (van Malssen, 1994)
I	γ	290.3	268–278
II	α	296.3	290–295
III	β'	298.5	293–300
IV	β'	300.5	
V	β	306.8	302–307
VI	β	309.3	

is desirable for high-quality applications of cocoa butter, due to its melting point and usual crystal morphology (e.g., Rousseau, 2007).

In order to achieve the desired product texture, polymorphic form and hence organoleptic properties, several manufacturing process steps – ingredients mixing, refining, conching, tempering and cooling – must be executed. During the refining step, the particle size is further reduced prior to the conching. In the conching step the solid particles of the dispersion are mixed with the liquid fat for an extended period of time, from 6 to 72 hours, in order to ensure coating of the solid particles with fat. This coating is an important parameter for the perceived creaminess of a chocolate (Rousseau, 2007). In the tempering step the liquid mixture is cooled to around 27 °C in order to initiate the nucleation of the metastable polymorphic forms (III, IV and V). Once sufficient solid material is present, the temperature is raised to approximately 31 °C, depending on the actual product formulation, so that the lower-melting metastable polymorphs become unstable and only the polymorphic forms V and VI remain to be supersaturated. After tempering, the liquid mass can be poured into moulds and is set to crystallise under precisely controlled cooling conditions. Once crystallisation has finished, the products can be demoulded easily, due to their contraction.

As with other fat-based products, chocolate does not reach its final state after manufacturing and packing. Since there are many products in which chocolate is in contact with another fat-containing phase, such as filled chocolates and pralines, diffusion processes will occur (Maleky *et al.*, 2012; Ziegleder *et al.*, 1996, 2001). Depending on the nature of the product structure and the driving forces, this can lead to significant changes in the product (Tietz & Hartel, 2000). After high temperatures, cracks and large pores are the main factors that promote oil and fat migration in such products (Norberg, 2007). Another product defect is the formation of fat bloom, a whitish, sometimes artisanal-looking structure on the surface of the chocolate, which looks like mildew and is certainly undesirable for consumers. The occurrence of chocolate bloom over the shelf life is related to the polymorphic transition from form V to form VI and is promoted by temperature fluctuations (Rousseau, 2007). High storage temperatures promote this solid–solid transition, which is to some extent due to imperfect tempering (Smith & Dahlman, 2005). Such a transformation can also be promoted by the presence of liquid oil, which can be generated through either partial melting due to temperature fluctuations or the previously mentioned oil migration in filled products (Kinta & Hatta, 2007).

8.8 Conclusion

The field of application of fats and oils has undergone significantly changes in the past few decades. The main driver for this process has been the pressure to practically eliminate partially hydrogenated fats from food products. There is no genuine substitute for partially hydrogenated fats, because they are characterised by a combination of a high structuring efficiency, a quick crystallisation process, a high oxidative stability and good organoleptic properties. Consequently, *trans* fatty acid-free solutions have been developed to be application-specific. Since the pressure on partial hydrogenation has also made full hydrogenation less popular, the use of palm oil and its fractions as a general source of saturated fatty acids has increased dramatically. Palm oil-based fat compositions do not satisfy all product needs, due to palm oil's specific crystallisation behaviour, its slow polymorphic transition and the risk of recrystallisation. Consequently, the use of interesterified fats as substitutes for *trans*-containing fats has increased significantly.

In order to successfully formulate a fat composition for an application, it is necessary to understand the role of the fat phase in the product under consideration. This boils down to understanding the nutritional constraints and needs with respect to chemical stability, organoleptic delivery, processability and product structure. The latter three aspects are strongly influenced by the structuring fat and the TAG profile present in a formulation, while the chemical stability and nutritional delivery are primarily based on the fatty acid profile of the total fat composition and mainly relate to liquid oils.

With the increasing demand by consumers for reduced chemical processing and preferably unprocessed ingredients, the use of IEE and the application of straight fats are growing at the expense of IEC. This opens the compositional space because, due to the nature of the process, IEE can deliver a much broader array of TAG profiles. Unfortunately, our understanding of these emerging compositions is currently insufficient to allow the deliberate utilisation of this new IEE compositional space. The variations originating from IEE are currently comparable to the natural variations of straight materials: a nuisance. To cater for this variability, robust product and process design is necessary. Although partially hydrogenated fats are not yet completely eliminated from our diets, the conversion from IEC to IEE is in full swing and the toolbox of the 'fat technologist' continuous to evolve. Next to the exploration of IEE beyond simple substitution for the chemical process, one can expect that new raw material sources of fats and oils will emerge; examples are tropical oils beyond cocoa butter, coconut, palm and palmkernel oil as natural sources of stearic acid, and novel oils originating from genetically modified organisms – either algae or seed oils.

In addition to these developments within the domain of product structuring with TAGs – which could soon find their way into product applications – the technological area of oil structuring (also referred to as 'organogelation') is receiving increasing attention. In essence, this deals with the search for alternative means to TAGs by which to structure the liquid oil phase. The status of the search for an analogue for oils to what gelatine and other biopolymers are to water has been comprehensively described by Marangoni & Garti (2011) and by Co & Marangoni (2012). A recent publication on the application of ethylcellulose as a structuring ingredient of the lipid phase in frankfurters (Zetzl *et al.*, 2012) indicates that these alternative structurants are on their way from the laboratory bench to product application. Although these structuring systems should still be considered in their infancy, discussions on the taxation of saturated fats (as has already been introduced in Denmark) will certainly stimulate development of the oil-gelling area.

References

Berger, K.G. (1997) Ice cream. In: S.E. Friberg & K. Larsson (eds) *Food Emulsions*, 3rd edn. New York, NY, USA: Marcel Dekker, pp. 413–490.

Bockisch, M. (1998) *Fats and Oils Handbook*. Champaign, IL, USA: AOCS Press.

Bot, A., Flöter, E., Lammers, J.G. & Pelan, E.G. (2003) Controlling the texture of spreads. In: B.M. McKenna (ed.) *Texture in Foods, Volume 1: Semi-solid Foods*. Cambridge, UK: Woodhead Publishing, pp. 350–372.

Co, E.D. & Marangoni A.G. (2012) Organogels: an alternative edible oil-structuring method. *J. Am. Oil. Chem. Soc.*, **89**, 749–780.

de Bruijne, D.W. & Bot, A. (1999) Fabricated fat-based foods. In: A.J. Rosenthal (ed.) *Food Texture: Measurement and Perception*. Aspen, CO, USA: Gaithersburg, pp. 185–227.

de Bruijne, D.W., Hendrickx, H.C.A.M., Alderliesten, L. & de Looff, J. (1997) Mouthfeel of foods. In: E. Dickinson & P. Walstra (eds) *Food Colloids and Polymers: Stability and Mechanical*. Cambridge, UK: Royal Society of Chemistry, pp. 204–213.

Deffense, E. (1985) Fractionation of palm oil. *J. Am. Oil. Chem. Soc.*, **62**, 376–385.

de Hoog, E.H.A., Prinz, J.F., Huntjens, L., Dresselhuis, D.M. & van Aken, G.A. (2006) Lubrication of oral surfaces by food emulsions: the importance of surface characteristics. *J. Food Sci.*, **71**, 337–341.

Eisner, M.D., Wildmoser, H. & Windhab, E.J. (2005) Air cell microstructuring in a high viscous ice cream matrix. *Colloids and Surfaces A: Physicochemical and Engineering Aspects*, **263**, 390–399.

Foubert, I., Dewettinck, K., Van de Walle, D., Dijkstra, A.J. & Quinn, P.J. (2007) Physical properties: structural and physical characteristics. In: F.D. Gunstone, J.L. Harwood & A.J. Dijkstra (eds) *The Lipid Handbook*, 3rd edn. Boca Raton, FL, USA: CRC Press, pp. 471–534.

Garti, N. & Sato, K. (1988) *Crystallization and Polymorphism of Fats and Fatty Acids*. New York, NY, USA: Marcel Dekker.

Garti, N. & Sato, K. (2001) *Crystallization Processes in Fats and Lipid Systems*. New York, NY, USA: Marcel Dekker.

Gavazotti, A. (2007) A solid-state chemist's view of the crystal polymorphism of organic compound. *J. Pharm. Sci.*, **96**, 2232–2241.

Gibon, V. (2006) Fractionation of lipids for use in food. In: F.D. Gunstone (ed.) *Modifying Lipids for Use in Food*. Cambridge, UK: Woodhead Publishing, pp. 201–229.

Goff, D. (2006) Ice cream. In: P.F. Fox & P.L.H. McSweeney (eds) *Advanced Dairy Chemistry, Volume 2: Lipids*, 3rd edn. New York, NY, USA: Springer, pp. 441–450.

Golding, M. & Pelan, E. (2008) Application of emulsifiers to reduce fat and enhance nutritional quality. In: G.L. Hasenhuettl & R.W. Hartel (eds) *Food Emulsifiers and their Applications*, 2nd edn. New York, NY, USA: Springer, pp. 327–348.

Gribnau, M.C.M. (1992) Determination of solid/liquid ratios of fats and oils by low-resolution NMR. *Trends in Food Science and Technology*, **3**, 186–190.

Hagemann, J.W. (1988) Thermal behaviour and polymorphism of acylglycerides. In: N. Garti & K. Sato (eds) *Crystallization and Polymorphism of Fats and Fatty Acids*. New York, NY, USA: Marcel Dekker.

Haighton, A.J. (1965) Worksoftening of margarine and shortening. *J. Am. Oil. Chem. Soc.*, **42**, 27–30.

Hartel, R. (2001) *Crystallization in Foods*. Aspen, CO, USA: Gaithersburg.

Johansson, D. (1995) The influence of temperature on the interactions and structures in semi-solid fat. *J. Am. Oil. Chem. Soc.*, **72**, 1091–1099.

Johansson, D. & Bergenståhl, B. (1995a) Sintering of fat crystal networks in oil during post-crystallisation processes. *J. Am. Oil. Chem. Soc.*, **72**, 911–920.

Johansson, D. & Bergenståhl, B. (1995b) Wetting of fat crystals by triglyceride oil and water. 2. Adhesion to the oil/water interface. *J. Am. Oil. Chem. Soc.*, 933–938.

Johansson, D., Bergenståhl, B. & Lundgren E. (1995a) Water-in-glyceride oil emulsions; effect of fat crystals on stability. *J. Am. Oil. Chem. Soc.*, **72**, 921–932.

Johansson, D., Bergenståhl, B. & Lundgren, E. (1995b) Wetting of fat crystals by triglyceride oil and water. 1. The effect of additives. *J. Am. Oil. Chem. Soc.*, **72**, 939–950.

Judd, J.T., Clevidence, B.A., Muesing, R.A., Wittes, J., Sunkin, M.E. & Podczasy, J.J. (1994) Dietary trans fatty acids: effects on plasma lipids and lipoproteins of healthy men and women. *Am. J. Clin. Nutr.*, **59**, 861–868.

Kellens, M., Gibon, V., Hendrix, M. & De Greyt, W. (2007) Palm oil fractionation. *Eur. J. Lipid Sci. Technol.*, **109**, 336–349.

Kinta, Y. & Hatta, T. (2007) Composition, structure, and color of fat bloom due to the partial liquefaction of fat in dark chocolate. *J. Am. Oil. Chem. Soc.*, **84**(2), 107–115.

Kloek, W. (1998) Mechanical properties of fats in relation to their crystallization. PhD thesis, University of Wageningen, Netherlands.

Korver, O. & Katan, M.B. (2006) The elimination of trans fats from spreads: how science helped to turn an industry around. *Nutr. Rev.*, **64**, 275–279.

Larsson, K. (1966) Classification of glyceride crystal forms. *Act. Chem. Scand.*, **20**, 2255–2260.

Lillford, P.J. (2000) The material science of eating and food breakdown. *Mat. Res. Soc. Bul.*, **12**, 38–43.

Lonchampt, P. & Hartel, R.W. (2004) Fat bloom in chocolate and compound coatings. *EJLST*, **106**(4), 241–274.

Maleky, F., McCarthy, K.L., McCarthy, M.J. & Marangoni, A.G. (2012) Effect of cocoa butter structure on oil migration. *J. Food Sci.*, **77**, 74–79.

Marangoni, A.G. (2004) *Fat Crystal Networks*. New York, NY, USA: Marcel Dekker.

Marangoni, A.G. & Garti, N. (2011) *Edible Oleogels: Structure and Health Implications*. Champaign, IL, USA: AOCS Press.

Marangoni, A.G. & Narine, S.S. (2002) *Physical Properties of Lipids*. New York, NY, USA: Marcel Dekker.

Marangoni, A.G., Acevedo, N., Maleky, F., Co, E.D., Peyronel, F., Mazzanti, G., Quinn, B. & Pink, D. (2012) Structure and functionality of edible fats. *Soft Matter*, **8**, 1275–1300.

Mason, T.G., Bibette, J. & Weitz, D.A. (1995) Elasticity of compressed emulsions. *Phy. Rev. Lett.*, **75**, 2051–2054.

Mensink, R.P. & Katan, M.B. (1990) Effect of dietary trans fatty acids on high-density and low-density lipoprotein cholesterol levels in healthy subjects. *N. Engl. J. Med.*, **323**, 439–445.

Micha, M. & Mozaffarian, D. (2010) Saturated fat and cardiometabolic risk factors, coronary heart disease, stroke, and diabetes: a fresh look at the evidence. *Lipids*, **45**, 893–905.

Norberg, S. (2007) Chocolate and confectionary fats. In: D.J. McClemens (ed.) *Understanding and Controlling the Microstructure of Complex Foods*. Cambridge, UK: Woodhead Publishing, pp. 648–687.

Ostwald, W. (1897) Studien über die Bildung und Umwandlung fester Körper. *Z. der Phys. Chem.*, **22**, 289–293.

Pickering, S.U. (1907) Emulsions. *J. Chem. Soc. Trans.*, **91**, 2001–2021.

Princen, H.M. (1979) Highly concentrated emulsions. I. Cylindrical systems. *J. Colloid Interface Sci.*, **71**, 55–66.

Prinz, J.F. & Lucas, P.W. (1997) Mastication and swallowing: an optimisation model. *Proc. of the Royal Soc. of London B*, **264**, 1715–1721.

Reddy, S.Y. & Jeyaran, T. (2001) *Trans*-free bakery shortenings from mango kernel and mahua fats by fractionation and blending, *J. Am. Oil. Chem. Soc.*, **78**, 635–640.

Rousseau, D. (2007) The microstructure of chocolate. In: D.J. McClemens (ed.) *Understanding and Controlling the Microstructure of Complex Foods*. Cambridge, UK: Woodhead Publishing, pp. 648–687.

Sato, K. (1999) Solidification and phase transformation behaviour of food fats – a review. *Fett/Lipid*, **101**, 467–474.

Sato, K. (2001a) Crystallization behaviour of fats and lipids – a review. *Chem. Eng. Sci.*, **56**, 2255–2265.

Sato, K. (2001b) Uncovering the structures of β′ fat crystals: what do the molecules tell us? *Lipid Technology*, **3**, 36–40.

Sato, K., Ueno, S. & Yano, J. (1999) Molecular interactions and kinetic properties of fats. *Prog. Lipid Res.*, **38**, 91–116.

Smith, K.W. (2001) Cocoa butter and cocoa butter equivalents. In: F.D. Gunstone (ed.) *Structured and Modified Lipids*. New York, NY, USA: Marcel Dekker, pp. 401–422.

Smith, R.P. & Dahlman, A. (2005) The use of atomic force microscopy to measure the formation and development of chocolate bloom in pralines. *J. Am. Oil. Chem. Soc.*, **82**, 165–168.

Smith, K.W., Bhaggan, K., Talbot, G. & van Malssen, K.F. (2011) Crystallization of fats: influence of minor components and additives. *J. Am. Oil. Chem. Soc.*, **88**, 1085–1101.

Sofjan, R.P. & Hartel, R.W. (2004) Effects of overrun on structural and physical characteristics of ice cream. *Int. Dairy Journal*, **14**, 255–262.

Talbot, G. (2009) Vegetable fats. In: S.T. Beckett (ed.) *Industrial Chocolate Manufacture and Use*, 4th edn. Weinheim, Germany: Wiley.
Tanaka, L., Tanaka, K., Yamato, S., Ueno, S. & Sato, K. (2009) Microbeam x-ray diffraction study of granular crystals formed in water-in-oil emulsion. *Food Biophysics*, **4**, 331–339.
Tietz, R.A. & Hartel, R.W. (2000) Effects of minor lipids on crystallization of milk fat-cocoa butter blends and bloom formation in chocolate. *J. Am. Oil. Chem. Soc.*, **77**, 763–771.
Timms, R. (2003) *Confectionery Fats Handbook: Properties, Production and Application*. Bridgwater, UK: The Oily Press.
Turner, W.R. (1971) Normal alkanes. *Ind. Eng. Chem., Product Res. & Dev.*, **10**, 238–260.
Vaclavik, V.A. & Christian, E.A. (2008) *Essentials of Food Science*. New York, NY, USA: Springer.
van de Streek, J. (2001) Molecular modelling of the crystal structures of long-chain compounds. PhD thesis, University of Nijmegen, Netherlands.
van Malssen, K.F. (1994) *Real* time x-ray diffraction applied to cocoa butter and graphite intercalates. PhD thesis, University of Amsterdam, Netherlands.
Walstra, P. (2001) *Physical Chemistry in Foods*. New York, NY, USA: Marcel Dekker.
Watanabe, A., Tashima, I., Matsuzakia, N., Kurashige, J. & Sato, K. (1992) On the formation of granular crystals in fat blends containing palm oil. *J. Am. Oil. Chem. Soc.*, **69**, 1007–1080.
Wesdorp, L.H., van Meeteren, J.A., de Jong, S., Giessen, R.V.D., Overbosch, P., Grootscholten, P.A.M., Struik, M., Royers, E., Don, A., de Loos, T.H., Peters, C. & Gandasasmita, I. (2004) Liquid-multiple solid phase equilibria in fats: theory and experiments. In: A.G. Marangoni (ed.) *Fat Crystal Networks*. New York, NY, USA: Marcel Dekker, pp. 481–710.
WHO (2003) Diet, nutrition and the prevention of chronic diseases: report of a joint WHO/FAO expert consultation. World Health Organization Tech Report **916**(i–viii), 1–149. Geneva, Switzerland: World Health Organization.
Wille, R.L. & Lutton, E.S. (1966) Polymorphism of cocoa butter. *J. Am. Oil. Chem. Soc.*, **43**, 491–496.
Willett, W.C., Stampfer, M.J., Manson, J.E., Colditz, G.A., Speizer, F.E. & Rosner, B.A. (1993) Intake of *trans* fatty acids and risk of coronary heart disease among women. *Lancet*, **341**, 581–585.
Xu, X., Guo, Z., Zhang, H., Vikbjerg, A.F. & Damstrup, M.L. (2006) Chemical and enzymatic interesterification of lipids for use in food. In: F.D. Gunstone (ed.) *Modifying Lipids for Use in Food*. Boca Raton, FL, USA: CRC Press, pp. 234–272.
Zéberg-Mikkelsen, C.K. & Stenby, E.H. (1999) Predicting the melting points and the enthalpies of fusion of saturated triglycerides by a group contribution method. *Fluid Phase Equilibr.*, **162**, 7–17.
Zetzl, A.K., Marangoni A.G. & Barut, S. (2012) Mechanical properties of ethylcellulose oleogels and their potential for saturated fat reduction in frankfurters. *Food Funct.*, **3**, 327–337.
Zevenbergen, H., de Bree, A., Zeelenberg, M., Laitinen, K., van Duijn, G. & Flöter, E. (2009) Foods with a high fat quality are essential for healthy diets. *Ann. Nutr. Metab.*, **54**, 15–24.
Ziegleder, G., Mikle, H. & Petz, A. (2001) Fat migration in filled chocolates – the dominant influences. *Zucker- und Süßwarenwirtsch.*, **55**, 23–25.
Ziegleder, G., Moser, C. & Geier-Greguska, J. (1996) Kinetics of fat migration within chocolate products. 2. Influence of storage temperature, diffusion coefficient, solid fat content. *Fett-Lipid*, **98**, 253–255.

9

Quality and Food Safety Assurance and Control

Mar Verhoeff[1] and Gerrit van Duijn[2]

[1]*Laboratory Dr A. Verwey B.V., Rotterdam, The Netherlands*
[2]*Maas Refinery, Rotterdam, The Netherlands*

9.1 Introduction

The main constituents of oils and fats are triacylglycerol (TAG) molecules. The composition of these molecules determines the type of oil or fat and its main physical and chemical properties. Besides TAGs, a wide range of minor components are also present in unrefined oils and fats. These can be residues of the oil crop (seed or pulp) remaining in the oil after oil extraction, products of oil degradation or supply chain contaminants. Some minor components will affect the product quality, while others will (also) have a negative health effect.

This chapter will give an overview of the methods used to measure oil and fat compositions, followed by an examination of the minor components and contaminants, looking at each of the following points:

- origin;
- analytical technique;
- level in the crude oil;
- removal by refining;
- specification and/or legal level.

This information will be summarised in the quality and food safety assurance system.

Edible Oil Processing, Second Edition. Edited by Wolf Hamm, Richard J. Hamilton and Gijs Calliauw.

9.2 Analytical methods for measuring oil and fat composition

Oils and fats can be characterised for a large part by their fatty acid composition. Fatty acids can vary in length and in the number of unsaturated double bonds. The double bonds can occur in two geometrical forms: the *cis* and the *trans* form. In the *cis* configuration, the hydrogen atoms linked to the carbon atoms are on the same side of the double bond. In the *trans* configuration, they are on opposite sides.

A fatty acid analysis, including a *trans* fatty analysis, is performed using gas chromatography with flame ionisation detection (FID). The sample is esterified to methyl esters, dissolved in an organic solvent and injected into the gas chromatograph. In general, calculation of the different fatty acids is based on area percentages, giving relative percentages; however, in specific cases absolute values can be calculated using an internal standard.

TAG or triglyceride composition can be analysed using gas chromatography without the methyl-esterification step. Equally, high-performance liquid chromatography (HPLC) is used to determine the distribution of diacylglycerols and TAGs in vegetable oils. The samples need no chemical preparation, but call for a careful selection of the mobile phase and detector system. Separation of positional isomers of TAGs (i.e. different positions of the fatty acids on the glycerol backbone) typically requires silver-ion HPLC, however.

The type of fatty acid, and to a much lesser extent its position on the glycerol, determines the melting point of a TAG. Natural fats are complex mixtures of TAGs, containing many different fatty acids. As a result, fats normally have a gradual melting curve. This means that between a low temperature, where all components are solid, and a high temperature, where all components are liquid, there is a gradual increase of the liquid fraction and decrease of the solid fraction. The line characterising the fraction of solid material at different stabilisation temperatures is called the solid phase line.

Solid fraction analysis is performed using pulsed-wavelength nuclear magnetic resonance (pNMR). The quantities are calculated by comparison using standard materials. The sample is melted and undergoes several temperature treatments before being tempered at the desired temperature, at which point measurement is performed. The result is shown as the solid fat content (SFC) curve, plotted against the temperature at which the measurement is carried out.

9.3 Quality analyses

9.3.1 Free fatty acids

Free fatty acids (FFA) are fatty acids in unesterified form. FFA are formed after harvesting of a seed or fruit (in fruit they can even be formed during ripening) by enzymatic or microbiological hydrolysis of oils and fats. Additional FFA can be formed by chemical hydrolysis during the storage and transport of the crude oils. The formation mechanisms are:

$$\text{TAG} + \text{water} \rightarrow \text{diacylglycerol} + \text{FFA} \quad (9.1)$$

and:

$$\text{diacylglycerol} + \text{water} \rightarrow \text{monoacylglycerol} + \text{FFA} \quad (9.2)$$

The amount of FFA is determined by neutralising the acids of a known quantity of oil with an alkaline solution of known strength in an organic solvent. The calculation is based on the average molecular weight of the oil/fat that is analysed. In general, oleic acid (282) is used. In the case of palm oil or palm oil products, palmitic acid (256) is used instead, or in the case of coconut or palm kernel oil, lauric acid (200).

FFA levels are relatively high in crude tropical oils (palm, palm kernel and coconut oil), medium in crude soft-seed oils (sunflower and maize germ oil) and low in crude hard-seed oils (soybean and rapeseed oil) (see Table 9.1).

In chemical refining, FFA are largely removed by reaction with sodium hydroxide to form soap and by subsequent separation by gravity or in a centrifuge. In physical refining, FFA are reduced by stripping in a deodoriser. The terms 'chemical' and 'physical' refining refer to the way in which FFA are removed: by chemical reaction or by physical stripping.

Relatively high levels of FFA in refined oils may result in a soapy product off-flavour (depending on FFA chain length and product type) or in smoke formation during frying. The specified residual level after refining is therefore relatively low (0.04–0.10%; see Table 9.2).

Table 9.1 Standard contractual quality specifications for major crude and semirefined oils and fats.

Oil	FFA max. (%)	M&I max. (%)	P max. (ppm)	Colour	Others
Crude degummed soybean oil	1.25	0.38	250	Yes	
Crude degummed rapeseed oil	1.75	0.40	300	–	C22 : 1 <2 %
Crude sunflower oil	3.00	0.50	–	–	
Crude maize oil	4.00	0.50	500	–	
Crude palm oil	5.00	Pure	–	–	
RBD palm oil	0.10	0.10	–	Yes	MP 33–39 °C
RBD palm olein	0.10	0.10	–	Yes	MP < 24 °C, IV > 56
RBD palm stearin	0.20	0.15	–	Yes	MP > 44 °C, IV < 48
Crude palm kernel oil	5.00	0.50	–	–	IV < 19
Crude coconut oil	4.00	1.00	–	–	

FFA, free fatty acid; M&I, moisture and impurities; P, phosphorus; ppm, parts per million; RBD, refined, bleached, deodorised; MP, slip melting point; IV, iodine value.

Table 9.2 Standard industry specifications for refined vegetable oils, excluding olive oil.

Component	Unit	Maximum level in refined oils
Free fatty acids (FFA)	%	0.05–0.10
Peroxide value	meq/kg	1
Phosphorous	mg/kg	5
Soap	mg/kg	10
Dirt	–	Not visible
Moisture	%	0.05
Iron	mg/kg	0.1–0.5
Copper	mg/kg	0.01–0.05

9.3.2 Peroxides

Peroxides are the first oxidation products of fatty acids (free as well as esterified fatty acids). They are mainly formed during storage, handling and transport of crude, semiprocessed and refined oils and fats in contact with air.

Individual fatty acids oxidise to so-called hydroperoxides, which in turn form volatile aldehydes and ketones. The initial hydroperoxides are tasteless, but the secondary reaction products, the aldehydes and ketones, give the oil an off-flavour.

In order to analyse peroxides, the sample is dissolved in an organic solvent. Potassium iodate solution is added and the mixture is stored in the dark for 10 minutes. During the reaction of the peroxides with the iodate, iodine is formed. The amount of iodine is determined by titration with sodium thiosulfate solution of known strength, using a starch solution as indicator. The result (peroxide value, POV) is expressed as:

$$\mathrm{POV} = (\mathrm{S} - \mathrm{B}).\mathrm{N}.1000/\mathrm{W} \tag{9.3}$$

where S is the titration of the sample (ml), B is the titration of the blank (ml), N is the normality of the thiosulfate solution and W is the weight of the sample.

The POV of oil is often used as a measure for the amount of oxidation. Fully refined oil with a POV < 1 is considered to be of good quality, while a higher POV may indicate secondary oxidation and the formation of rancid off-flavour.

9.3.3 Phosphorus

Phosphorus in a vegetable oil stems from the plant tissue and is present as a variety of phospholipids, generally labelled as lecithin. During oil extraction, part of the lecithin will disperse in the oil. Lecithin is (deliberately) largely removed during degumming and neutralisation (especially during centrifuge neutralisation with acid pretreatment). Bleaching with acid pretreatment will only remove relative low-level residues of lecithin. Before deodorisation, the phosphorus level of oil needs to be low (<5 ppm) in order to avoid fouling of the heating coils and brown colouring of the oil by 'burnt' lecithin.

Several methods can be used for this determination, of which the following is an example. A small sample of the oil (0.1–10.0 g) is weighed into a quartz crucible, and magnesium carbonate is added. This mixture is combusted until a white ash is formed, which is dissolved in hydrochloric acid. When a sulfate molybdate solution is added and the mixture is left in the dark, a blue colouration forms, which can be measured at 720 nm. The absorption value thus obtained can be converted to a concentration with the aid of a calibration curve derived from a series of dipotassium hydrogen sulfate solutions. In a more automated fashion, inductively coupled plasma (ICP) linked with atomic emission spectroscopy (AES) is also commonly used to measure levels of phosphorus (and other elements such as Na, Ca, Mg etc.) at <1 ppm concentrations.

The phosphorus level in fully refined oils and fats will be low, in order to satisfy the low level required before deodorisation. Also, in hot applications such as frying, phospholipid levels should be low to avoid 'burning' of the product. The phosphorus specification for fully refined oil varies between a maximum 5 ppm and less than 2 ppm. Note however that this analysis only measures the P-content and doesn't differentiate between organic and anorganic phosphorus (e.g. phosphoric acid).

9.3.4 Moisture and dirt

Moisture and dirt in crude oils and fats can be caused by oil crop residues remaining in the oil after the extraction process or else can originate from the supply chain (dirty tanks, condensation of moisture in air, steam blowing etc.). Moisture and dirt are removed in the first refining step: neutralisation in chemical refining or bleaching in physical refining. Bleaching earth residues or spots of polymerised material from a fouled deodoriser can also be the cause of dirt in fully refined oils and fats.

The moisture (and volatile matter) content is measured by determining the loss of weight during heating at 105 °C on a hot plate or in an oven. Another option, for determining moisture only, is the Karl Fischer method. This is a dead-stop endpoint titration, using special titration equipment. The impurities are determined by dissolving the sample in an organic solvent such as petroleum ether, filtering the mixture through a paper filter and finding the weight of residues on the filter.

There should be no presence of visible dirt or moisture in a fully refined oil or fat. Dirt should be absent for quality reasons, while free moisture can be the cause of microbiological contamination.

9.3.5 Colour

Crude vegetable oils will have a colour varying from red to green via yellow and brown. The main contributors to this colour are carotenoids (red) and chlorophyll (green). Colour is mainly removed during bleaching by adsorption or by interaction with the chemical active sites on the surface of the bleaching earth. Carotenoids are also decomposed during high-temperature deodorisation.

Colour measurements are based on comparison with standard colour glasses. Several colour standards are used in the oils and fats industry: Lovibond 1 and 5¼ inch cell, Gardner, FAC and iodine scale.

In general, refined oils and fats should be more or less colourless. The 'natural' oil or fat colour does not always match the desired colour of the food product containing this oil or fat. In addition, some colour particles promote oil deterioration or contribute to off-flavours.

9.3.6 Metals

Some metals, such as iron and copper, are catalysts for oxidation. Metals can originate from the original oil crop or by pickup in the crude oil supply chain. Copper is such a strong oxidation catalyst that its presence, and that of its alloys, should be avoided in the entire oils and fats supply chain. Metals are largely removed by neutralisation in chemical refining or by acid pretreatment followed by bleaching in physical refining.

Several techniques are available for the measurement of metals content. The most common is atomic absorbance spectroscopy, either after digestion of the sample (flame AAS) or after dilution in an organic solvent (flameless AAS). Another method is the use of ICP after dilution of the sample in an organic solvent.

The metal levels in fully refined oils and fats should be made as low as possible using standard refining techniques. Specifications for iron in refined oil vary from a maximum 0.5 ppm for frying fats to a maximum 0.1 ppm for oils and fats used as ingredients in sensitive products. Copper specifications are in general a factor 10 below iron specifications.

9.3.7 Deterioration of Bleachability Index

The Deterioration of Bleachability Index (DOBI) is an indicator for the bleachability of crude palm oil. An oil's DOBI value is calculated by dividing the extinction value at 446 nm by the extinction value at 269 nm of a sample dissolved in cyclohexane. The higher the value, the better the bleachability of the palm oil. The palm oil industry uses a range from above 3.5 (very good) to below 1.5 (very bad) (Wai-Lin & Ping-Tou, 2001).

9.3.8 Tocopherols

Tocopherols are natural constituents in vegetable/animal oils and fats. The compositions and levels of natural tocopherol mixes vary from source to source. Some tocopherols are active as antioxidants, while others have a vitamin E activity. Tocopherol levels are reduced during refining (mainly at high temperature); tocopherols lost this way can be partly recovered from the deodoriser distillate.

Tocopherol composition is analysed by normal-phase HPLC and compared with a standard solution containing known concentrations of the tocopherols.

9.4 Supply chain contaminants

9.4.1 Polycyclic aromatic hydrocarbons

Polycyclic aromatic hydrocarbons (PAHs) constitute a large class of organic compounds that are composed of two or more fused aromatic rings. They are primarily formed by incomplete combustion or pyrolysis of organic matter (Alexander *et al.*, 2008). Humans are exposed to PAHs by inhalation, if they are smokers, and by consumption of contaminated food. Oil crop can be contaminated with PAHs by absorption of these components from exhaust gases, when these gases are in direct contact with the crop during drying. The generally used marker for PAHs in crude oil is benz(a)pyrene (BaP). High levels of BaP have been found in coconut oil and sunflower oil from specific regions (van Duijn & den Dekker, 2010).

PAHs are removed by active carbon dosing in the bleaching process. Volatile PAHs will additionally be reduced during high-temperature deodorisation. The volatility depends on the number of aromatic groups in the PAH compound; four or less is referred to as light PAH, while five or more is called heavy PAH. The tracer compound BaP has five aromatic groups and is a heavy PAH.

Several methods of analysis are available, using gas chromatography–mass spectrometry (GCMS) and HPLC. A widely accepted option in Europe is the donor accepted column chromatographic (DACC) method. This provides an online HPLC cleanup procedure, using a preparative column combined with backflush on another, analytical, HPLC column. Calculation is performed using standard materials. Other HPLC methods and the GCMS methods consist of an offline cleanup by column chromatography and/or liquid–liquid partition, followed by the actual analysis.

PAHs in food have a proven carcinogenic and/or genotoxic effect associated with long-term exposure to a relative low dosage. In such a case, the limit will be set by application of the ALARA (As Low As Reasonably Achievable) principle. Good industry practices result in a BaP level after refining of around 1 ppb. Current EU regulation sets a limit for BaP and a limit for the sum of four PAHs for products that involve drying/roasting in processing. The four PAHs included in this regulation are: benz(a)pyrene, benzanthracene, benzofluoranthene and chrysene. The limits for oils and fats intended for direct consumption or as ingredients in food are 2 ppb BaP and 10 ppb for the sum of the four PAHs. An exception is made for coconut oil, for which the limits are 2 ppb BaP and 20 ppb for the sum of the four PAHs .

9.4.2 Pesticide residues

Plant protection products or pesticides can be used to protect an oil crop during growth, to reduce weeds and to protect oilseeds during storage and transport. For permitted applications of pesticides, so-called maximum residue limits (MRLs) are introduced. These limits are pesticide- and crop-specific. They are defined on the basis of residues found after pesticide use according to good agricultural practices. Generally, these levels are much lower than the harmful toxicological thresholds (see Figure 9.1). Pesticide residue levels may initially increase during oil extraction (if they are oil- or hexane-soluble), but most pesticides are largely reduced during oil refining (van Duijn, 2008).

In general, pesticide residues are analysed using gas chromatography for the nonpolar types and liquid chromatography for the polar types. In oils and fats, the most common pesticides are of the nonpolar type. Gas chromatographic detection methods use electron capture detection (ECD) techniques. Such detectors are very sensitive to chlorinated compounds. Nitrogen phosphorus detection (NPD) methods are used for phosphorus-containing and nitrogenated compounds. Today, mass spectrometry (MS) detection is the most used method, including single-quad MS and triple-quad MS/MS. Cleanup is done by column chromatography, liquid–liquid partition or gel permeation chromatography (GPC), or a combination of these techniques.

In general, pesticide residues are related to the oil crop, and it is assumed that no further change of concentration occurs during processing. This is not necessarily the case for pesticide levels in refined oil as compared to the levels in the oil crop as harvested. To date, pesticide residues have only been found in crude seed oils. The types of pesticide residue (organophosphorus

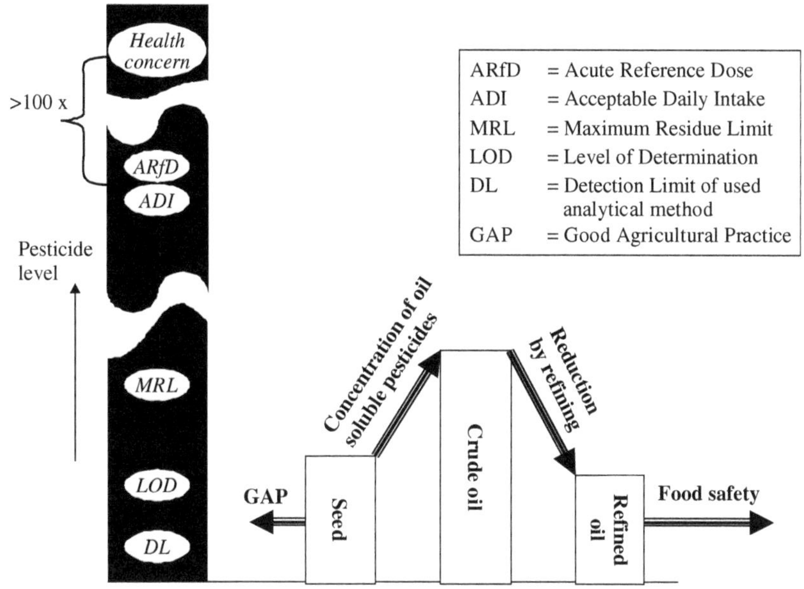

Figure 9.1 The changing pesticide level from seed to crude oil (increased by oil extraction) to refined oil (reduced by refining). The vertical column shows a possible relationship with various health, legal and detection limits.

insecticides) found indicate that the pesticides were used for post-harvest treatment of seeds or that they originate from cross-contamination in silos. These organophosphorus insecticides are relatively volatile and are largely reduced during deodorisation.

9.4.3 Hydrocarbons of mineral origin

Mineral oil products can be present in crude edible oils due to contamination during processing (lubricants and hydraulic oils), residues from previous cargoes during transport and storage, fraudulent adulteration and environmental pollution (Neukom *et al.*, 2002). Their presence can also be the result of their permitted use as a processing aid – like hexane in solvent extraction – as a pesticide solvent and as an antidusting agent in oilseed storage. Many plants and animals synthesise long-chain alkanes, resulting in considerable levels of naturally occurring alkanes in crude oils.

For hydrocarbon analyses, gas chromatographic techniques are used: either gas chromatography–flame ionisation detection (GCFID) or GCMS. The sample is cleaned using column chromatography with aluminum oxide or silicon oxide. After evaporation of the solvent, the sample is injected into the gas chromatography system. As oils and fats contain naturally present hydrocarbons, interpretation of the chromatogram is rather difficult and requires a lot of experience on the part of the analyst.

Crude oils and fats must be free from contamination by hydrocarbons of mineral origin. Long-chain hydrocarbons will not be removed by the refining process, and most hydrocarbons will not be present in a pure form and may contain additives or other impurities.

The following limits are industry standards for crude oils based on good agricultural and manufacturing practice:

- Short-chain hydrocarbons (<C10) are volatile and contractually limited by the flashpoint of the crude oil.
- Diesel (C10–C24) levels in crude palm oil and palm products should be below 25 mg/kg (agreement by the Dutch, Malaysian and Indonesian governments).
- The total hydrocarbon level in crude sunflower oil should be below 50 mg/kg (after correction for natural alkanes).
- For all other vegetable oils, further investigation is required if the level of total hydrocarbons exceeds 300 mg/kg.

9.4.4 Mycotoxins

Two types of mycotoxin are found in crude edible oils:

(1) Aflatoxin in coconut and groundnut oil. Aflatoxins are carcinogenic and their legal limits in foodstuffs are very low.
(2) Zearalenone in crude maize germ oil. Various studies have reported a negative effect of zearalenone on the fertility of pigs; the effect on humans is currently unknown (Kuiper-Goodman *et al.*, 1987).

In general, mycotoxins are analysed using HPLC, following a cleanup procedure. The fluorescence detector is the most sensitive for most toxins. A very modern way of analysing mycotoxins is through liquid chromatography–mass spectrometry (LCMS). With the LCMS method, it is possible to detect more mycotoxins in one analysis run. Some mycotoxins can be detected using gas chromatographic methods.

Both the chemical and the physical refining process reduce observed levels of aflatoxin in crude oil to below the detection limit in the refined oil.

Chemical refining will remove more than 80% of zearalenone, while reduction by physical refining varies between 70 and 80%. The EU limit for Zearalenone in refined oil is 400 μg/kg, based on ALARA principles (European Commission, 2006).

9.4.5 Other contaminants

Monitoring programmes for dioxins, furans and dioxin-like polychlorinated biphenyls (PCBs) have shown levels well below those permitted for oils and fats intended for human consumption. Only crude fish oils sometimes contain relatively high dioxin levels, arising from their concentration in the fish feed chain.

Heavy metals are seldom present at detectable levels in crude edible oils.

9.5 Quality and food safety assurance

9.5.1 Crude oil analyses

Crude oils and fats are analysed for the following reasons:

(1) To verify that the product is delivered according to contract. This verification is carried out on samples taken by an independent superintendent at the point mentioned in the contract (normally the port of loading). The following parameters or characteristics are checked:
   (i) that the oil or fat is of good merchantable quality (GMQ) (i.e. it is not adulterated or contaminated);
   (ii) that, as a principle, the transport of oils and fats is only permitted in conveyances which are dedicated to foodstuffs; for sea transport, some exceptions can be made (see Federation of Oils, Seeds and Fats Associations (FOSFA) or EU List of Acceptable Previous Cargo, see also chapter 2);
   (iii) the quality parameters mentioned in the contract (see Table 9.1).

(2) To check on contaminants which are legally not permitted in crude oils (e.g. pesticides above MRL), or which cannot be removed by refining (long-chain hydrocarbons of mineral origin). This analysis is normally done after unloading into land tanks.

(3) To check on quality parameters as an input to adjust the refiner's process conditions. The main parameters to be analysed are FFA and phosphorus. This analysis is normally done in the process laboratory on samples taken from the input of the oil into the refinery.

(4) To check levels of contaminants that have a limit in refined oil and can be removed by refining. These contaminant levels can be an input to adjust the refiner's process conditions or a check on whether they are lower than the levels used during the refiner's process validation (see Section 9.5.3). Contaminant analyses are costly and time-consuming; therefore, samples should be taken from the largest possible batch (e.g. the crude oil tank park) some time before processing. An overview of the risk of contaminant presence and the frequency of analyses is given in Section 9.5.2.

Table 9.3 Crude oil risk matrix. This shows the risk classification for contaminant presence in a crude oil. It also shows the recommended frequency of analysis if an oil is of unknown origin. The full colour version can be found on the plates.

	Pesticides	PAH	Mineral oil in edible oil	Dioxins and PCBs	Aflatoxins	Zearalenone
Limit	MRL or LOD	BaP < 2 ppb				
Soybean oil	xx	xx	x	x		
Sunflower oil	xxx	xxx	xx	x		
Rapeseed oil	xx	xx	x	x		
Corn oil	xx	xx	x	x		xxx
Palm oil	x	x	xx	x		
Palm kernel oil	x	xx	xx	x		
Coconut oil	x	xxx	xx	x	xx	
Groundnut oil	x	x	x	x	xxx	
Fish oil	x	xx	x	xxx		
Linseed oil	xx	xx	x	x		
Cottonseed	xx	xx	x	x		
Grape seed	x	xxx	x	x		
Olive	xx	xx	x	x		

PAH, polycyclic aromatic hydrocarbon; PCB, polychlorinated biphenyl; MRL, maximum residue limit; LOD, level of determination (see Figure 9.1); BaP, benzopyrene. xxx, high risk (regular occurrence (> once a year), monitor every batch); xx, medium risk (occasional occurrence (every 1–5 years), monitor at least once a quarter); x, low risk (infrequent occurrence (< once every 5 years), monitor once a quarter at most).

9.5.2 Crude oil risk matrix

Section 9.4 gave an overview of the origin and occurrence of contaminants in oils and fats. This experience, together with published analytical results (Pages *et al.*, 2010; van Duijn & den Dekker, 2010), is the basis of the so-called crude oil risk matrix. The crude oil risk matrix, shown in Table 9.3, gives the risk classification (high, medium or low) for the presence of a contaminant in a crude oil, where the origin of this crude oil is unknown. Knowledge of practices or procedures in dedicated supply chains may further reduce the risk classification if these practices reduce contamination risk. The crude oil risk matrix can be used to determine the required frequency of contaminant analyses in crude oils. The proposed frequencies are:

- *high risk*: check every delivery;
- *medium risk*: quarterly monitoring;
- *low risk*: annual monitoring.

Crude oils and fats limits are set for pesticides, hydrocarbons of mineral origin and previous cargoes:

- The pesticide level in the crude oil should not exceed the MRL for the pesticide/oilseed combination.
- The level of hydrocarbons of mineral origin should not exceed the limits defined by the industry (see Section 9.4.3).
- Previous cargoes are checked by comparing a ship's logbook with the EU or FOSFA List of Acceptable Previous Cargo, taking into account the construction material of the ship's tanks. This activity is normally performed by an independent superintendent.

Other contaminants have no legal or industry limits in crude oil, but are regulated in the fully refined product.

9.5.3 Process validation contaminant removal

The refinery process validation for contaminant removal will ensure that the contaminant level in the fully refined oil is below the regulated or defined limit, even for the crude oil feedstock with the highest observed contaminant level to date. The validation process is as follows:

(1) The refinery is informed of a crude oil delivery with a contaminant level higher than the highest level used in previous process validations and the contaminated lot is blocked.
(2) A minimum batch of contaminated oil is processed in the refinery using the standard refining recipe. The contaminant levels are analysed in deodorised end product (and preferably also after the intermediate refining steps).
(3) The crude oil is deblocked and the whole lot is processed if the contaminant level in the deodorised oil is below the regulated or defined limit. The validation process must be repeated with modified process conditions if the contaminant level in the deodorised oil is still too high. Alternatively, the crude oil can be sold for non-food application (feed or biofuel) if removal is technically or economically not feasible.
(4) This validation process must be repeated for every delivery of crude oil with a contaminant level higher than the level used in previous process validations.

9.5.4 Oil processing link tables

The purification of an oil or fat occurs through a reduction of minor components and contaminants in the various refining steps. For chemical refining, these steps are combined degumming and neutralisation, bleaching and

Table 9.4 Refining link table for quality-related minor components. This table summarises the process validation experience of minor component reduction.

	Free fatty acids	Peroxides	Phosphorous	Dirt	Metals	Taste	Colour
Degumming			P	P	P		
Neutralisation	C		C	C	C		
Bleaching			P	P	P	+	+
Deodorisation	P	+			+	+	
Refined oil storage		+				+	

P, physical refining; C, chemical refining; +, both physical and chemical refining.

Table 9.5 Refining link table for food safety-related contaminants. This table summarises the process validation experience of contaminant reduction. Long-chain hydrocarbons cannot be removed by refining; crude oils contaminated with long-chain hydrocarbons should therefore be rejected. Crude oils containing pesticide levels above the maximum residue limit (MRL) (or level of determination (LOD) for unallowed pesticides) are not permitted by pesticide legislation and should be rejected.

	Hydrocarbons $<$ C20	Hydrocarbons $>$ C20	PAH (BaP)	Pesticides	Aflatoxin B1	Zearalenone
Crude oil reception		+				
Degumming						
Neutralisation				+	C	C (93%)
Bleaching			+	+	P	P (77%)
Deodorisation	+		+	+		

P, physical refining; C, chemical refining; +, both physical and chemical refining.

deodorisation. For physical refining, these steps are degumming, bleaching (with acid pretreatment) and high-temperature deodorisation. The process step which reduces the level of a specific minor component or contaminant is known from theory or process validation experience. This knowhow can be summarised in so-called refining link tables. Table 9.4 shows the links between the process steps and the reduction of quality related minor components. Table 9.5 shows the links with food safety-related contaminants. These link tables can be used as a quick reference for process optimisation and troubleshooting. The contaminant link table can also be the basis for a hazard analysis critical control points (HACCP) of the refining process.

9.5.5 Food safety control points

Each food production site must perform an HACCP analysis in order to secure the food safety of its products. Such an analysis must be based on the seven HACCP principles:

(1) Conduct a hazard analysis.
(2) Identify critical control points.
(3) Establish critical limits for each critical control point.
(4) Establish critical monitoring requirements.
(5) Establish corrective actions.
(6) Establish record-keeping procedures.
(7) Establish procedures for ensuring the HACCP system is working as intended.

It is not the objective of this section to present a complete HACCP analysis of the refining process – such an analysis is the responsibility of company management. Instead, the most critical hazards, their critical limits and the necessary corrective actions are summarised in Table 9.6.

A site HACCP may result in additional hazards related to actual processing procedures and equipment being identified, such as a risk of grease oil

Table 9.6 Sample HAACP analysis.

Hazard	**Critical limit**	**Corrective action**
Crude oil intake		
Nonpermitted previous cargo	Cargo not on EU or FOSFA positive list	Block and reject for food use
Mineral oil contamination	25–300 ppm (see Section 9.4.3)	Block and reject for food use
Residue of nonpermitted pesticide	LOD (level of determination)	Block and reject for food use
Too-high residue of permitted pesticide	MRL (maximum residue limit)	Block and reject for food use
Processing		
Too-high level of PAHs in crude oil	Legal limit in refined oil	Apply validated reduction process
Hexane residue in crude seed oil	LOD (level of determination)	Apply validated reduction process
Aflatoxin in crude coconut or groundnut oil	Legal limit in refined oil	Apply validated reduction process
Zearalenone in crude maize germ oil	Legal limit in refined maize germ oil	Apply validated reduction process, or reject for food use if removal is not feasible

contamination, foreign bodies in the final product and so on. These should lead to appropriate corrective actions, following the HACCP principles.

References

Alexander, J., Benford, D., Cockburn, A., Cravedi, J.-P., Dogliotti, E., Di Domenico, A., Fernández-Cruz, M.L., Fink-Gremmels, J., Fürst, P., Galli, C., Grandjean, P., Gzyl, J., Heinemeyer, G., Johansson, N., Mutti, A., Schlatter, J., van Leeuwen, R., Van Peteghem, C. & Verger, P. (2008). Scientific opinion of the panel on contaminants in the food chain on a request from the European Commission on polycyclic aromatic hydrocarbons in food. *The EFSA Journal*, **724**, 1–114.

European Commission (2006). Setting maximum levels for certain contaminants in food-stuffs. Regulation no. 1881. *Official Journal of the European Union*, **L364**, 5–24.

Kuiper-Goodman, T., Scott, P.M. & Watanabe, H.W. (1987) Risk assessment of mycotoxin zearalenone. *Toxicology and Pharmacology*, **7**(3), 253–306.

Neukom, H-P., Grob, K., Biedermann, M. & Noti, A. (2002) Food contamination by C20–C50 mineral paraffins from the atmosphere. *Atmospheric Environment*, **36**, 4839–4847.

Pages, X., Morin, O., Birot, C., Gaud, M., Fazeuilh, S. & Gouband, M. (2010). Raffinage des huiles et des corps gras et élimination des contaminants. *OCL*, **17**, 86–99.

van Duijn, G. (2008). Industrial experiences with pesticide removal during edible oil refining. *European Journal of Lipid Science and Technology*, **110**, 982–989.

van Duijn, G. & den Dekker, G. (2010). Unilever food safety assurance system for refined vegetable oils and fats. *OCL*, **17**, 100–103.

Wai-Lin, S. & Ping-Tou, G. (2001) Deterioration of bleachability index (DOBI). *INFORM*, **12**(12), 1183ff.

10 Oil Processing Design Basics

Gerrit van Duijn[1] and Gerrit den Dekker[2]

[1]*Maas Refinery, Rotterdam, The Netherlands*

[2]*Retired, Unilever R&D Vlaardingen, Vlaardingen, The Netherlands*

10.1 Introduction

A project to build a new refinery or to upgrade an existing one starts with the design of the block diagram. The block diagram is the basis for the initial investment and operational costs estimates required to support the business case. This simplified process flowsheet does not contain any details of the process steps but does specify the process flow, required capacities, first indications of storage capacities and utility requirements.

The design of the block diagram starts with a market study to determine required product volumes and the estimated development of these volumes over the coming years. Flexibility towards longer-term market changes should also be considered, since an oil processing site is built to last for at least 30 years. A systematic approach designing the block diagram and obtaining the information needed in the initial phase of the project includes the following steps:

(1) Selection of the refining and modification process routes that will deliver the products demanded, beginning with the available raw materials.
(2) Design of the oil processing block diagram, based on selected process routes. This will require decisions on working pattern, flexibility (batch or continuous) and so on.
(3) Calculation of the required capacities of the equipment used in the various process steps on the basis of the effective operational time.

Edible Oil Processing, Second Edition. Edited by Wolf Hamm, Richard J. Hamilton and Gijs Calliauw.

(4) Determination of the design basics of crude, intermediate and refined oil storage from internal and external logistics.
(5) Estimation of utilities consumptions and effluent productions, including the capacities of the utility units in the block diagram and a first input for the operational costs calculation.
(6) Integration of occupational safety-based design, as early as possible in the project.

The six sections of this chapter will lead the reader through these six steps. Some sections will give more information than is strictly needed for primary design purposes, but these details are not given elsewhere in this book.

10.2 Refining and modification process routes for most common oil types

10.2.1 Process step definitions

Each processed oil and fat will have an optimal sequence of process steps (process route) for reaching the final required quality at the lowest cost. The process routes given in this chapter are based on the Unilever best-practice process routes (Unilever Oil Processing Recommendations, 1988).

The costs of oil processing for a given oil or fat are dependent upon a number of cost items, such as oil losses, labour, processing aids, energy, repair and maintenance, investments and so on.

The final quality is dependent on the crude oil quality, the treatment given and the quality measures taken during treatment. Product safety and occupational safety issues also play a role.

The optimum process route will minimise costs and produce a product of good quality, according to market standards. This process route will consist of a sequence of process steps as outlined in this section.

10.2.1.1 Degumming or water degumming (degummed)

A pretreatment process applied to seed oils to remove impurities and partly remove phospholipids. The crude oil is treated with water, which leads to hydration of hydratable phospholipids. The hydrated material is removed by centrifugation.

10.2.1.2 Deep degumming (ddg)

A pretreatment process applied to seed oils to reduce the phosphorus content below 30 ppm. It is normally a two-step process, with the addition of an acid (typically citric acid) to remove both hydratable and nonhydratable phospholipids. The phospholipids are removed by centrifugation. The process conditions can be chosen in such a way that part of the waxes is removed as well.

10.2.1.3 Neutralisation (n)

The purpose of neutralisation is to reduce the concentration of free fatty acids (FFA) to a maximum 0.10% through the use of a diluted alkali solution, typically sodium hydroxide. This process can be applied batchwise in vessels (bn) and continuously by means of centrifuges (cn). After alkali treatment, the oil is washed with hot water or treated with silica to reduce its residual soap level.

Soda silicate boil can be used for batch neutralisation of crude or water-degummed seed oils.

10.2.1.4 One-step bleaching (osb)

This is usually applied to chemical-refined and interesterified oils and blends, which have low residual concentrations of nonvolatile impurities. The main purpose is to remove residual soap, pigments and oxidised components. This process is applied under vacuum with the addition of bleaching earth. Sometimes other absorption agents, such as active carbon or silicate, are added to promote better removal of unwanted minor components.

10.2.1.5 Two-step bleaching (tsb)

This process is typically applied as pretreatment for physical refining and as post-treatment of hydrogenated oils and fats, in order to facilitate removal of nonvolatile impurities such as phosphatides and metals. Unlike in one-step bleaching, the oil is pretreated with acid and water before addition of the bleaching earth.

10.2.1.6 Deodorisation (d)

Under high vacuum, the oil is heated to 180–240 °C and brought in contact with steam to remove volatile components and to create an oil with a bland taste and increased storage stability.

10.2.1.7 Deodorisation/stripping (ds)

Stripping is an integral part of physical refining, in which the aim is to reduce the content of FFA to a maximum 0.1%, to reduce the concentration of volatile components and to produce an oil with a bland taste. This is achieved by operating under high vacuum at temperatures between 220 and 270 °C and in contact with steam.

10.2.1.8 Hydrogenation (h)

In the presence of a catalyst, typically nickel, the addition of hydrogen increases the melting point of the oil. This increase occurs because the double bonds are either partly or fully saturated. At the same time, a more oxidation-stable fat is produced, due to both the lower saturation level and the reduction of oxidation agents (aldehydes, peroxides etc.). Hydrogenation can be partial (only some of the double bonds are saturated) or full (all double bonds are saturated). In the case of partial hydrogenation, some of the remaining double bonds will be in the *trans* configuration; the *trans* fatty acid level of full hydrogenation will be close to zero.

10.2.1.9 Interesterification (ie)

Interesterification is a modification process in which the fatty acids of the various triglycerides are randomly rearranged to create a new fat with a new triglyceride distribution. This process is carried out under vacuum with the presence of a catalyst, typically sodium methylate (chemical interesterification, IEC), or with the help of enzymes (enzymatic interesterification or rearrangement, IEE)

10.2.1.10 Dewaxing/winterisation (wi)

This process is applied to certain seed oils, typically sunflower oil, to achieve an oil that will remain clear at lower temperatures. Dewaxing is achieved by cooling the oil to below the crystallisation temperature of its waxes. The crystallised waxes are then removed by filtration.

10.2.1.11 Dry fractionation (df)

The aim of dry fractionation is to separate high-melting triglycerides from low-melting triglycerides. This is achieved by cooling the oil to below the crystallisation temperature of the high-melting triglycerides. In a subsequent filtration, the crystallised triglycerides are removed from the noncrystallised triglycerides, creating the stearin and olein fractions, respectively.

10.2.1.12 Soapstock splitting (ss)

The soapstock that is formed during the neutralisation and/or interesterification process is split into acid oil and acid water. This process can be carried out either batchwise or continuously. A strong acid is used as the splitting reagent; normally sulfuric acid. The acid oil is separated from the water by gravity. The acid water is further treated and neutralised to obtain an effluent, which is discharged in line with local regulations.

10.2.2 Process routes for straight refined oils and fats

The recommended refining routes used to obtain straight fully refined oils and fats for the most common oil types are given in Table 10.1. The process

Table 10.1 Process routes for straight refined oils and fats.

		ddg	n	osb	tsb	d	ds
Soybean oil	crude	x			x		x
	degummed		x	x		x	
	ddg				x		x
Sunflower oil	crude	x			x		x
	degummed		x	x		x	
	ddg				x		x
Rapeseed oil	crude	x			x		x
	degummed		x	x		x	
	ddg				x		x
Maize germ oil	Crude	x			x		x
	Ddg				x		x
Cottonseed oil	crude		x	x		x	
	neutralised				x		x
Olive oil	crude, <5% FFA				x		x
	crude, >5% FFA		x	x		x	
Groundnut oil	crude		x	x		x	
	ddg				x		x
Palm oil	crude				x		x
	rbd				x		x
Palm kernel oil	crude				x		x
	rbd				x		x
Coconut oil	crude				x		x
	rbd				x		x

routes can be divided into chemical refining and physical refining. The main differences between the two processes are:

- During chemical refining, FFA are removed by saponification with the help of NaOH (caustic soda) in the neutralisation (n) step. Soapstock is removed by settling or centrifugation.
- During physical refining, FFA are removed by distillation at high temperature under deep vacuum during the deodorisation/stripping (ds) step.

The starting oils that are to be treated can take different qualities. The main ones are:

- *Crude*: the oil/fat is not treated after the oil milling step.
- *Degummed*: the oil is simply water degummed to remove a part of the phospholipids (P < 200 ppm).
- *Deep degummed (ddg)*: the oil is deep degummed to remove the major part of the phospholipids (P < 30 ppm).

- *Semiprocessed (rbd)*: the oil/fat is refined before the process, but requires further refining due to its storage, transport and handling history.
- *Neutralised*: mainly for cottonseed oil, which needs to be neutralised immediately after milling to remove unwanted components (gossypol).

10.2.3 Process routes pre- and post-hydrogenation

Oils to be hydrogenated need to be free from catalyst poisons. The main catalyst poisons are: phospholipids, soap, FFA and sulfuric components present in some seed oils, such as rapeseed and soybean oil. In Tables 10.2 and 10.3, the recommended process routes for oils and fats prior to hydrogenation are shown. For a description of the different crude oil qualities, see Section 10.2.2.

After hydrogenation, residual nickel still present in the oil following the main catalyst filtration needs to be removed by a two-step bleaching process.

Table 10.2 Process routes prehydrogenation.

		n	osb[b]	tsb	d	ds
Soybean oil	crude	x	x			
	degummed	x	x			
	ddg[a]			x		x
Sunflower oil	crude	x	x			
	degummed	x	x			
	ddg[a]			x		x
Rapeseed oil	crude	x	x			
	degummed	x	x			
	ddg[a]			x		x
Maize germ oil	crude	x	x			
	ddg[a]			x		x
Cottonseed oil	crude	x	x			
	neutralised			x		x
Groundnut oil	crude	x	x			
	ddg[a]			x		x
Palm oil	crude			x		x
	rbd			no pretreatment[c]		
Palm kernel oil	crude			x		x
	rbd			no pretreatment[c]		
Coconut oil	crude			x		x
	rbd			no pretreatment[c]		

[a]ddg oils can alternatively be neutralised before hydrogenation, depending on equipment availability.
[b]Bleaching is only necessary when the soap content after neutralisation is above 250 ppm.
[c]If FFA $> 0.5\%$, neutralisation is required.

Table 10.3 Process route post-hydrogenation.

	n	osb	tsb	d	ds
All fats			x		x

Typical hydrogenation off-flavours and FFA formed by reaction with water require a deodorisation/stripping step.

10.2.4 Process routes pre- and post-IEC

The main catalysts used for IEC are sodium methylate and sodium ethylate. These catalysts are largely deactivated by the following minor components in oil: water, FFA and hydroperoxides. Therefore, oils need to be chemically neutralised or physically refined before interesterification. Refined, bleached, deodorised (RBD) oils can be used without pretreatment if they are not oxidised or hydrolysed during transportation and handling. Drying of the oils to a very low water level ($<0.01\,\%$) is considered part of the interesterification process.

The recommended process routes for oils and fats undergoing IEC are given in Tables 10.4 and 10.5. In many cases, a mixture of oils and fats is used for interesterification. The mixture can be made before pretreatment or after,

Table 10.4 Process routes preinteresterification (IEC).

		n	osb[a]	tsb	d	ds
Soybean oil	crude	x				
Sunflower oil	degummed	x				
Rapeseed oil	hydrogenated			no pretreatment[b]		
Coconut oil	crude			x		x
	rbd			no pretreatment[b]		
	hydrogenated			no pretreatment[b]		
Palm kernel oil	crude			x		x
	rbd			no pretreatment[b]		
	hydrogenated			no pretreatment[b]		
Palm oil and fractions	crude			x		x
	rbd			no pretreatment[b]		
	hydrogenated			no pretreatment[b]		

[a]Bleaching is only necessary when the soap content after neutralisation is above 250 ppm.
[b]If FFA > 0.1%, neutralisation is required.

Table 10.5 Process route post-interesterification (IEC).

	n	osb	tsb	d	ds
All fats		x			x

depending on equipment availability, quantities, the quality of an individual component and so on. The process route for a mixture is mainly determined by its largest component. For a description of the different crude oil qualities, see Section 10.2.2.

Deactivation of the catalyst by water washing and drying is considered to be part of the interesterification process. Bleaching and deodorisation are a must after IEC, in order to remove reaction products such as soap and methyl esters.

10.2.5 Process routes pre- and post-IEE

To protect the enzymes from degradation and activity loss, all oils entering the process should be of a fully refined quality and free from inorganic acid residues. After IEE, the oils require deodorisation to remove off taste components.

10.2.6 Process routes in dry fractionation and dewaxing

The concentration of phosphatides should be as low as possible before dewaxing by filtration. This is achieved by neutralisation and bleaching in chemical refining and deep degumming and two-step bleaching in physical refining.

Full physical refining is recommended before fractionation. High-temperature deodorisation of the olein may cause a change of melting point due to interesterification at high temperatures in the deodoriser.

Tables 10.6 and 10.7 show the recommended process routes for oils and fats that are dry fractionated or dewaxed.

10.3 Oil processing block diagram design

10.3.1 Standard oil processing block diagrams

The design of the oil processing block diagram will be based on the processing sequence recommended in Section 10.2. Additional information required to detail the design includes:

Table 10.6 Process route pre fractionation and dewaxing.

		n	osb	tsb	d	ds
Palm oil	crude			x		x
Sunflower oil	crude	x	x			
	degummed	x	x			
	ddg			x		

Table 10.7 Process route post fractionation and dewaxing.

	n	osb	tsb	d	ds
Palm oil (fractions)			no post treatment[a]		
Sunflower oil			d or ds		

[a]Post-treatment is necessary if during the storage and transport of palm oil fractions the FFA content increases to >0.1%.

- production volumes;
- equipment design capacities;
- working pattern and operational time;
- required flexibility (batch or continuous lines);
- crude and refined oil storage requirements.

This section will propose standard oil processing block diagrams based on the process routes recommendations and required flexibility. These standard block diagrams are selected based on the most common working practices; different local circumstances may lead to deviating block diagram designs. Later sections will deal with effective equipment capacity and tank park design rules.

10.3.2 Batch and continuous processes

The refining and oil modification processes can be carried out in batch or continuous equipment, while deodorisation can be carried out semicontinuously in an intermediate form.

Batch processes are mainly performed within a single piece of equipment; the process steps are sequential in time. The input is discontinuous at the start of the sequence, the output discontinuous at the end.

In continuous processes, the process steps are simultaneous in different pieces of equipment; both input and output are more or less constant over time.

Comparison of batch and continuous processes gives the following main advantages and disadvantages.

10.3.2.1 Batch processes

Main advantages:

- short product changeover time;
- suited for small production lots;
- flexible recipe;
- simple maintenance;
- can be operated manually.

Main disadvantages:

- limited scope for heat recovery;
- requires sequence control;
- many parallel lines for high capacity (space requirement).

10.3.2.2 Continuous processes

Main advantages:

- suited for high-capacity lines (low space requirement);
- input/output heat recovery;
- simple automation and control;
- low manning level.

Main disadvantages:

- long product changeover time;
- complex and costly maintenance;
- high electrical energy consumption.

Due to economical constraints, only small or specialised refineries will operate mainly with batch equipment (batch size: 5–20 tonnes). Fully continuous plants are used for large runs of the same products: continuous refining of seed oils on the same site as the seed oil extraction plant, for example. Product changeover in a continuous plant means a loss of production time; this loss depends on the average residence time of the oil in the equipment (oil content divided by capacity) and the product intermixing specifications. High-capacity flexible refineries consist of continuous neutralisation lines, continuous and batch bleachers (batch for post-treatment after hydrogenation and/or IEC, for example) and semicontinuous deodorisers. Hydrogenation, IEC and fractionation are almost always batch processes. IEE is a continuous process.

10.3.3 Refining of straight oils and fats

10.3.3.1 Chemical refining

Chemical refining is mainly applied for degummed seed oils (see Section 10.2.2 and Table 10.1). It can also be applied for crude seed oils in stand-alone refineries (not connected to an onsite extraction plant), which do not have the ability to add the gums from the deep degumming process to the extracted meal.

Chemical refining has the following disadvantages compared to the physical refining route:

(1) Higher oil loss due to entrainment of neutral oil in the soapstock.
(2) The need for a soapsplitting facility.
(3) Higher liquid effluent discharge (the acid water out of the soapsplitting plant).

The main advantage of the chemical refining process is its flexibility in obtaining a high-quality product, more or less independently of feedstock pretreatment and quality.

Figure 10.1 shows the schematic outline of a continuous chemical refining process for seed oils. Crude or degummed oil storage is followed by continuous neutralisation, continuous bleaching and continuous deodorisation, and finally the fully refined oils are stored in the refined oil storage. Figure 10.1 does not show intermediate buffer storage tanks, since these are considered part of the continuous lines. The following intermediate buffer storage tanks are normally installed:

- A feed tank for the continuous neutralisation line, to allow a feed of a constant quality.
- Buffer tanks between continuous neutralisation, continuous bleaching and continuous deodorisation, to buffer during stoppage or breakdown of one of the lines.

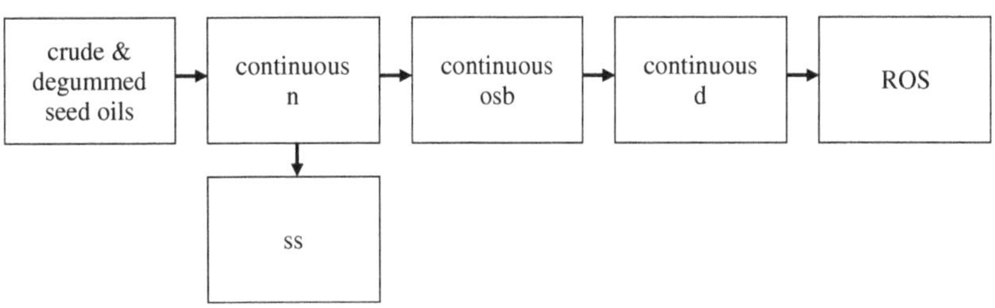

Figure 10.1 Continuous chemical refining of seed oils. ROS, refined oil storage. For other abbreviations, see Section 10.2.

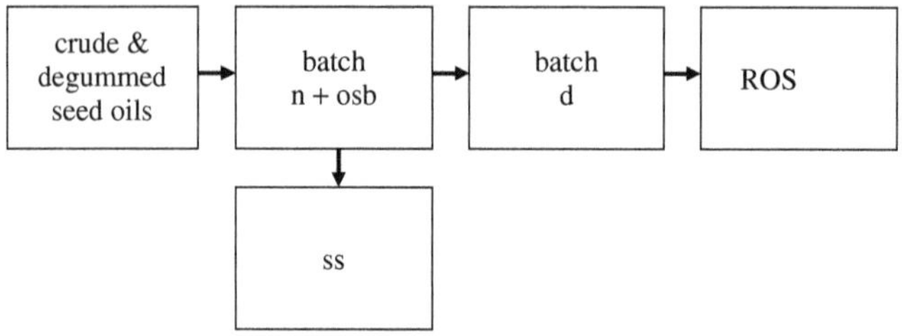

Figure 10.2 Batch chemical refining of seed oils. ROS, refined oil storage. For other abbreviations, see Section 10.2.

Figure 10.2 shows the schematic outline of a batch chemical process. The batch neutralisation and batch bleaching are normally done in the same vessel, followed by batch deodorisation. A buffer tank is installed between the bleaching earth filter and batch deodoriser to improve capacity utilisation of both processes.

10.3.3.2 Physical refining

Physical refining can be applied for seed oils after deep degumming of the crude oil. The phosphorus level following the deep degumming process should be below 30 ppm, in order to avoid uneconomically high bleaching earth levels in the bleaching step. The deep degumming process is preferably carried out in the seed oil extraction plant. In that case, the gums can be recycled to the meal. A deep degumming process for standalone refineries is less attractive, since the gums are microbiologically unstable and should be dried or utilised immediately after production.

Figure 10.3 gives an outline of the physical refining process for seed oils (for process buffers, see Section 10.3.3.1). The bleaching step following deep degumming should be two-step bleaching (see Section 10.2.2), in order to hydrate and agglomerate residual phospholipids with acid and water before their removal with bleaching earth. The deodorisation/stripping temperature should be such that FFA are removed and the *trans* fatty acid level remains below the specified limit (see van Duijn *et al.*, 2006). Figure 10.4 shows the block diagram of the physical refining process for tropical oils.

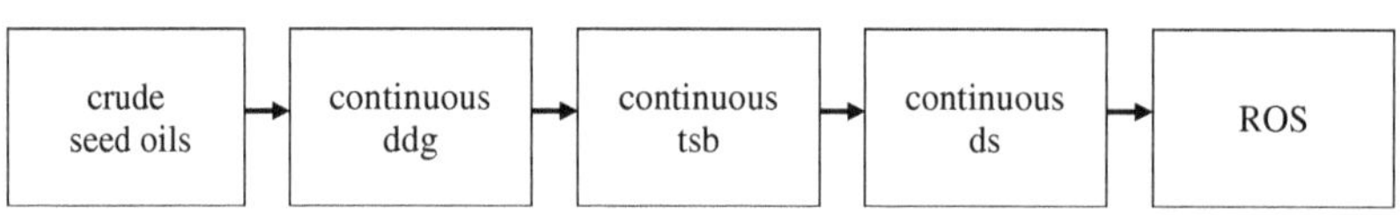

Figure 10.3 Physical refining of seed oils. ROS, refined oil storage. For other abbreviations, see Section 10.2.

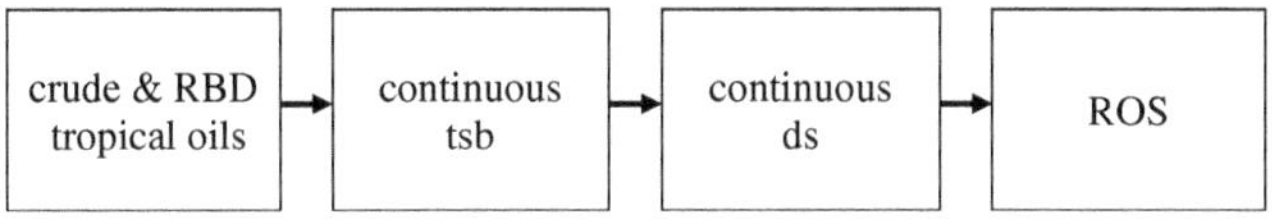

Figure 10.4 Physical refining of tropical oils. ROS, refined oil storage. For other abbreviations, see Section 10.2.

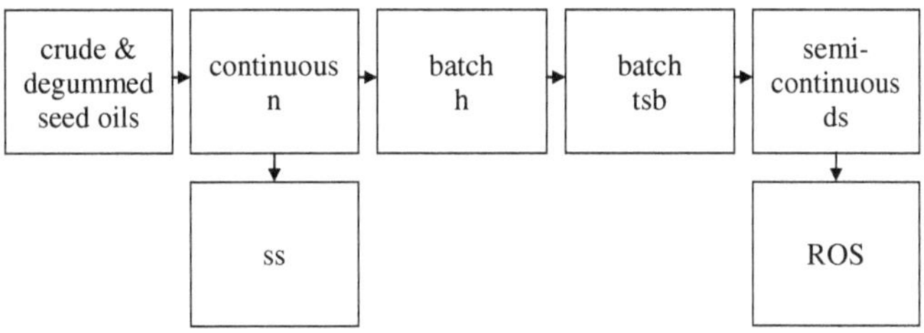

Figure 10.5 Chemical refining and hydrogenation of seed oils. ROS, refined oil storage. For other abbreviations, see Section 10.2.

10.3.4 Refining combined with hydrogenation

In the case of batch refining, crude and degummed seed oils require neutralisation and bleaching before hydrogenation (see Figure 10.5), resulting in low FFA, phosphorus and soap levels. In continuous refining, bleaching can be omitted where there is good-quality centrifuge neutralisation, which results in low soap levels. In this latter case, the catalyst dosing will be slightly higher, but this on-cost will be largely compensated for by the omission of the bleaching step.

Batch two-step bleaching (or an acid treatment followed by filtration) and semicontinuous deodorisation/stripping are mostly applied after hydrogenation. Starting with seed oils, a range of products with different melting points can be produced by the process steps sequence of centrifuge neutralisation, partial hydrogenation to various melting points, batch two-step bleaching and semicontinuous deodorisation/stripping. This sequence is largely applied in the soybean oil-based refineries in the USA, for example.

In physical refining of deep degummed seed oils and tropical oils, the oils need to be two-step bleached and deodorised/stripped before hydrogenation (Figure 10.6). The post-refining sequence is identical with the chemical refining route.

10.3.5 Refining combined with interesterification

Figure 10.7 gives the total process sequence for chemical refining, IEC and post-bleaching followed by deodorisation/stripping. IEC is a batch process.

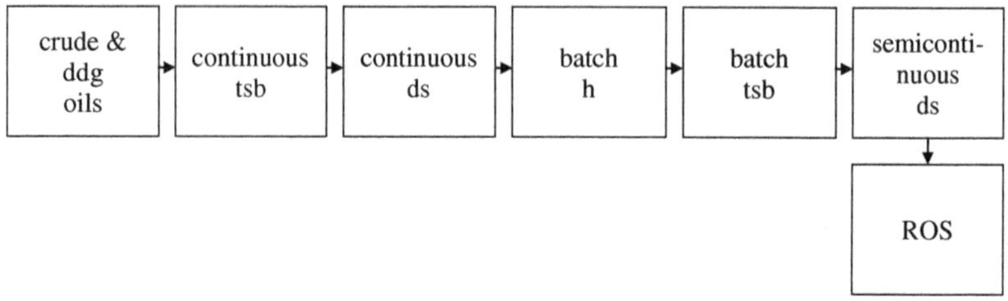

Figure 10.6 Physical refining and hydrogenation of ddg seed oils and tropical oils. ROS, refined oil storage. For other abbreviations, see Section 10.2.

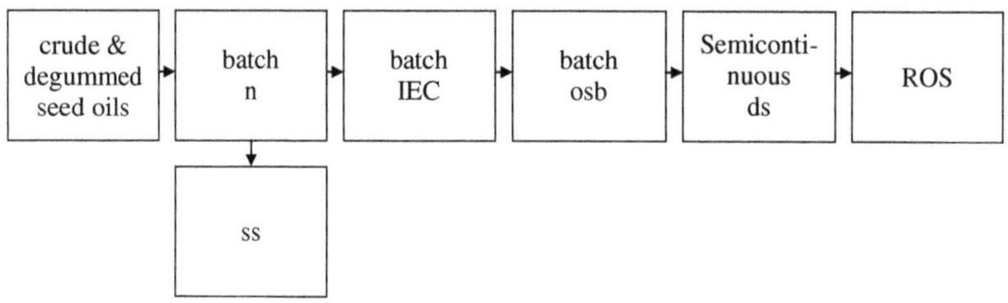

Figure 10.7 Chemical refining and IEC of seed oils. ROS, refined oil storage. For other abbreviations, see Section 10.2.

Preneutralisation and post-bleaching are therefore often performed in the same batch vessel. This vessel should be equipped with a caustic and water dosing facility, soapstock draining, a recirculation loop and deep vacuum for drying, a catalyst dosing unit and a dosing system for bleaching earth (multipurpose vessel).

All oil components are already bleached and deodorised before they enter the interesterification vessel in combined physical pretreatment/IEC (see Figure 10.8). Post-bleaching can still be performed in the multipurpose vessel.

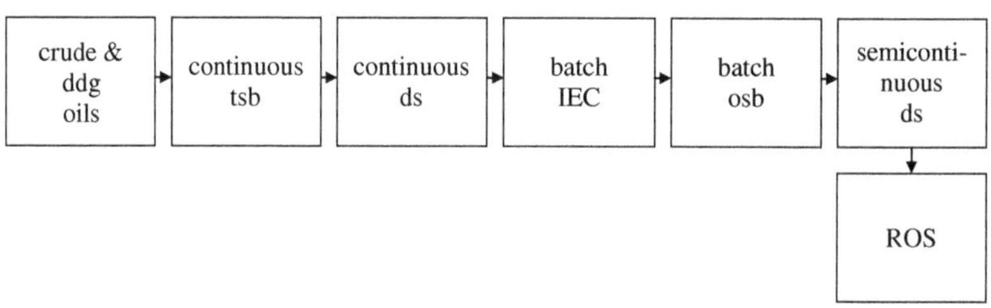

Figure 10.8 Physical refining and IEC of ddg seed oils and tropical oils. ROS, refined oil storage. For other abbreviations, see Section 10.2.

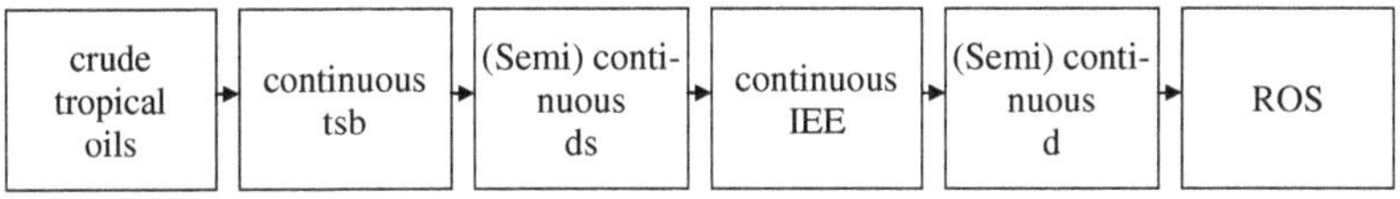

Figure 10.9 Physical refining and IEE of tropical oils. ROS, refined oil storage. For other abbreviations, see Section 10.2.

IEE is a continuous process; the feedstocks are fully refined materials, though without citric acid dosing after deodorisation. Enzymatically interesterified oils need only post-deodorisation/stripping to remove FFA and odour/taste components. This deodorisation can be continuous if it is linked with one IEE line, or semicontinuous if it is linked with several lines. Figure 10.9 shows the block diagram for the combination of physical refining and enzymatic interesterification of tropical oils.

10.3.6 Refining and dewaxing

Predewaxing can be combined with deep degumming or neutralisation and is not discussed in this section. Final dewaxing and filtration with filter aid is a continuous process following bleaching. Input/output heat exchange is required to reduce heating and cooling energy input.

10.3.7 Refining and fractionation

Fractionation of palm oil is normally done after full physical refining of the oil (see Section 10.2.6). In most fractionation filters, the olein resides in closed pipes while the stearin is in open contact with the environment. Special care should be taken to avoid contamination of the stearin by dirt from the environment (high-care zone).

10.3.8 Production of *trans*-free hard fats

Until the mid 1990s, partial hydrogenation was the most widely applied technique for increasing the melting point of unsaturated oils and fats. Next to saturation, hydrogenation inevitably involves *cis*/*trans* isomerisation of some of the double bonds. Publications in the early 1990s indicated that *trans* fatty acids have a negative effect on blood cholesterol and, hence, on coronary heart disease risk. This resulted in a gradual elimination of partial hydrogenation as an oil modification technique.

The combination of full hydrogenation (iodine value, IV < 2), interesterification and fractionation on a wide variety of feedstocks will produce hard fats with a wide range of melting performances (van Duijn *et al.*, 2006).

Liquid seed oils first need full hydrogenation to generate solids. These fully hydrogenated oils then require interesterification with nonhydrogenated oils in order to reduce solid levels at high temperatures. These solid levels can be further reduced by fractionation (see Figure 10.10).

The presence of relatively high solid levels in natural tropical oils creates more flexibility in process routes. Fractionation of palm oil alone will produce a relatively soft stearin with a relatively long tail of the solid phase line. Stearin is as such not optimal for structuring products such as margarine. Fractionation of palm oil (or double fractionation) followed by interesterification with palm kernel oil will produce nonhydrogenated fat phase components with more suitable solid phase lines. Full hydrogenation of palm oil and palm kernel oil followed by interesterification is an alternative way of obtaining fat components without fractionation.

This combination of techniques (see Figure 10.10) creates a tool for producing optimal fat phase products. This tool is almost as flexible as partial hydrogenation.

A multipurpose refinery will combine the refining and processing steps outlined in the previous sections. The combination of chemical refining for degummed seed oils, physical refining for crude seed oils (including a deep degumming process step) and tropical oils with the *trans*-free hard fat modification techniques just mentioned is illustrated in Figure 10.11. This figure shows the increase of complexity compared to the combination of

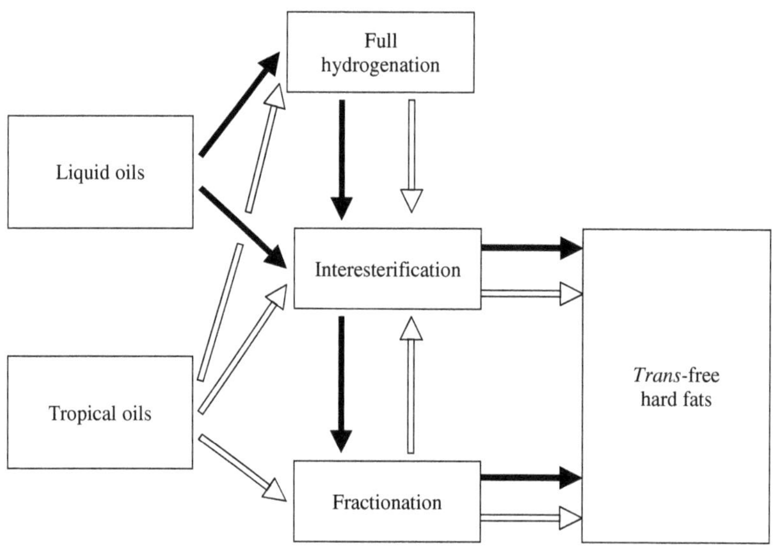

Figure 10.10 Virtual *trans*-free modification techniques.

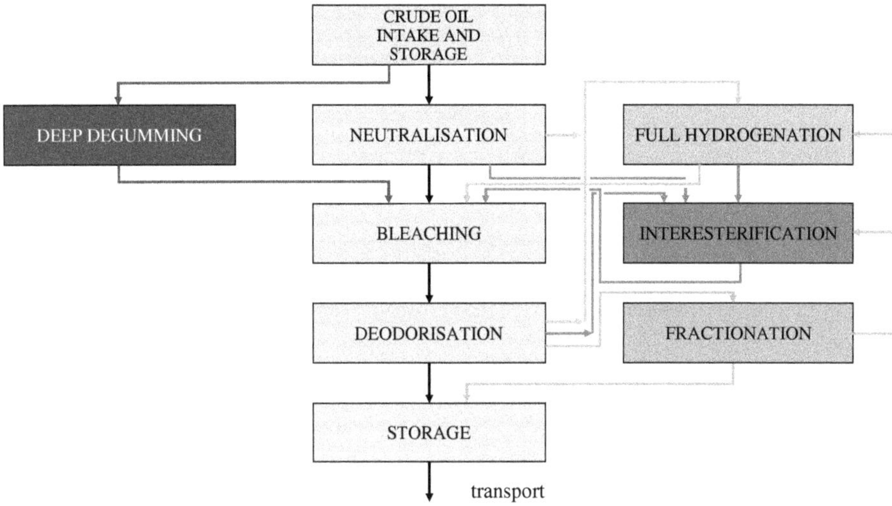

Figure 10.11 An oil processing flowsheet, combining the virtual *trans*-free modification techniques with both chemical and physical refining.

refining and partial hydrogenation mentioned in Section 10.3.4 and illustrated by Figure 10.5.

10.4 Effective equipment capacity

The design of an oil processing block diagram is in general based on an annual volume estimate, corrected for seasonal variations. Equipment manufacturers specify the capacity of their equipment as an hourly or 24-hours production volume. This section gives an estimation of annual production capacities, starting from the hourly capacity as specified by the equipment manufacturer.

First, the effective production time is estimated by the following analysis:

(1) The total time in a year (see Figure 10.12, T) is $365 \times 24 = 8760$ hours (the extra day every 4 years is not taken into account).
(2) The available production time (A) is total time minus unavailable time. Unavailable time is statutory and religious holidays, weekends, shifts not worked and enforced factory shutdowns.
(3) The used production time (U) is available time minus available unused time. Available unused time is time lost because there are no production orders.
(4) The operational time (O) is used time minus planned non-operational time. The planned non-operational time is scheduled maintenance and planned tests without production.

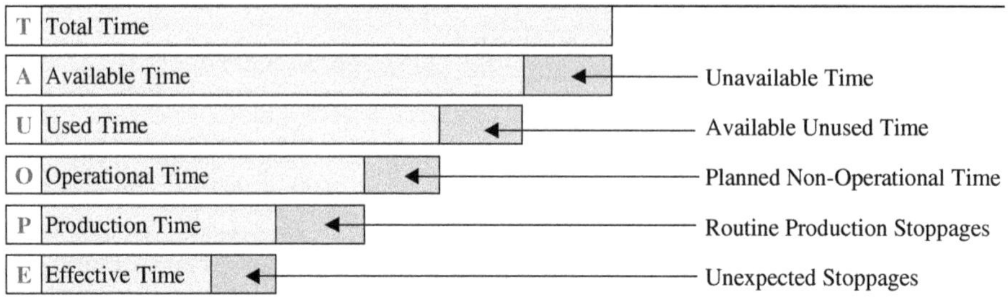

Figure 10.12 Effective machine time analysis.

(5) The production time (P) is operational time minus routine production stoppages. These stoppages are startups and shutdowns, cleaning and breaks (meals, rest etc.). Startups and shutdowns are important in 5-days-a-week operations; breaks are not applicable in fully automated plants.
(6) The effective time (E) is operational time minus unexpected stoppages. Unexpected stoppages are unplanned repairs or breakdowns, failures of material supply, full or empty tanks, absences of operators and so on.

The time difference between total and operational time can be estimated from events that can be planned in advance. The time difference between operational and effective time contains unpredictable events (cleaning, breakdowns etc.). Thus the operational efficiency (EO) has been introduced:

$$EO = E/O \times 100\% \tag{10.1}$$

EO is an average over years of experience and depends on type of equipment, operating mode (batch or continuous) and weekly working pattern (5 × 24 hours or 7 × 24 hours a week).

The EO given in Table 10.8 can be used as guideline for oil processing design.

The estimated annual capacity can now be calculated from the hourly capacity specified by the equipment manufacturer:

$$\text{annual capacity} = \text{hourly capacity} \times EO \times \text{operational time (tonnes/annum)} \tag{10.2}$$

Table 10.8 Guideline EO for oil processing design.

EO (%)	5 days a week	7 days a week
Batch equipment	85	85
(Semi)continuous equipment	85	95

10.4.1 Example: calculation of effective times for 5- and 7-days-a-week operations

10.4.1.1 5 days a week

- A year has 52 weekends of 48 hours.
- There are three additional stops of 24 hours for public holidays.
- The preventive maintenance stop is 5 days of 24 hours.
- Calculation:
  - total time = 8760 h/a;
  - unavailable unused time = 2368 h/a;
  - planned non-operational time = 120 h/a;
  - operational time = 6072 h/a;
  - effective time all equipment (EO = 85%) = 5161 h/a.

10.4.1.2 7 days a week

- Stops for public holidays: 5 days a year.
- Preventive maintenance stop: 6 days of 24 hours.
- Calculation:
  - total time = 8760 h/a;
  - unavailable unused time = 120 h/a;
  - planned non-operational time = 144 h/a;
  - operational time = 8496 h/a;
  - effective time batch equipment (EO = 85%) = 7222 h/a;
  - effective time continuous equipment (EO = 95%) = 8071 h/a.

Note that a continuous refinery running 5 days a week has only 64% of the annual capacity of a continuous refinery running 7 days a week, while the operational time at 5 days a week is 71% of that at 7 days a week. This difference is caused by the weekly startups and shutdowns due to weekend stoppages.

10.5 Tank park design rules

10.5.1 Storage capacity

The oil processing plant will store crude or semiprocessed oils and fats unloaded in large volumes from ships, trains and trucks. The crude oil tank park should have sufficient capacity to store the received volume in separate empty tanks. Topping up of partly full tanks can only be allowed if the oil in the tank and the arriving oil are of equal type and quality, and if the arriving oil has been checked and approved for quality and food safety. Ship deliveries will have an uncertainty around their arrival date and time, leading to additional buffer storage requirement. At arrival, ships need to be unloaded within a restricted time interval in order to avoid penalties. Train deliveries are normally well planned and have defined time arrival and departure times. Truck deliveries are more flexible but may still have restrictions due to traffic conditions, limited circulation at weekends and so on.

Oil processing plants on the same site as the oil mill will store crude oils to match differences in capacities and to buffer for upstream breakdowns.

Oil Processing plants will store oils in between processing steps to match differences in capacities and to buffer for upstream and downstream breakdowns. The storage capacity for breakdown buffering should be sufficient to buffer the volume of one production shift (8 hours' production).

Fully refined oils are stored after deodorisation to match differences in capacities or working patterns of the oil processing plant and its customers, to buffer for upstream breakdowns, to deliver full tank-car loads and to store components in case of blends delivery. Fully refined oils and fats should be stored for as short an amount of time as possible in order to minimise quality downgrading.

A minimum of two tanks per oil type should be used for continuous production lines; this allows one tank to receive the deodorised oil while trucks are loaded from the other one, and makes complete emptying during changeover from one tank to the other possible.

Multiproduct sites will normally make flexible use of the available tanks. Spare tanks will allow regular emptying. Topping up of a remaining quantity of oil in a storage tank with a new batch of identical product is only possible if the following topping-up rules are applied:

- The previous batch in the tank should not be too old (more than a few days).
- The previous batch should have a bland taste.
- The chemical and physical parameters of the previous batch should be within specifications.
- The tank must be regularly emptied, so as to ensure traceability and avoid residues with a very long residence time.

10.5.2 Degradation during storage

Oils and fats will slowly deteriorate during storage due to chemical reactions. Storage also introduces the risk of downgrading through intermixing or contamination. This results in a loss of quality and/or value of the oil. The main causes of degradation are:

- *Hydrolysis*: decomposition of triacylglycerol to FFA and diacylglycerol further to monoacylglycerol.
- *Oxidation*: reaction with oxygen from the air.
- *Intermixing with other types or qualities of oil.*
- *Contamination by chemicals or impurities.*

10.5.2.1 Hydrolysis

This chemical reaction requires dissolved water. The rate increases with temperature and the reaction is autocatalysed by FFA. Figure 10.13 shows the hydrolysis of palm oil, saturated with dissolved water, as a function of time, temperature and the initial FFA concentration (Hilder, 1968). This graph shows that minimum oil loss by hydrolysis is obtained at low temperature (storage temperature should be at least 5–10 °C above melting point in order to avoid fractionation), low initial FFA content and short storage time. Also, the dissolved water concentration in the oil should be kept as low as possible by avoiding water addition from steam blowing of pipes, rainwater through open manholes, leaking heating coils and so on.

10.5.2.2 Oxidation

Contact of oils and fats with oxygen present in the atmosphere causes chemical changes in the product, which downgrades its quality (through formation of hydroperoxides, which further react to aldehydes and ketones). Some of the effects of oxidation may be rectified by refining, which involves some extra processing, and therefore extra cost. However, the deterioration can be so severe that it can never be fully rectified. Much is to be gained by reducing

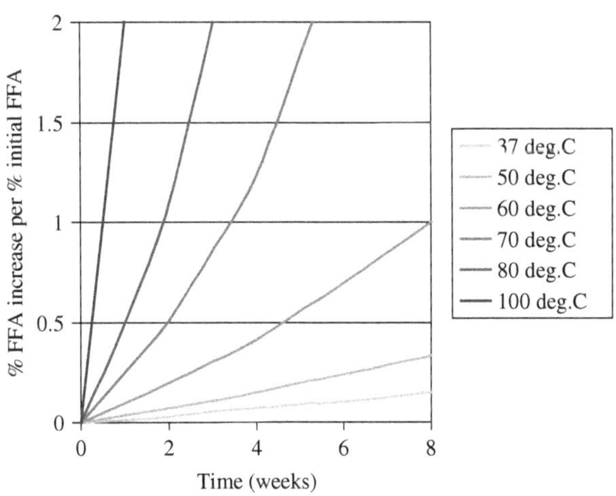

Figure 10.13 Free fatty acid (FFA) increase over time as a function of temperature and initial FFA concentration (Hilder, 1968).

the air contact, and this principle is the basis of several tank design rules (see Section 10.5.3). Oxidation proceeds more rapidly as temperature increases, so each operation must be carried out at the lowest practicable temperature. The rate of oxidation is greatly increased by the catalytic action of copper or copper alloys, even when trace amounts are present. Because of this, copper and copper alloys must be rigorously excluded from the systems. Other metals, such as iron, also have a catalytic effect, although it is less than that of copper.

The oxidation sensitivity depends on the level of unsaturation of the oil: highly unsaturated oils like soybean and rapeseed oil oxidise more easily than more saturated tropical oils like palm and coconut oil.

10.5.2.3 Intermixing with other oils

Intermixing with other types or different qualities of oil may lead to a serious value loss. In most cases, the mixture obtains the value of the cheapest component, if the intermixing level is above a generally accepted technically unavoidable limit. In some cases, the mixture may even downgrade to the value of rework. The risk of intermixing can be minimised by tank dedication, zoning of piping, adequate cleaning of tanks and pipes before product changeover and proper topping-up procedures.

10.5.2.4 Contamination by chemicals or impurities

Contamination by stored chemicals can be avoided through full separation of oil storage and chemicals (including fatty acids) storage. Cleaning with food-grade chemicals followed by flushing with water and oil should be applied for new tanks and following tank repairs. Pumps and tank stirrers should be placed in such a way that lubricant oil cannot come in contact with the product. Tanks should be heated with low-pressure steam or hot water; thermal heating oil should be excluded.

All openings to the tanks environment (like manholes, unloading connections, hoses etc.) must be kept closed when not in use, so as to avoid the entrance of impurities such as sand, bird droppings or even animals.

Impurities in the crude oil may settle with water to form tank bottoms. When processed, these tank bottoms result in high refining losses. In some cases, they cannot be processed and must be discharged as chemical waste. Tank bottoms can be avoided by a proper design of the oil outlet, combined with controlled agitation.

10.5.3 Tank design rules

10.5.3.1 Tank shape and material of construction

The most suitable shape is the vertical, circular cross-section tank with self-supporting fixed roof, the latter preferably conical in shape (see Figure 10.14).

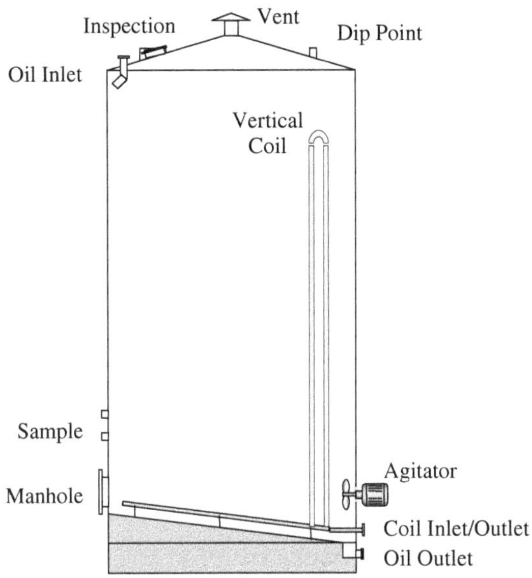

Figure 10.14 The basic design of a crude oil storage tank.

Tall, narrow tanks are preferred as they minimise contact between the oils or fats and oxygen from the air.

The bottoms of large crude oil tanks should be sloped (with a sump with a separate outlet) to facilitate draining. The bottoms of refined oil storage tanks should be conical, to allow complete emptying.

All openings (manholes, inlets, outlets etc.) should be made such that they can be locked and/or effectively sealed.

All materials used in the construction of tanks and for ancillary equipment (including heating facilities) should be inert to oils and fats and should be suitable for use in contact with food.

Stainless steel is the preferred material for the construction of tanks. Stainless steel will never cause an iron contamination and stainless-steel tanks can be used for all types of oils and fats.

Mild steel tanks can be used for the storage of crude oils and of fully refined liquid oils at low temperature, provided that the protective layer formed by polymerised oil is not removed or damaged by water cleaning.

Copper and its alloys (brass, bronze, gun metal etc.) must never be used in contact with oils and fats.

Temperature gauges containing mercury should not be used, since they may break and lead to mercury contamination of the oil.

Glass equipment and glass sample bottles should be avoided in situations where breakage might lead to contamination.

10.5.3.2 Tank heating

All tanks for solid or semisolid oils and fats (including liquid oils in low-temperature climates) should be provided with heating coils so that the product is liquid and homogenous when transferred or unloaded. Heating coils should be made of stainless steel.

Heating by hot water (about 90 °C) is the generally preferred system, as it is the least likely to cause local overheating.

If steam heating is used, the steam pressure should not exceed 2.5 bar (maximum temperature 127 °C). The heating coils' position should enable easy cleaning of the tank bottom. At least one vertical coil should be provided, to create a channel of melted fat from the top to the bottom of the tank and thus prevent any overpressure caused by the volume increase during melting of solid fats or by steam pressure from a leaking coil. This also prevents the collapse of the tank through pumping out of oil while the top of the tank content is still solid.

As a guideline: for insulated tanks with hot water, a coil area of between 0.1 m^2 per tonne of tank capacity (for small tanks) to 0.05 m^2 per tonne of tank capacity (for large tanks) is sufficient to heat the contents of the tank at 5 °C per day or to melt solidified fat at 5% per day.

For the melting solidified of fats, hot water heating requires more time than steam heating.

10.5.3.3 Tank insulation

Storage tanks should preferably be insulated (including the roof), particularly in moderate and cold climates. Heated tanks should always be insulated. Insulation must be designed to avoid the absorption of oil or water in the insulation material.

10.5.3.4 Avoiding air contact

Filling can be done from the bottom or via a dip pipe from the top to the bottom. When pipe blowing is applied, filling should be from a pipe at the top, with the opening pointing to the tank wall, so as to obtain an oil flow along the tank wall downwards. This avoids splashing and aeration of the oil.

Nitrogen blanketing is generally not recommended for crude and partly processed oils, but can be applied for fully refined oils in longer (>3 days) storage.

It is recommended that storage tanks for crude oils and fats be provided with proper agitation (e.g. side-entry stirrers or ejector mixers) in order to prevent settling of impurities during storage. The stirrers have to be designed and operated in such a way that air entrainment and the formation of tank bottoms are effectively avoided during storage. This can be achieved by a side-entrance stirrer positioned near the bottom of the tank running intermittently. The stirrer should be switched off if the oil reaches a low level close to the top of the stirrer to avoid air entrainment.

Tanks for fully refined oils and fats should not be stirred.

10.5.4 Piping design rules

10.5.4.1 Materials

Mild steel is acceptable for all crude and semirefined oils and fats pipelines, though stainless steel is preferable.

Flexible hoses that are used to connect pipelines during loading and unloading must be of inert material, be suitably reinforced and be of such a length and design as to make cleaning easy. Exposed ends should be capped when not in use. Couplings for flexible hoses should be of stainless steel or another inert material.

10.5.4.2 Insulation and heating

Pipelines used for oils and fats which may solidify at ambient temperatures should preferably be lagged and also be provided with heating, for example by steam tracing lines or electrical heating tape. The pipe temperature should not exceed 110 °C, in order to avoid overheating of the oil or fat.

10.5.4.3 Layout

It is preferred to have dedicated lines for groups of oils of a similar fatty acid composition when different types of oils and fats are stored (as e.g. palm oil, lauric oils and liquid oils).

The layout of the pipelines should be such that draining by gravity is aided and not hindered. Horizontal pipes should have a slight slope downwards in the direction of the flow. Blowing of pipes should always be from high to low; if needed, a small-diameter pipe can be used to route the blown oil upwards.

A pigging system can be used for line clearing when different types of oils and fats are transported via one pipeline.

10.6 Design estimates for utilities consumptions and effluent production

10.6.1 Introduction

Oil processing processes require such utilities as:

- Energy in the form of steam, electricity and hot water.
- Water for cooling and cleaning.

- Compressed gases such as air and nitrogen for control purposes, oil protection and blowing.

The processes produce waste in the form of:

- Liquid effluent from neutralisation/soapsplitting and deodorisation.
- Solid waste from bleaching and dewaxing.
- Exhaust gases from deodorisation.

The quantification of utilities consumptions and effluent production requires detailed equipment characteristics, environmental parameters (such as cooling water temperature) and oil quality data. These will not be available in the early phase of the design process. Estimated utilities and effluent data may be required to help justify a project and the design principles of a utility installation. This chapter will supply estimates based on best-practice data which can be used for first-design purposes.

10.6.2 Utilities

10.6.2.1 Heating

During processing, oil needs to be heated at the following stages:

(1) *At arrival*: heating to storage temperature (for oils which are stored in heated tanks), in case of a temperature drop during transport.
(2) *In the tank park*: to compensate for heat losses during storage in heated tanks.
(3) *From storage temperature to first process temperature* (for neutralisation, 95 °C; for bleaching, 70–120 °C): continuous processes may have heat recovery; the oil output has a temperature lower than the process temperature.
(4) *In chemical refining*: from neutralised oil storage temperature to bleaching temperature.
(5) *In hydrogenation*: oil needs to be heated to the temperature at which the reaction starts (120–150 °C); the reaction itself produces heat.
(6) *In IEC*: the oil must be heated to drying temperature (max. 110 °C); the IEE temperature is relatively low (70 °C) and no additional heating is required.
(7) *In dewaxing after bleaching*: the oil needs to be heated from dewaxing temperature to buffer temperature, or directly to deodorisation temperature.
(8) *From bleached oil buffer temperature to deodorisation temperature*: this is by far the largest temperature difference; heat recovery is normally applied.

(9) *In fractionation*: the stearin needs to be melted and heated to 15 °C above melting point.
(10) *In refined oil storage*: the temperature needs to be maintained at 15 °C above melting point.

The heating in steps 1–7 is or by condensing steam or hot water. The high temperature in deodorisation is reached through a closed high-pressure steam loop heated by a light fuel oil, a gas-fired boiler or high-pressure steam. Hot water is preferred for heating after deodorisation (steps 9 and 10).

The heat loss of a storage tank can be calculated by means of the overall heat transfer coefficient:

$$\Phi = U.A.(\Delta T) \tag{10.3}$$

where Φ is heat loss (W), U is the overall heat transfer coefficient (W/m^2 °C), A is the surface area through which the heat is transmitted (m^2) and ΔT is the temperature difference between the oil in the tank and the ambient air (°C).

For a first estimate, the area of the tank walls plus roof can be taken as:

$$A = 4.8\,V^{2/3} \tag{10.4}$$

where V is the total tank volume.

The overall heat transfer coefficient for this area can be taken as:

$$U\ (\text{unlagged}) = 7\,\text{W/m}^2\,{}^\circ\text{C} \tag{10.5}$$

$$U\ (\text{lagged}) = 1\,\text{W/m}^2\,{}^\circ\text{C}\ (\text{for a lagging thickness of } 50\,\text{mm}) \tag{10.6}$$

This assumes that the tank is in regular use (on average 50% full). The temperature difference ΔT is the difference between the average tank contents temperature and the average ambient temperature in the tank farm area.

The mass of steam required to heat 1 tonne of oil is given by the relation:

$$M_{steam} = C_{oil}.\Delta T.1000.(1 + X/100)/C_{steam} \tag{10.7}$$

where M_{steam} is the mass of steam per tonne of oil in kg/tonne, C_{oil} is the specific heat of the oil in kJ/(kg °C), ΔT is the temperature difference in °C, X is the relative heat loss to the environment in % and C_{steam} is the specific heat of steam in kJ/kg.

The specific heat of oil is temperature dependent and varies from 2.0 kJ/(kg °C) at 20 °C to 2.5 kJ/(kg °C) at 200 °C.

The specific heat of steam is the heat released by condensation and further cooling of the condensate. For low-pressure steam, this is around 2200 kJ/kg.

The light fuel oil consumption to heat oil in a deodoriser via a closed high-pressure steam circuit is given by:

$$M_{fuel} = C_{oil}.\Delta T.(1-Y/100).(1+X/100)/(H_{fuel}.E_{boiler}/100) \qquad (10.8)$$

where M_{fuel} is the mass of fuel required to heat 1 tonne of oil in kg/t, C_{oil} is the specific heat of oil in kJ/(kg °C), ΔT is the difference between deodorisation temperature and oil inlet temperature in °C, Y is the heat recovery in %, X is the relative heat loss to the environment in %, H_{fuel} is the heat of combustion of light fuel oil in kJ/kg and E_{boiler} is the boiler efficiency in %. The heat of combustion of light fuel oil (H_{fuel}) is 42 000 kJ/kg.

10.6.2.2 Open steam and vacuum

Open steam is used in the deodorisation process to remove volatile components by stripping. The mass of open steam per tonne of deodorised oil depends mainly on the oil temperature and vacuum pressure. Deodoriser configuration will have a secondary effect.

In the refining process, vacuum is used for drying (after neutralisation, during bleaching, before interesterification) and degassing (to evacuate hydrogen from a hydrogenation vessel and for deaeration before deodorisation), and to enhance the stripping effect of open steam in deodorisation. The vacuum pressure for drying after neutralisation, during bleaching and during degassing is relatively high (100 mbar). This can be generated by the combination of a condenser and an ejector set or vacuum pump for the noncondensable gases. Drying before IEC requires a pressure of around 20 mbar, which necessitates a booster between the interesterification vessel and condenser to compress the vapours from 20 mbar to condenser pressure.

The pressure in deodorisation is normally between 2 and 6 mbar. Such a pressure can be obtained in one of two ways:

(1) By compressing the deodoriser vapours to condenser pressure using one booster or two boosters in series. The motive steam flow of these boosters is between 2 and 4 times the open steam flow of the deodoriser.
(2) By condensing at a very low temperature (around –20 °C). The deodoriser steam freezes on the coils of the condenser (dry condensing). The very low temperature of the coolant inside these coils is produced by a cooling compressor driven by electrical energy.

The booster system consumes much more steam, the dry condensing system slightly more electricity. The total energy balance is in favour of the dry condensing system.

10.6.2.3 Electrical energy

General electricity use is through consumption by pumps, stirrers, electrical tracing and heating, lighting and the process control system.

Specific machines with relatively high electricity consumptions are:

- Centrifuges in deep degumming and neutralisation.
- Cooling compressors in dewaxing, fractionation and dry condensing.
- Cooling water circulation pumps and cooling towers.
- Vacuum pumps.

10.6.2.4 Cooling water

In oil refining, the refined oil will leave the site at more or less the same temperature as that at which it entered the site. Hence the net effect of heating to process temperatures (after heat recovery) should be compensated for by cooling.

Cooling to temperatures slightly above ambient can be achieved by circulating the cooling water that passes the process heat exchangers directly over a cooling tower.

Cooling to temperatures slightly above the nearby open-water (river, lake or sea) temperature can be achieved by circulation of open water directly through the process heat exchangers. To minimise the risk of pollution and corrosion of the process heat exchangers, it is preferable to cool indirectly. A closed loop that passes the process heat exchangers will exchange heat with open water in a secondary heat exchanger.

In oil modification, the temperature of the product leaving the site may be different (in general higher) than that of the product entering the site.

10.6.2.5 Gases

Compressed air is mainly used for control purposes as a driving gas for controlling valves. Compressed air is sometimes used to blow filter cake at low temperature after dewaxing. It is not recommended to use air blowing at high temperatures after bleaching, because of the fire risk.

Compressed nitrogen has several applications:

- As a gas blanket in storage tanks of sensitive oils, to reduce oxidation.
- To blow filter cake after bleaching and catalyst filter cake after hydrogenation.
- As sparging gas in the outlet pipeline of a deodoriser, to reduce the risk of air entrainment.
- To blow pipelines.
- As a safety gas in hydrogenation (Section 10.7.4.1) and deodorisation (Section 10.7.3.3).

Nitrogen can be delivered as liquid nitrogen in containers, or else can be produced onsite by pressure swing adsorption or fractional distillation of liquid air. A reduced oxygen level in air due to the leakage of nitrogen inside closed areas may lead to oxygen depletion, resulting in loss of consciousness and even death (see Section 10.7.5.2).

Hydrogen is used in the hydrogenation process. Hydrogen can be produced onsite by gas reforming or electrolysis. It can also be transported to the site as high-pressure gas (200 bar) and reduced to working pressure in a pressure-reduction station.

10.6.3 Effluent

10.6.3.1 Liquid effluent

The following more or less constant liquid effluent flows are the result of oil processing operations:

- *Acid water from soapsplitting of soapstock from neutralisation.* This low-pH acid water is, before discharge or further treatment, neutralised with sodium hydroxide to obtain a pH around 7. Washwater from neutralisation or IEC can be added to the flow either before soapsplitting or after soapsplitting but before neutralisation. The neutralised effluent flow contains sodium sulfate, phosphates, some residual fatty matter, metals or metal compounds and organic residues, depending on the processed oil type. The total effluent flow is more or less equal to the sum of all process input flows (dilution water of chemicals, washwater, bowl flush water of centrifuges etc.). Sometimes, cooling water is added to the neutralised effluent flow to reduce the sodium sulfate concentration to below a critical limit, in order to avoid degradation of concrete sewer pipes.
- *The bleed of the alkaline recirculation system or the melted condensate of a dry condensing system.* The bleed of the alkaline recirculation system requires acidification to recover the fatty matter. The total effluent mass flow is more or less equal to the sum of the open steam flows and the booster/ejector motive steam flows.

Incidental liquid effluent discharges can occur due to:

- floor cleaning;
- equipment or tank cleaning;
- rainwater from the inside of closed tank farm bunds.

Fat-containing flows should pass a fat trap, to avoid blocking of sewer pipes by accumulated fat.

10.6.3.2 Solid waste

The main waste discharged in an oil processing operation is spent bleaching earth. The volume and composition of spent bleaching earth depends strongly on the type and quality of the bleached oil. The following components may be found in spent earth:

- *Moisture*: after steam blowing of the cake with dry steam, the moisture level will be limited (<10%). Higher moisture levels may occur in the case of hot water treatment (up to 50%) or water spraying to prevent autoignition.
- *Oil/fat*: this will vary between 20% for well-blown vertical leaf filters to 50% for nonblown cakes. In further treatment, the fatty matter may be valued as acid oil or as a source of energy. It may also create inconveniences such as self-ignition, and mineral nitrogen consumption in the case of biological degradation.
- *pH*: the (water phase) pH of spent earth will normally be low, due to the acid activation of the fresh earth and the residues of phosphoric or citric acid from acid pretreatment.
- *Active carbon*: bleaching earth may contain active carbon, used for the adsorption of polycyclic aromatic hydrocarbons (PAHs). Disposing of spent earth that contains active carbon is in general not a problem; in fact, in energy recovery applications the carbon will make a positive contribution. However, in some applications carbon may have a negative influence. Therefore, the waste user should be informed when carbon is present.
- *Polycyclic aromatic hydrocarbons (PAHs)*: assuming a crude coconut oil that contained 40 ppb heavy PAH has been treated with 0.2% carbon and 0.8% earth, and the spent earth from coconut oil bleaching is 20% of the total spent earth mixture, the final PAH content of the spent earth will be about 800 ppb.
- *Pesticides*: pesticides are hardly adsorbed on bleaching earth (with the exception of pirimiphos-methyl); the pesticide content of spent earth is normally negligible.
- *Impurities*: spent earth will contain colour components, soaps, salts and so on, on the parts-per-million level.
- *Nickel*: after the removal of the Ni catalyst from the hardened oil by filtration, the oil will contain a maximum 5 ppm nickel. Adsorption on to bleaching earth may lead to nickel contents of up to 2000 ppm.
- *Other heavy metals*: after agglomeration, other heavy metals present in the oil (Cr, Cu, Pb, Zn etc.) will be concentrated on the earth. The total level of heavy metals (excluding Ni) will normally not exceed 100 ppm.
- *Calorific value*: the calorific value of a normally blown filter cake, including moisture, is relatively low (around 10 000–15 000 kJ/kg for a steam-blown cake).

In the past, disposal of spent bleaching earth was only a minor problem in oil refining. After defatting by caustic or hexane extraction to recover acid oil, the residue could be dumped on public discharges or used as landfill. Sometimes the earth disposal even created financial revenue, due to the value of the recovered acid oil. The oil refiner ceased to be liable for damage caused by the spent earth once the waste had been removed from their site.

Today, disposal of spent earth is an important problem in oil refining, as a result of the increased interest in and concern for environmental matters. Spent bleaching earth is considered to be an industrial waste, which means that transport and treatment are strictly regulated and controlled. The restriction in disposal outlets, the increase of administration costs and the use of more costly transport and treatment methods have increased the disposal costs. The main disposal outlets are:

- *Animal feed*: this application is limited to earth that is low in contaminant levels (PAH adsorbed by active carbon, nickel from bleaching after hydrogenation etc.) and is restricted to refineries that are fully linked to an oil mill (processing the same oil).
- *Agricultural recycling*: direct recycling on agricultural land is not recommended. The preferred route is to add spent earth to other organic waste and then compost it.
- *Incorporation in building materials*: in this application, the inert part of the spent bleaching earth is incorporated in cement, bricks or expanded clay. During the fabrication process, the organic part of the earth is burnt, contributing to the energy needed for the process.

Spent filter aid from dewaxing and inactive hydrogenation catalyst will normally have a residual value, due to the sunflower oil or nickel content.

10.6.3.3 Exhaust gases

The noncondensable gas outlet of the vacuum unit of deodorisation is the main source of odorous gases. Other sources are the soapslitting area, acid oil storage and crude oil storage of smelly oils (coconut oil and fish oil). Reduction of odours can be obtained by passing the gasses through a water scrubber or a biological filter.

10.6.4 Utility consumption and effluent data per process

10.6.4.1 Storage

The steam consumption of storage tanks of different capacities is calculated using Equations 10.3–10.8 and the following assumptions:

- The tank is lagged and the heat transfer coefficient is 1.0 W/m^2 °C.
- The tank is on average half full with oil during the effective time.
- The temperature of the oil stored in the tank is 50 °C (e.g. palm oil). The oil enters the tank at storage temperature. The yearly average ambient temperature in the tank park area is 10 °C.
- The tank is effectively used for 8000 hours per year. It is empty and not heated for the remainder of the year.
- The tanks are completely filled 12 times per year.

Table 10.9 shows the average steam consumption per tonne of stored product required to keep oil at storage temperature. The calculations are made for tanks with storage capacities of between 100 and 5000 tonnes. Storing oil in a single large tank always results in a lower steam use per tonne of product than storing the same amount of oil in two smaller tanks. For example, storing 60 000 tonnes per annum of palm oil in one 5000 tonne tank uses 60% of the heating energy required to store the same amount of oil in five tanks of 1000 tonnes each.

The electricity consumption per tonne of stored oil varies between 5 and 10 kWh, depending on the frequency of pumping and the use of stirrers. An additional 5 kWh/tonne should be added if electrical tracing of pipes is applied.

10.6.4.2 Refining processes

Table 10.10 shows, for each refining process step, the estimated utility consumptions and liquid effluent or solid waste productions per tonne

Table 10.9 Heating energy requirement to keep oil in storage tanks at storage temperature.

Tin	50.0 C	Latent heat of steam	2250 kJ/kg
Tout	10.0 C	1 kW =	1.60 kg/h steam
DT	40.0 C	Effective time	8000 h/year
HTC =	1.0 W/m2C	No batches per year	12

Oil content tons	Tank volume m3	Effective surface m2	Heat loss kW	Heating steam kg/h	Annual steam use t/y	Steam use per t product kg/t
100	122	118	4.73	7.6	61	50
200	244	188	7.51	12.0	96	40
500	611	346	13.83	22.1	177	29
1000	1222	549	21.95	35.1	281	23
2000	2444	871	34.84	55.7	446	19
5000	6111	1604	64.18	102.7	821	14

Table 10.10 Estimated utilities consumptions and liquid effuent or solid waste productions per tonne of processed oil.

Process	**Utilities and effluent per ton oil processed**							
	Steam kg	light fuel oil kg	Electricity kWh	total energy Mj	cooling water m3	Nitrogen m3	liquid effluent m3	solid waste kg
Deep degumming (ddg)	55		15	175	2			
Centrifuge neutralisation + soapsplitting (cn + ss)	100		10	256	3		0.2	
Batch neutralisation (bn)	150		6	352	4		0.4	
Bleaching (osb and tsb)	60		6	154	2	4		5-30
Dewaxing/winterisation (wi)	30		15	120	3			10-20
Batch deodorisation (d)	280		20	688	6	1	0.14	
Semi conti deodorisation (d and ds)								
-with alkaline recirculation	140	8	18	709	16	1	0.14	
-with dry condensing	30	8	20	474	5	1	0.03	
Conti deodorisation with dry condensing (d and ds)	25	3	17	242	5	1	0.03	

of processed oil. The following inputs are used for the deodoriser calculations:

- batch deodorisation:
  - deodorisation temperature = 180 °C;
  - no heat recovery;
  - deodorisation pressure = 5 mbar;
  - condenser water outlet temperature = 32 °C.
- semicontinuous deodorisation:
  - deodorisation temperature = 240 °C;
  - heat recovery = 40%;
  - deodorisation pressure alkaline recirculation = 5 mbar;
  - condenser water outlet temperature = 32 °C;
  - deodorisation pressure dry condensing = 3 mbar.
- continuous deodorisation:
  - deodorisation temperature = 240 °C;
  - heat recovery = 80%;
  - deodorisation pressure dry condensing = 1.5 mbar.

The calculations show that the total energy for semi continuous deodorisation with alkaline recirculation is almost 3 times as high as the total energy for continuous deodorisation with dry condensing.

10.6.4.3 Modification processes

The estimated utilities consumptions and liquid effluent and solid waste productions for the oil modification process are given in Table 10.11. This table shows a significant lower total energy use for IEE than for IEC. The inactive spent catalyst is not given as a solid waste because of its relative high value due to its nickel content.

10.7 Occupational safety by design

10.7.1 Introduction

Oil processing operations are carried out under conditions that may introduce occupational safety hazards: high temperature, high pressure or vacuum, in high buildings, sometimes using aggressive chemicals, with a risk of fire or explosions and so on. The design of the installation should address these occupational safety risks and contribute to a zero-accidents working environment. A structured and systematic examination of the existing or designed process should be carried out to identify and evaluate problems that

Table 10.11 Estimated utilities consumptions and liquid effuent or solid waste productions per tonne of processed oil.

Process	**Utilities and effluent per ton oil processed**							
	Steam Kg	Light fuel oil kg	Electricity kWh	Total energy Mj	cooling water m3	Nitrogen m3	Liquid effluent m3	Solid waste kg
Hydrogenation (h)	55		15	175	2	3		
Chemical interesterification including post bleaching and soapsoplitting (iec + b + ss)	150		6	352	4	3	0.2	10
Enzymatic interesterification (iee)	22		1	52	0	0.5		0.66
Fractionation	80		25	266	3			

may represent risks to personnel or equipment. Hazard and operational study (HAZOP) is such a structured approach (Tyler *et al.*, 2008). In HAZOP, a process flow diagram is examined in detail, specifying its design intention, possible deviations, feasible causes and likely consequences. HAZOP studies should be carried out by local teams on actual design details or in existing plants. This chapter cannot and will not replace these local HAZOP studies; rather, it provides an overview of process design-related hazards, which can be used as one of the inputs for these local studies.

The basis of all safety systems is documentation, education and continuous improvement. All safety procedures should be well documented, operators and contractors should be trained and accidents and near accidents should be evaluated in order to produce improved procedures.

10.7.2 General hazards

The following hazards are not directly process-related but do belong to the main causes of accidents in oil processing operations:

- *Slippery floors*: spillage of oils and fats or fatty vapours will result in an oil or fat film on the surfaces of floors, gangways and staircases inside the refinery buildings, as well as on roads in the direct surroundings. Slipping and falling by operators, contractors and visitors may lead to serious injuries. Proper selection of surface material, frequent cleaning, antislip safety shoes and mandatory use of handrails will reduce this risk. Open floor grating will minimise slipping but can lead to more serious injuries, and frequent cleaning of the grating is required to avoid fouling. Fouled grating will rapidly transfer flames between floors if there is a fire.
- *Burns from hot water and steam*: pipes transporting hot water and steam are normally well insulated. Insulation should be properly reinstalled after pipe repairs. The following provide risk of contact with hot water and steam:
  - exposure during deblocking of pipes using steam or hot water;
  - cleaning of floors or equipment using hot water (risk of getting hot water in shoes or boots);
  - steam or hot water from steam traps, overpressure releases and overflow of closed containers.
- *Falling from height*: refinery buildings, structures around equipment (deodorisers) and tanks are all high. Gangways are normally protected by handrails, but operators or contractors may have to work on unprotected high surfaces during maintenance or construction work. Fall-prevention equipment (such as safety harnesses) should be used if secure fencing is not practical. Also, objects may fall from height and cause serious injuries.

Removal of loose objects and the use of helmets to protect operators, contractors and visitors will reduce this risk.

- *Noise*: high noise levels may occur close to rotating machines, during (steam) blowing operations and at the discharge of self desludging centrifuges. As far as possible, noise should be reduced at the source through better design and sound-protection walls. In normal practice, noise cannot be fully eliminated, and operators, contractors and visitors should wear ear protection, such as earplugs or earcaps.
- *Eye protection*: spillage of chemicals, dust and aggressive vapours may cause eye damage. Use of proper safety glasses will protect against this.

10.7.3 Main occupational hazards of oil refining

10.7.3.1 Neutralisation and soapsplitting

- *Centrifuge vibration*: strong vibrations must be prevented in order to avoid damage to the centrifuge bearings. Excessive vibrations can eventually lead to the centrifuge becoming disconnected from its support and starting to 'walk'. Centrifuge disc stacks rotate at a high rotation speed (>4000 rpm), resulting in a high rotation energy. They will continue to rotate for a relative long period (around 1 hour) after a power shutdown or disruption, moving the centrifuge through the building and causing serious damage. Regular maintenance and vibration control will eliminate this risk.
- *Chemicals*: strong chemicals are used in the neutralisation and soapsplitting process. Direct contact with these chemicals should be avoided by following proper procedures and using personnel protection. Caustic soda (sodium hydroxide) is a strong base. It reacts violently with acids and is highly corrosive to some metals (like zinc and aluminum), causing possible formation of hydrogen gas. Sulfuric acid is a strong mineral acid. It reacts violently with alkalis and water and is highly corrosive to metals, again with possible formation of hydrogen gas. The addition of caustic soda and sulfuric acid to water results in the evolution of large quantities of heat, which can cause local boiling. Caustic and sulfuric acid solutions are dangerous to human tissue. Contact with the skin will cause serious chemical burns. A few seconds of contact can result in the loss of an eye or in impaired vision due to scarring. Ingestion causes severe damage to the throat and deeper tissues. Exposure reduction through good design (critical couplings behind transparent plates) and proper handling procedures should be the basis for occupational safety. Goggles and chemical-resistant clothing, boots and gloves should be used as additional measures. Showers/eye showers followed by immediate medical attention is the correct procedure if accidents occur.

- *Risk of fire*: a potential fire hazard exists at the outlet of the air evacuation system of the vacuum set of the dryer, due to possible traces of hexane in crude seed oils. This risk can be reduced by proper positioning of the outlet opening and by marking the area around the opening as an explosion risk area.

10.7.3.2 Autoignition of spent bleaching earth

Autoignition of spent bleaching earth can occur at much lower temperatures than the flash point of edible oils (300–350 °C). It is probably caused by the oxidation of residual oil with the liberation of the heat of reaction and the formation of oxidation products. The heat generated by oxidation is only slowly transferred by the surrounding fatty bleaching earth, which may result in hot spots in the spent bleaching earth; oxidation products in these hot spots can reach their autoignition temperature, which is much lower than the edible oil flash point.

The oil–oxygen contact area is small, and therefore the risk of autooxidation is low, when the spent bleaching earth is completely saturated with oil. The heating effect is limited when there is less than 5% fatty matter in the spent earth. The critical fatty matter range for autoignition of spent bleaching earth is:

$$5\% < \text{fatty matter in spent bleaching earth} < 40\% \tag{10.9}$$

After blowing with nitrogen and/or dry steam, the spent bleaching earth contains a maximum 10% moisture and 20–50% fatty matter. The risk of autoignition increases with increasing oxygen–oil contact time, and also with increasing unsaturation of the oil; high-risk oils are linseed oil, soybean oil, rapeseed oil and sunflower oil.

The following principles should be applied to minimise the risk and/or damage of autoignition:

- Spent bleaching earth should remain inside the refinery building for as short a length of time as possible. This can best be achieved if the spent bleaching earth outlet of the filters is outside the building. Alternative solutions are discharge in a small container that is directly transported outside the building and transport of the spent earth outside the building by a closed slow-speed chain conveyor.
- Containers of spent bleaching earth should be regularly removed from the spent bleaching earth area and transported to spent bleaching earth treatment/disposal.
- The spent earth containers should be regularly inspected for starting fires (inspection every shift; a starting fire can be detected by its smell or by carbon monoxide detectors).

- Spent bleaching earth should never be left onsite if personnel is not present (e.g. during the weekend, in the case of a 5-days-a-week operation).
- The transporter and end user of the spent earth must be informed by writing of its self-ignition properties.
- After emptying, the containers must be free from spent bleaching earth residues.

10.7.3.3 Deodoriser safety

The deodoriser is operated at the highest temperature of the refining plant. The oil is kept under noncombustible conditions, due to the absence of air in the deodoriser. A fire hazard is possible when air is entrained in the deodoriser at high temperature. The following preventative measures should be used to avoid this situation:

- Properly maintained equipment, to avoid air leaks.
- Trained personnel, to identify air leaks and react to starting fires.
- Well-established procedures for startup, shutdown and maintenance.

A nitrogen supply line, activated by remote control, must be connected to the deodoriser so that the oxygen level can be decreased in the event of fire.

A fire hazard potentially exists at the outlet of the air evacuation system of the vacuum set, due to possible traces of hexane in crude seed oils.

The high-pressure boiler develops up to 90 bar pressure. Its condition and safety system integrity must be periodically checked.

A deodoriser must be protected from overpressure by a bursting disc and a pressure relief valve combination opening at 1.5 bar.

During repair or maintenance of the deodoriser, the required vessel entry procedures (see Section 10.7.5.1) should be applied.

10.7.4 Main occupational hazards of oil modification

10.7.4.1 Hydrogenation safety hazards

Hydrogenation is a reaction of hot oil with highly inflammable hydrogen under pressure using a metallic catalyst with a high specific surface area. Operators work in the direct environment of the hydrogenation equipment to inspect, take samples, operate valves, clean and do repairs. This combination of risk factors categorises hydrogenation as a high-safety-risk process.

- *Safety related to pressure, temperature and working environment*: vessels, coils and all connections should be regularly tested for leaks. The exothermic reaction at high reaction temperature requires an effective temperature control system and proper insulation of vessels and pipes. Clear safety procedures, including cooling, isolation and safe evacuation of hydrogen, are needed in case of repairs and maintenance.

- *Safety related to hydrogen*: hydrogen is highly inflammable, and mixtures of hydrogen with oxygen (air) are highly explosive even at relative low temperatures or ignition energies. Hydrogen escaping from a pressurised vessel or pipe into air may autoignite (static electricity) and burn with an invisible high-temperature flame.

Precautions:

- Avoid hydrogen–air mixtures at any time and in any part of the total hydrogenation system, including buildings, storage and so on.
- Two methods can be applied to avoid hydrogen–air mixtures during filling and emptying of the hydrogenation reactor:
  - *the hydrogen blanket system*: outside the reaction, the vessel is always kept under a slight hydrogen overpressure by connecting it with a hydrogen gasholder;
  - *the hydrogen evacuation system*: during filling and emptying, the vessel is filled with air; it is evacuated by an explosion-proof vacuum system when changing from air to hydrogen or vice versa.
- *Minimise leaks from the hydrogen system to air*: avoid flanges and regularly inspect all couplings, especially those at high pressure. Start from the assumption that all leaking/escaping hydrogen will autoignite.

Hydrogen leak = hydrogen fire.

- Avoid all ignition sources and high-temperature spots in the complete hydrogenation system and its surroundings (protected electrical and control equipment, special tools etc.).
- Special attention should be paid to the hydrogen unloading system where hydrogen is delivered by high-pressure (200 bar) road tankers. A leaking coupling or disrupted hose, pipe or flange in the high-pressure area will result in an explosion followed by a several-metres-long jet flame. The unloading station must protect operators and equipment against the effects of possible leaks (explosion and jet flame). However, it should be open enough for easy access by fire-fighting services and for the escape of personnel in case of an emergency.
- Always purge lines and installations with nitrogen before any maintenance activities are carried out.
- *Safety related to catalyst handling*: the active catalyst surface (mostly nickel) will rapidly oxidise in contact with air. This will lead to high temperatures and the risk of fire. Therefore, the catalyst particles should always be surrounded by fatty material. Fresh catalyst is suspended in a fat with high melting point during its manufacture and is distributed as fatty flakes or

pills. Cake blowing after catalyst filtration should be well controlled, in order to avoid a too 'dry' cake. Spent catalyst has to be stored and disposed of while immersed in solid fat.

10.7.4.2 Safety of IEC

The interesterification catalysts sodium methylate (NaOMe) and sodium ethylate (NaOEt) are hazardous materials and should be handled according to the appropriate preventative measures. NaOMe/Et with moisture (even humid air) produces NaOH (caustic, corrosive) and (m)ethanol (highly flammable). The NaOH part causes severe injuries on bodily contact (with the skin or eyes, or by inhalation or ingestion). The (m)ethanol part ignites when in contact with hot surfaces or ignition sources; the autoignition temperature is 40–70 °C.

NaOMe/Et is normally delivered in closed plastic bags of 5–10 kg; the bags are transported in drums. The following measures minimise the risk of fire and contact with dust from bag handling or catalyst dosing:

- Store the drums in closed dry areas. The drums should be electrically connected to earth.
- Transport the bags from drum to dosing system inside a container (such as a bucket).
- Use the complete contents of one or more bags; there should be no leftovers in the bags after use.
- Use protective clothing, goggles and mask or a glove box to open the bag.
- Submerge empty bags in a drum filled with water.
- Do not unblock the dosing pipes with steam; instead, prevent pipe blocking by regularly flushing the dosing pipe with water.

In case of fire, dry NaOMe/Et burns with small blue flames. However, violent eruptions of several-metres-long flames occur when water is sprayed on this fire. Therefore, never use water to fight NaOMe/Et fires! Gently covering fires with a blanket or sand or letting them burn seems to be the best solution.

10.7.5 Main occupational hazards of oil storage and handling

10.7.5.1 Access to tanks and processing vessels

Entry into tanks or processing vessels is a high-risk operation; the entry is relatively small (a manhole), the tank may still be hot, the atmosphere in the tank may be low in oxygen, there may be residues of oil, oil may be pumped in the tank, stirrers may start to rotate and so on. Therefore, access to tanks must always be covered by a Permit-to-Work: a set of procedures designed to prepare the tank before entry. This Permit-to-Work should be confirmed by a safety certificate signed by production, maintenance and the contractor (if applicable), posted close to the point of entry. The certificate should be properly cancelled when the work inside the tank is completed and all persons have left the tank.

The Permit-to-Work should include a check of internal conditions and the complete isolation of the tank:

- A visual check that the tank is empty.
- Closing of all connections to the tank (including heating and oil in- and outlets) by blind flanges.
- A gas test to check the oxygen level in the tank.
- A temperature check of the tank walls and inside piping.
- Isolation of any motor drive by the removal of electrical fuses.

There should always be a person standing close to the manhole, in constant contact with the person inside while work is in progress. A safety harness is required for entry via a top manhole or for work at height inside the tank. The person entering the tank must apply hygienic measures like hand washing, boot protection, a hairnet and so on.

10.7.5.2 Top access to tank cars

Top access to tank cars like rail cars or road tankers is necessary for visual inspection of the interior of the tanks, for the opening, closing and sealing of manholes and sometimes for sample taking. Falling from the top of the tanker (around 3 m high) may lead to serious injuries and can even be fatal. The following safety measures should be in place (see Figure 10.15 for an example of bad practices):

- Safe access to the top of the tank by a ladder connected to the tanker or via the stairway of the loading station.
- A gangway fixed to the top of the tank, with open floor grating to minimise slipping.
- Guarding to protect the operator or driver from falling: either safety rails connected to the loading bay or handrails on top of the tanker.
- Use of nonslip footwear and open floor grating on stairs and gangways in the loading bay.
- Daily cleaning of floor areas, including roads and the ground, to minimise oily surfaces.
- Use of a safety harness in case proper guarding cannot be guaranteed.

An increased level of nitrogen may be present inside the tanks if the tanker is flushed with nitrogen before loading or if nitrogen is sparged during

Figure 10.15 Unsafe top access to a road tanker.

loading. A reduced oxygen level in the air due to the presence of excess nitrogen may lead to oxygen depletion, resulting in loss of consciousness and even death. Leaning into the manhole presents a significant risk if increased nitrogen levels are present in the tank (empty or in the headspace above the oil). Operators and drivers must be made aware of the risk related to increased levels of nitrogen. The use of nitrogen should be clearly marked on the manhole cover and specified in the vendor certificate and vehicle load documents.

References

Hilder, M.A. (1968) The solubility of water in edible oils and fats. *J. Am. Chem. Soc.*, **45**, 703–707.

Tyler, B., Crawley, F. & Preston, M. (2008) *HAZOP: Guide to Best Practice*, 2nd edn. Rugby, UK: IChemE.

Unilever (1988) Unilever Oil Processing Recommendations: E3 – Best Proven Practice Process Routes.

van Duijn, G., Dumelin, E.E. & Trautwein, E.A. (2006) Virtually *trans* free oils and modified fats. In: C. Williams & J. Buttriss (eds) Improving the fat content of foods. Cambridge, UK: Woodhead Publishing, pp. 490–507.

Further Reading

Chapter 1

Lai, O.-M., Tan, C.-P. & Akoh, C. (2012) *Palm Oil: Production, Processing, Characterization, and Uses*. Champaign, IL, USA: AOCS Press.

Mittaine, J.-F. & Mielke, T. (2012) The globalisation of international oilseeds trade. *OCL*, **19**(5), 249–260.

Chapter 2

Cuvelier, M.-E. & Maillard, M.-N. (2012) Stability of oils during storage [in French]. *OCL*, **19**(2), 125–132.

Chapter 3

Dijkstra, A.J. (2013) *Edible Oil Processing from a Patent Perspective*. New York, NY, USA: Springer.

Erickson, D. (1995) *Practical Handbook of Soybean Processing and Utilization*. Champaign, IL, USA: AOCS Press.

Heimann, M. (2008) Advantages and disadvantages in particle reduction techniques. 16th Annual ASA-IM SEA Feed Technology and Nutrition Workshop, 26–30 May 2008, The Regent, Singapore.

Jung, S., Mahfuz, A. & Maurer, D. (2009) Structure, protein interactions and in vitro protease accessibility of extruded and pressurised full-fat soybean flakes. *J. Am. Oil Chem. Soc.*, **86**(5), 475–483.

Lamsal, B.P., Murphy, P.A. & Johsnon, L.A. (2006) Flaking and extrusion as mechanical treatments for enzyme-assisted aqueous extraction of oils from soyabeans. *J. Am. Oil Chem. Soc.*, **83**, 973–979.

Edible Oil Processing, Second Edition. Edited by Wolf Hamm, Richard J. Hamilton and Gijs Calliauw.

Prezusio, S.M., Di Serio, M.G., Biasone, A., Viato, R., Mucciarella, M.G. & Giovacchino, L. (2010) Influence of olive crushing methods on the yields and oil characteristics. *Eur. J. Lipid Sci. Technol.*, **112**, 1345–1355.

Wang, L. (2009) *Energy Efficiency and Management in Food Processing Facilities*. Boca Raton, FL, USA: CRC Press.

Chapter 4

Baümler, E.R., Crapiste, G.H. & Carelli, A.A. (2010) Solvent extraction: kinetic study of major and minor components. *J. Am. Oil Chem. Soc.*, **87**, 1489–1495.

Lebovka, N., Vorobiev, E. & Chemat, F. (2012) *Enhancing Extraction Processes in the Food Industry*. Boca Raton, FL, USA: CRC Press.

Van Hoed, V., Ben Ali, C., Slah, M. & Verhé, R. (2010) Quality differences between pre-pressed and solvent extracted rapeseed oil. *Eur. J. Lipid Sci. Technol.*, **112**(11), 1241–1247.

Chapter 5

De Greyt, W. (2012) Efficient recovery of tocopherols from vegetable oils. *INFORM*, **23**(9), 557–561.

Chapter 6

List, G.R. & King, J.W. (2011) *Hydrogenation of Fats and Oils: Theory and Practice*. Champaign, IL, USA: AOCS Press.

Chapter 8

Leal-Calderon, F. (2012) Emulsified lipids : formulation and control of end-use properties. *OCL*, **19**(2), 111–119.

Marangoni, A. (2012) Organogels: an alternative edible oil-structuring method. *J. Am. Oil Chem. Soc.*, **89**(5), 749–780.

Mohos, F. (2010) *Confectionery and Chocolate Engineering*. Chichester, UK: Wiley-Blackwell.

Talbot, G. (2011) *Reducing Saturated Fats in Foods*. Cambridge, UK: Woodhead Publishing.

Chapter 10

Rigas, F. & Amyotte, P. (2013) *Hydrogen Safety*. Boca Raton, FL, USA: CRC Press.

Index

Edible Oil Processing, Second Edition. Edited by Wolf Hamm, Richard J. Hamilton and Gijs Calliauw.

www.ingramcontent.com/pod-product-compliance
Ingram Content Group UK Ltd.
Pitfield, Milton Keynes, MK11 3LW, UK
UKHW051035091026
15624UKWH00011B/437

9 781444 336849